STUDY GUIDE
STUDENT SOLUTIONS MANUAL

BO LOU

COLLEGE
PHYSICS

JERRY D. WILSON/ANTHONY J. BUFFA

THIRD EDITION

PRENTICE HALL, UPPER SADDLE RIVER, NJ 07458

Acquisition Editor: ***Wendy A. Rivers***
Production Editor: ***James Buckley***
Special Projects Manager: ***Barbara A. Murray***
Production Coordinator: ***Ben D. Smith***
Supplement Cover Manager: ***Paul Gourhan***

Printed in the United States of America

10 9 8 7 6 5 4

ISBN 0-13-505116-9

Prentice-Hall International (UK) Limited, *London*
Prentice-Hall of Australia Pty. Limited, *Sydney*
Prentice-Hall Canada Inc., *Toronto*
Prentice Hall Hispanoamericana, S.A., *Mexico*
Prentice-Hall of India Private Limited, *New Delhi*
Prentice-Hall of Japan, Inc., *Tokyo*
Simon & Schuster Asia Pte. Ltd., *Singapore*
Editoria Prentice-Hall do Brasil, *Ltda., Rio de Janero*

TABLE OF CONTENTS

Preface

This **Study and Guide and Student Solutions Manual** for **College Physics**, third edition, was prepared to help students gain a greater understanding of the principles of their introductory physics courses. Most of us learn by summary and examples, and this manual has been organized along these lines. For each chapter you will find:

- **Chapter Objectives**
  states the learning goals for the chapter. The objectives tell you what you should know upon completion of chapter study. Your instructor may omit some topics.

- **Key Terms**
  lists the key terms for the chapter. The definitions and/or explanations of the most important key terms can be found in the next section.

- **Chapter Summary and Discussion**
  outlines the important concepts and provide a brief overview of the major chapter contents. Extra worked out examples are included to further strengthen the concepts and principles. This review allows you to check the thoroughness of your study and serves as a last minute quick review (about 30 minutes) before quizzes or tests. Common students' mistakes and misconceptions are noted.

- **Mathematical Summary**
  lists the important mathematical equations in the chapter. The purpose is for self-review. You should be identify each symbol in an equation and explain what relationship the equation describes. The equation number in the text is included for reference. A last glance of this section is helpful before taking a test or quiz.

- **Solutions to Selected Exercises and Paired Exercises**
  provides the worked out solutions of the even-numbered annotated (blue dot) end-of-chapter text exercises. The paired exercises are similar in nature. You should try to work out the even-numbered paired text exercise independently and then check your work with solutions in this manual. After working the odd-numbered exercise of a pair, you can check your answer in the Answers to Odd-

Numbered Exercises at the back of the text. Solutions for some additional end-of-the-chapter exercises are also included.

- **Practice Quiz**

  consists of multiple choice questions and problems of the most fundamental concepts and problem solving skills in the chapter. This allows you to self-check your knowledge of the chapter. The answers to the quizzes are also given.

As you can see, this manual provides a through review for each chapter. The conscientious student can make good use of the various sections to assist in understanding and mastering the course contents and preparing for exams. I hope you find it so.

ACKNOWLEDGMENT

I would like to thank the following people and organizations for their enormous help and support.

First to the authors, Jerry D. Wilson and Anthony J. Buffa, for their meticulous checking of my work and numerous constructive and helpful comments and discussions.

To John Kinard and Jerry Wilson for their solutions manual of the second edition Wilson book. Some of the solutions are from their work.

To the editors of Prentice Hall, Tim Bozik, Paul Corey, Alison Reeves, and Wendy Rivers, for their guidance, and Pam Holland-Moritz for assistance.

To Carolyn Gauntt for proof-reading this manual.

Last but not the least, to my family, Lingfei and Alina, for their essential and generous support and love. I dedicate this manual to them.

Bo Lou
Ferris State University
Big Rapids, Michigan 49307

CHAPTER 1

Units and Problem Solving

I. Chapter Objectives

Upon completion of this chapter, you should be able to:

1. distinguish standard units and system of units.
2. describe the SI and specify the references for the three main base quantities of this system.
3. use common metric prefixes and nonstandard metric units.
4. explain the advantage of and apply dimensional analysis and unit analysis.
5. explain conversion factor relationships and apply them in converting units within a system, or from one system of units to another.
6. determine the number of significant figures in a numerical value and report the proper number of significant figures after performing simple calculations.
7. establish a problem-solving procedure and apply it to typical problems.

II. Key Terms

Upon completion of this chapter, you should be able to define and/or explain the following key terms:

standard unit
system of units
International System of Units (SI)
SI base units
SI derived units
meter (m)
kilogram (kg)
second (s)
mks system
cgs system
fps system
liter (L)
dimensional analysis
unit analysis
density (ρ)
conversion factor
exact number
measured number
significant figures (sf).

The definitions and/or explanations of the most important key terms can be found in the following section: **III. Chapter Summary and Discussion.**

III. Chapter Summary and Discussion

1. International System of Units (SI)

Objects and phenomena are measured and described using **standard units**, a group of which makes up a **system of units**.

(1) The International System of Units (SI), or the metric system, has only seven (see Table 1.1) base units. The base units for length, mass, and time are the **meter** (m), the **kilogram** (kg), and the **second** (s), respectively. A derived unit is a combination of the base units. For example, the units of speed, **meters per second**, are a combination of **meter** and **second**. There are many derived units.

(2) The metric system is a base-10 (decimal) system, which is very convenient in changing measurements from one unit to another. Metric multiples are designated by prefixes, the most common of which are **kilo–** (1000), **centi–**(1/100), and **milli–**(1/1000). For example, a centimeter is 1/100 of a meter, etc. A complete list of the metric prefixes is given in Table 1.2. A unit of volume or capacity is the **liter** (L), and 1 L = 1000 mL = 1000 cm^3 (cubic centimeters).

2. Dimensional Analysis

Fundamental base quantities, such as length, mass, and time are called **dimensions**. These are commonly expressed by bracketed symbols [L], [M], and [T], respectively. **Dimensional analysis** is a procedure by which the dimensional correctness of an equation may be checked. Both sides of an equation must not only be equal in numerical value, but also in dimension; and dimensions can be treated like algebraic quantities. Units, instead of symbols, may be used in **unit analysis**.

Dimensional analysis can be used to

(1) check whether an equation is **dimensionally correct**, i.e., if an equation has the same dimension (unit) on both sides.

(2) find out dimension or units of derived quantities.

Example 1.1: Check whether the equation $v^2 = 2ax$ is dimensionally correct, where x is length, v is velocity, and a is acceleration.

Solution: The dimensions and units of x, v, and a, are [L], m; $\frac{[L]}{[T]}$, m/s; and $\frac{[L]}{[T]^2}$, m/s^2; respectively.

Dimensional analysis: Dimension of left side is $\left(\frac{[L]}{[T]}\right)^2 = \frac{[L]^2}{[T]^2}$.

Dimension of right side is $2\frac{[L]}{[T]^2}[L] = 2\frac{[L]^2}{[T]^2}$.

So the dimension of the left side is equal to the dimension of the right side and the equation is dimensionally correct. Note: dimensionally correct does not necessarily mean the equation is correct. For example, the "equation", 2 tables = 3 tables, is dimensionally correct, but not numerically correct.

Unit analysis: Units of the left side are $(m/s)^2 = m^2/s^2$.

Units of the right side are $2\,(m/s^2)(m) = 2\,m^2/s^2$.

So the units of the left side are equal to the units of the right side and the equation is dimensionally correct.

Example 1.2: Newton's second law states that the force on a body is equal to its mass times its acceleration. Determine the dimension and units for force.

Solution: Since force is equal to mass times acceleration, the dimension (units) of force must be equal to the dimension (units) of mass times the dimension (units) of acceleration.

So the dimension of force is $[M] \times \frac{[L]}{[T]^2} = \frac{[M]\cdot[L]}{[T]^2}$, and the units of force is $kg \times m/s^2 = kg\cdot m/s^2$, which is called newton (N).

---

3. Conversions of Units

A quantity may be expressed in other units through the use of **conversion factors** such as (1 mi/1609 m) or (1609 m/1 mi). Note that any conversion factor is equal to 1 and so they can be multiplied or divided to any quantity without altering the quantity. The appropriate form of a conversion factor is easily determined by dimensional (unit) analysis.

Example 1.3: Convert 3200 meters to miles.

Solution: Here we need to convert meters to miles. We can accomplish this by using the conversion factor (1 mi/1609 m). The result units will be $m \times (mi/m) = mi$. We cannot multiply (1609 m/1 mi) because the result units would be $m \times (m/mi) = m^2/mi$.

$$(3200\ \cancel{m}) \times \frac{1\ mi}{1609\ \cancel{m}} = 1.99\ mi \approx 2.0\ mi.$$ Note the cancellation of the units m.

We can also use the conversion factor (1609 m/1 mi). Then we have to divide 3200 m by (1609 m/1 mi) in order to get mi. $\dfrac{3200\ \cancel{m}}{\dfrac{1609\ \cancel{m}}{1\ mi}} \approx 2.0\ mi.$

Example 1.4: Convert 55 mi/h (miles per hour) to m/s.

Solution: Here we need to convert miles to meters *and* hours to seconds. We can use the conversion factor (1609 m/1 mi) to convert miles to meters and (1 h/3600s) to convert 1/h to 1/s [why can't we use (3600 s/1 h)?]

$$(55\ \cancel{mi}/\cancel{h}) \times \frac{1609\ m}{1\ \cancel{mi}} \times \frac{1\ \cancel{h}}{3600\ s} = 25\ m/s.$$

We can also use the direct conversion (1 mi/h = 0.447 m/s). $(55\ \cancel{mi}/\cancel{h}) \times \dfrac{0.447\ m/s}{1\ \cancel{mi}/\cancel{h}} = 25\ m/s.$

---

4. Significant Figures

The number of **significant figures** in a quantity is the number of reliably known digits it contains. In general, *the final result of a multiplication and/or division should have the same number of significant figures as the quantity with the least number of significant figures used in the calculation. The final result of an addition and/or subtraction should have the same number of decimal places as the quantity with the least number of decimal places used in the calculation.* The proper number of figures or digits is obtained by rounding off a result.

Example 1.5: Perform the following operations:

(a) $0.586 \times 3.4 =$

(b) $13.90 \div 0.580 =$

(c) $(13.59 \times 4.86) \div 2.1 =$

(d) $4.8 \times 10^{5} \div 4.0 \times 10^{-3} =$

(e) $(3.2 \times 10^{8})(4.0 \times 10^{4}) =$

Solution: The final result of the multiplication and/or division should have the same number of significant figures as the quantity with the least number of significant figures.

(a) $0.586 \times 3.4 = 2.0$.

(b) $13.90 \div 0.580 = 24.0$.

(c) $13.59 \times 4.86 \div 2.1 = 31$.

(d) $4.8\times10^5 \div 4.0\times10^{-3} = 1.2\times10^8$.

(e) $(3.2\times10^8)(4.0\times10^4) = 1.3\times10^{13}$.

Example 1.6: Perform the following operations:

(a) $23.1 + 45 + 0.68 + 100 =$

(b) $157 - 5.689 + 2 =$

(c) $23.5 + 0.567 + 0.85 =$

(d) $4.69\times10^{-6} - 2.5\times10^{-5} =$

(e) $8.9\times10^4 + 2.5\times10^5 =$

Solution: The final result of the addition and/or subtraction should have the same number of decimal places as the quantity with the least number of decimal places.

(a) $23.1 + 45 + 0.68 + 100 = 169$.

(b) $157 - 5.689 + 2 = 153$.

(c) $23.5 + 0.567 + 0.85 = 24.9$.

(d) $4.69\times10^{-6} - 2.5\times10^{-5} = 0.469\times10^{-5} - 2.5\times10^{-5} = -2.0\times10^{-5}$.

(e) $8.9\times10^4 + 2.5\times10^5 = 8.9\times10^4 + 25\times10^4 = 34\times10^4 = 3.4\times10^5$.

---

5. Problem Solving

Problem solving is a skill that has to be learned and accumulated gradually. We can not learn this skill in a lecture or overnight. It takes practice, lots of practice and the exact procedure you adopt will probably be unique to you. The point is to develop one that works for you. However, there are some suggested problem solving procedures that can be followed.

(1) Say it in words.

(2) Say it in pictures.

(3) Say it in equations.

(4) Simplify the equations.

(5) Check the units.

(6) Insert numbers and calculate; check significant figures.

(7) Check the answer: is it reasonable?

The details of these procedures can be found on page 20 in the text.

Example 1.7: Starting from city A, an airplane flies 250 miles east to city B, then 300 miles north to city C, and finally 700 miles west to city D. What is the distance from city A to D?

Solution: Given: the distances and directions of each trip.

Find: the distance from city A to D.

Following the problem statement, we draw a diagram. From the diagram, it is easy to see that the distance from A to D is the hypotenuse of the shaded right-angle triangle. The sides perpendicular to each other are 300 mi and (700 mi − 250 mi) = 350 mi.

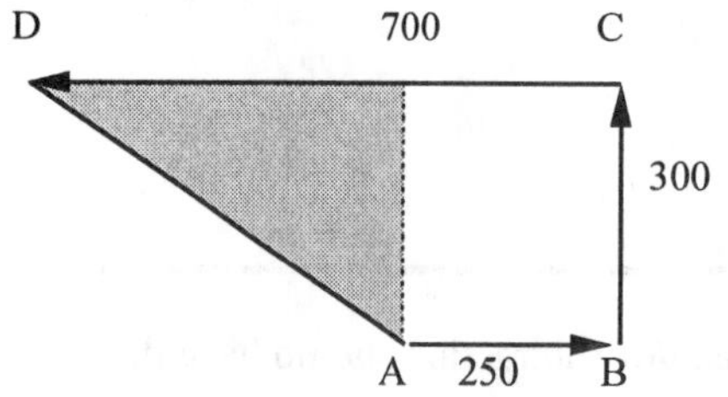

To find the hypotenuse of a right-angle triangle, we use Pythagorean theorem:

$c^2 = a^2 + b^2$, where a and b are the sides perpendicular to each other and c is the hypotenuse.

$$c = \sqrt{a^2 + b^2} = \sqrt{(300 \text{ mi})^2 + (350 \text{ mi})^2} = 461 \text{ mi}.$$

Obviously, the units, miles, are right; the answer, 461, is reasonable; and the number of significant figures, 3, in the final result, is the same as the least number of significant figures between 350 mi and 350 mi.

Example 1.8: The density of air is 1.29 kg/m^3. Find the mass of air in the room having the dimensions: 13 ft × 16 ft × 9.0 ft.

Solution: Given: the dimensions of the room and the density of air. Find: the mass of the air.

First find the volume of the room. Be aware of the different units in the problem; so we first convert the room dimensions to meters.

$L = 13 \text{ ft } (0.3048 \text{ m/ft}) = 4.0 \text{ m}$, $W = 16 \text{ ft } (0.3048 \text{ m/ft}) = 4.9 \text{ m}$,

$H = 9.0 \text{ ft } (0.3048 \text{ m/ft}) = 2.7 \text{ m}$. So the volume is $V = L \times W \times H = (4.0 \text{ m})(4.9 \text{ m})(2.7 \text{ m}) = 53 \text{ m}^3$.

Next find the mass of the air.

$\rho = \dfrac{m}{V}$. Multiplying V on both sides yields $m = \rho V$. So $m = (1.29 \text{ kg/m}^3)(53 \text{ m}^3) = 68 \text{ kg}$.

IV. Mathematical Summary

Density	$\rho = \frac{m}{V}\left(\frac{\text{mass}}{\text{volume}}\right)$ (1.1)	Define density in terms of mass and volume.

V. Solutions of Selected Exercises and Paired Exercises

4. (a) No, different ounces are used for volume and weight measurements.

   (b) Two different pound units are used. Avoirdupois lb = 16 oz, troy lb = 12 oz.

12. The units of the left side are m^2. The units of the right side are $(\text{m})^2 = \text{m}^2$.

    So the equation is dimensionally correct.

16. Since $x = \frac{gt^2}{2}$, $g = \frac{2x}{t^2}$. So units of g are m/s^2.

22. Since $mgx = -kx^2$, $k = -\frac{mgx}{x^2} = -\frac{mg}{x}$. So the units of k are $\frac{\text{kg(m/s}^2)}{\text{m}} = \text{kg/s}^2$.

40. 1 gal = 3.785 L. So $15 \text{ gal} = (15 \text{ gal}) \times \frac{3.785 \text{ L}}{1 \text{ gal}} = 57 \text{ L}$.

46. (a) The length of the tube is $x = (25 \text{ in.})\cos 37° = 20 \text{ in.}$

    and the width of the tube is $y = (25 \text{ in.})\sin 37° = 15 \text{ in.}$

    So the area $A = xy = (20 \text{ in.})(15 \text{ in.}) = 3.0\times 10^2 \text{ in.}^2$

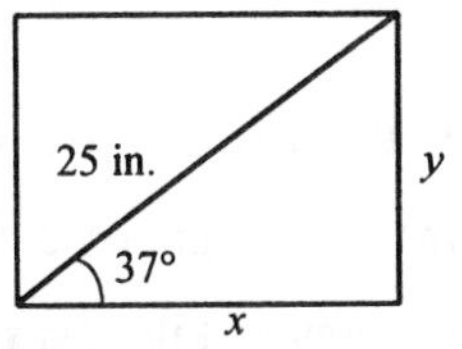

    (b) $A = 3.0\times 10^2 \text{ in.}^2 = (3.0\times 10^2 \text{ in.}^2) \times \left(\frac{2.54 \text{ cm}}{1 \text{ in.}}\right)^2 = 1.9\times 10^3 \text{ cm}^2$.

56. Since the last digit is estimated in measurements, the smallest division is 0.001 m or 1 mm.

58. (a) 4.

    (b) 3.

    (c) 5.

    (d) 2.

65. The area is the sum of that of the top, the bottom and the side. The side of the can is a rectangle with a length equal to the circumference and width equal to the height of the can.

So $A = \frac{\pi d^2}{4} + \frac{\pi d^2}{4} + C\,h = \frac{\pi d^2}{4} + \frac{\pi d^2}{4} + (\pi d)h$

$= \frac{\pi(12.559\text{ cm})^2}{4} + \frac{\pi(12.559\text{ cm})^2}{4} + \pi(12.559\text{ cm})(5.62\text{ cm}) = 470\text{ cm}^2.$

71. $\rho = \frac{m}{V} = \frac{6.0\times10^{24}\text{ kg}}{1.1\times10^{21}\text{ m}^3} = 5.5\times10^3\text{ kg/m}^3.$

78. A better buy gives you more AREA (more pepperoni) per dollar. Calculating the ratio of area to dollar.

For the 9.0-in.: $\frac{\pi(4.5\text{ in})^2}{\$7.95} = 8.0\text{ in.}^2\text{/dollar}$. For the 12-in.: $\frac{\pi(6.0\text{ in})^2}{\$13.50} = 8.4\text{ in.}^2\text{/dollar}$.

So the larger one is the better buy.

90. $V = AH = (\pi r^2)H = \pi(4.0\text{ cm})^2(12\text{ cm}) = 6.03\times10^2\text{ cm}^3 = (6.03\times10^2\text{ cm}^3)\times\frac{1\text{ L}}{1000\text{ cm}^3} = 0.60\text{ L}.$

94. $3.36\text{ g/cm}^3 = (3.36\text{ g/cm}^3)\times\frac{1\text{ kg}}{1000\text{ kg}}\times\left(\frac{100\text{ cm}}{1\text{ m}}\right)^3 = 3.36\times10^3\text{ kg/m}^3.$

Since $\rho = \frac{m}{V}$, $m = \rho V = (3.36\times10^3\text{ kg})\times\frac{4\pi}{3}\times[(1080\text{ mi})(1609\text{ m/mi})]^3 = 7.39\times10^{22}\text{ kg},$

which is very close to 7.4×10^{22} kg listed in the inside back cover.

VI. Practice Quiz

1. In the SI, the base units for length, mass, and time are

(a) meters, grams, seconds. (b) kilometers, kilograms, seconds. (c) centimeters, kilograms, seconds.
(d) meters, kilograms, seconds. (e) kilometers, grams, seconds.

2. If v has units of m/s and t has the units of s. What are the units of the quantity $a = v/t$?

(a) m (b) m/s^2 (c) s/m (d) s/m^2 (e) m/s^3

3. Which one of the following has the same dimension as time?

(x is length, v is velocity, and a is acceleration)

(a) $\frac{x}{a}$ (b) $\sqrt{\frac{2x}{a}}$ (c) $\sqrt{\frac{v}{x}}$ (d) vx (e) xa

4. Which one of the following is *not* equivalent to 2.50 miles?

(a) 1.32×10^4 ft (b) 1.58×10^5 in. (c) 4.02×10^3 km (d) 4.02×10^5 cm (e) 4.40×10^3 yd

5. When (3.5×10^4) is multiplied by (4.00×10^2), the product is which of the following expressed to the correct number of significant figures?

(a) 1.40×10^2 (b) 1.4×10^6 (c) 1.40×10^7 (d) 1.4×10^7 (e) 1.40×10^6

6. The density of water is 1.0×10^3 kg/m^3. Find the mass of water needed to fill a 2.0 L soft drink bottle.

(a) 0.020 kg (b) 0.20 kg (c) 2.0 kg (d) 20 kg (e) 200 kg

7. The area of a room floor is 15 m^2. How many cm^2 are there on the floor?

(a) 0.15 cm^2 (b) 1.5×10^{-3} cm^2 (c) 1.5×10^3 cm^2 (d) 1.5×10^4 cm^2 (e) 1.5×10^5 cm^2

8. An iron cube has a mass of 30 kg. What is the length of each side of the cube? (the density of iron is 7.86×10^3 kg/m^3)

(a) 3.8×10^{-3} m (b) 0.16 m (c) 0.062 m (d) 5.5×10^{-8} m (e) 2.6×10^2 m

9. A person stands 35.0 m from a flag pole. With a protractor at eye-level, he finds that the angle the top of the flag pole makes with the horizontal is 25.0°. How high is the flag pole? (the distance from his feet to his eyes is 1.7 m)

(a) 16.5 m (b) 33.4 m (c) 18.0 m (d) 76.8 m (e) 61.7 m

10. A rectangular garden measures 15 m long and 13.7 m wide. What is the length of a diagonal from one corner of the garden to the other?

(a) 29 m (b) 1 m (c) 18 m (d) 4.1×10^2 m (e) 20 m

Answers to Practice Quiz:

1. d 2. b 3. b 4. c 5. d 6. c 7. e 8. b 9. c 10. e

CHAPTER 2

Kinematics: Description of Motion

I. Chapter Objectives

Upon completion of this chapter, you should be able to:

1. distinguish between scalars and vectors and define distance and displacement.
2. define and calculate speed and velocity, and perform graphical analyses of velocity.
3. explain the relationship between velocity and acceleration and perform graphical analyses of acceleration.
4. explain kinematic equations and apply them in physical situations.
5. analyze free fall using the kinematic equations.

II. Key Terms

Upon completion of this chapter, you should be able to define and/or explain the following key terms:

mechanics
kinematics
dynamics
motion
distance
scalar (quantity)
vector (quantity)
displacement
speed
average speed
instantaneous speed
velocity
average velocity
instantaneous velocity
acceleration
average acceleration
instantaneous acceleration
acceleration due to gravity
free fall.

The definitions and/or explanations of the most important key terms can be found in the following section: **III. Chapter Summary and Discussion**.

III. Chapter Summary and Discussion

1. Distance and Displacement

Motion is related to change of position. The length traveled in changing position may be expressed in terms of **distance**, the actual path length between two points. Distance is a scalar quantity, which has only a magnitude with no direction. The direct straight line from the initial point to the final point is called **displacement** (change in position). It only measures the change in position, not the details involved in the change in position. Displacement is a vector quantity, which has both magnitude and direction. In the Figure shown, an object goes from point A to point C by following paths AB and BC. The distance traced is 3.0 m + 4.0 m = 7.0 m, and the displacement is 5.0 m in the direction of the arrow.

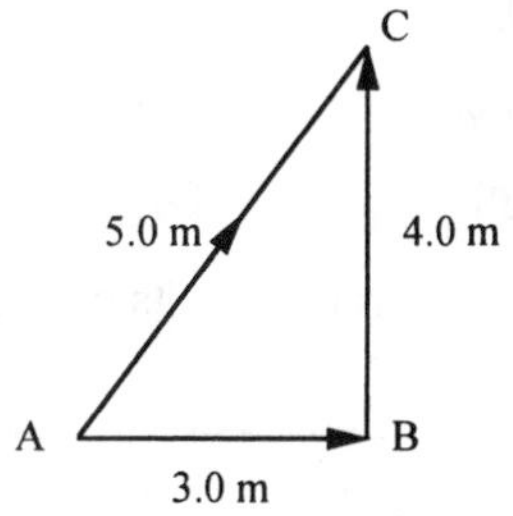

Example 2.1: In a soccer game, a midfielder kicks the ball back 10 yards to a goalkeeper. The goalkeeper then kicks the ball straight up the field 50 yards to a forward. What is the distance traveled by the football? What is the displacement of the football?

Solution: Sketch a diagram of the situation. For clarity, the arrows are laterally displaced.

It is obvious from the diagram that the soccer ball traveled 10 yards, then 50 yards. So the *distance traveled* is 10 yards + 50 yards = 60 yards.

Displacement is the straight line from the initial point to the final point. The ball displaced only 50 yards − 10 yards = 40 yards **straight up the field**.

Try to solve this problem without the diagram. You will find it is very difficult to do. That is why you are encouraged to try to draw a diagram for every problem.

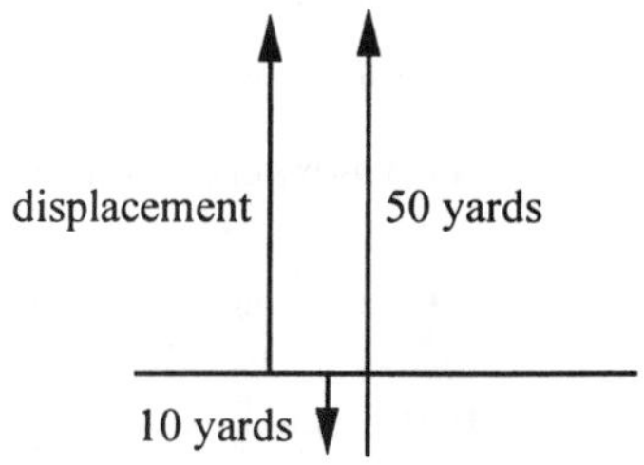

---

2. Speed and Velocity

In describing motion, the rate of change of position may be expressed in several ways.

Average speed is defined as the distance traveled divided by the time interval to travel that distance, $\bar{s} = \frac{\Delta d}{\Delta t}$, where $\bar{s}$ is average speed (the bar above s stands for average), Δd is distance traveled, and Δt is time interval (change in time). **Instantaneous speed**, s, is the speed at a particular time instant (Δt is close to zero). Since distance is a scalar quantity with no direction, so are average speed and instantaneous speed. Both tell us only how fast objects are moving.

Average velocity is defined as *displacement* divided by the time interval, $\bar{v} = \frac{\Delta x}{\Delta t}$, where $\bar{v}$ is average velocity, Δx is displacement (change in position), and Δt is time interval. (Direction of displacement indicated by sign, + or −.) **Instantaneous velocity**, v, is the velocity (magnitude and direction) at a particular instant of time (Δt is close to zero). Since displacement is a vector quantity, so are average velocity and instantaneous velocity. Both tell us not only how fast, but in which directions objects are moving. (Direction of velocity indicated by sign, + or − .) The metric units of speed and velocity are m/s.

Example 2.2: If the play described in Example 2.1 lasts 3.0 s, what is the average speed of the soccer ball? What is the average velocity of the soccer ball?

Solution: According to the definition, average speed $\bar{s} = \frac{\Delta d}{\Delta t} = \frac{60 \text{ yd}}{3.0 \text{ s}} = 20$ yd/s.

average velocity $\bar{v} = \frac{\Delta x}{\Delta t} = \frac{40 \text{ yd straight up the field}}{3.0 \text{ s}} = 13$ yd/s **straight up the field.**

---

3. Acceleration

Acceleration is the rate of change of velocity of time. Note: it is *velocity* (a vector), not speed (a scalar), hence acceleration is a vector.

Average acceleration is defined as the change in velocity divided by the time interval to make the change, $\bar{a} = \frac{\Delta v}{\Delta t}$, where $\bar{a}$ is average acceleration, Δv is change in velocity, and Δt is time interval.

Instantaneous acceleration is the acceleration at a particular instant of time (Δt is close to zero). As noted, velocity is a vector quantity, so are average acceleration and instantaneous acceleration. The metric units of acceleration are m/s/s or m/s^2.

A common misconception about velocity and acceleration has to do with their directions. Since velocity has both magnitude and direction, a change in either magnitude (speed) or direction will change velocity, resulting in an acceleration. We can accelerate objects by speeding up or down (change magnitude). We can also accelerate objects by changing their directions of travel. We often call the gas pedal of a car an accelerator. Can we call the brake pedal an accelerator? Can we call the steering wheel an accelerator? The answers are yes for both questions. (Why?)

But how about direction? For linear motion, when the velocity and acceleration of an object are in the same direction (they have the same directional signs), the velocity increases and the object speeds up (acceleration). When the velocity and acceleration are in opposite directions, the velocity decreases and the object slows down (deceleration).

Example 2.3: An object moving to the right has a decrease in velocity from 5.0 m/s to 1.0 m/s in 2.0 s. What is the average acceleration? What does your result mean?

Solution: Given: $v_o = +5.0$ m/s, $v = +1.0$ m/s, $t = 2.0$ s.

Find: $\bar{a}$.

According to the definition, average acceleration

$$\bar{a} = \frac{\Delta v}{\Delta t} = \frac{v - v_o}{t} = \frac{+1.0 \text{ m/s} - (+5.0 \text{ m/s})}{2.0 \text{ s}} = \frac{-4.0 \text{ m/s}}{2.0 \text{ s}} = -2.0 \text{ m/s}^2.$$

The negative sign means the acceleration is opposite to velocity (deceleration). The result means that the object *decreases* its velocity by 2.0 m/s every s or 2.0 m/s^2.

4. Kinematic Equations

Our discussion is restricted to **motions with constant accelerations.** In a motion with constant acceleration, the acceleration is not changing with time, in another word it is a constant. However, the constant could be zero, a negative or a positive non-zero constant. Zero acceleration simply means the velocity is a constant (no acceleration). For positive velocity, a negative acceleration means deceleration (speed decrease) and a positive acceleration means acceleration (speed increase). For negative velocity, a negative acceleration means acceleration (speed increase) and a positive acceleration means deceleration (speed decrease).

The symbols used in these equations are v_o: initial velocity; v: final velocity: a: acceleration; x: displacement; t: time interval. Be aware that the terms initial and final are relative. The end of one event is always the beginning of another. There are three general equations and two algebraic combinations of these equations that provide calculation convenience.

$x = \bar{v}\, t$, displacement = average velocity times × interval,

$\bar{v} = \dfrac{v + v_o}{2}$, average velocity = $\dfrac{\text{final velocity + initial velocity}}{2}$,

$v = v_o + a\,t$, final velocity = initial velocity + acceleration × time interval,

$x = v_o\,t + \frac{1}{2}a\,t^2$, displacement = initial velocity × time interval + $\frac{1}{2}$ × acceleration × time interval squared,

$v^2 = v_o^2 + 2\,a\,x$, final velocity squared = initial velocity squared + 2 × acceleration × displacement.

Which equation should you use in problem solving? The equation you select must have the unknown in it and everything else must be given. We can only solve for one unknown in one equation.

Example 2.4: An object starts from rest and accelerates with a constant acceleration of 5.0 m/s². Find its velocity and displacement at a time of 3.4 s.

Solution: Given: $v_o = 0$ (starts from rest), $a = 5.0\ \text{m/s}^2$, $t = 3.4$ s.

Find: v and x.

From $v = v_o + a\,t$, we have $v = 0 + (5.0\ \text{m/s}^2)(3.4\ \text{s}) = 17$ m/s.

$x = v_o\,t + \frac{1}{2}a\,t^2 = (0)(3.4\ \text{s}) + \frac{1}{2}(5.0\ \text{m/s}^2)(3.4\ \text{s})^2 = 29$ m.

Both velocity and displacement are in the direction of motion.

Example 2.5: An automobile brakes from 25 m/s to stop while traveling 120 m. What is the acceleration?

Solution: Given: $v_o = 25$ m/s, $v = 0$ (to stop), $x = 120$ m.

Find: a.

From $v^2 = v_o^2 + 2\,a\,x$, we have $a = \dfrac{v^2 - v_o^2}{2\,x} = \dfrac{(0)^2 - (25\ \text{m/s})^2}{2(120\ \text{m})} = -2.6\ \text{m/s}^2$.

The negative sign indicates that the acceleration is opposite to velocity (deceleration).

---

5. Free Fall

Objects in motion solely under the influence of gravity are said to be in **free fall**. A free fall does not necessarily mean a falling object. A vertically rising object is also said to be in free fall. The magnitude of **acceleration due to gravity** is often expressed with a symbol g. Near the surface of the Earth, the acceleration due to gravity is $g = 9.80\ \text{m/s}^2$ (downward) and near the surface of the Moon, it is $g = 1.7\ \text{m/s}^2$.

Since free fall is in the vertical direction and we often choose the upward direction as the $+y$ axis, we replace the x's by y's and a's by $-g$'s in the kinematic equation. The results are

$y = \bar{v}\,t,\quad \bar{v} = \dfrac{v + v_o}{2},\quad v = v_o - g\,t,\quad y = v_o\,t - \frac{1}{2}g\,t^2,\quad v^2 = v_o^2 - 2\,g\,y,$ where $g = 9.80\ \text{m/s}^2$.

Example 2.6: A ball is thrown upward with an initial velocity near the surface of the Earth. When it reaches the highest point

(a) its velocity is zero and its acceleration is non-zero,

(b) its velocity is zero and its acceleration is zero,

(c) its velocity is non-zero and its acceleration is zero,

(d) its velocity is non-zero and its acceleration is non-zero.

Solution: The answer is (a) , not (b) as you might think. The velocity has to change its direction at the highest point (goes from positive to negative) and so it is zero. However, the acceleration is not zero there. The acceleration is a constant 9.80 m/s^2 downward, independent of velocity. Stop and think, what if both the velocity and acceleration are zero? Will the ball fall down after it reaches the highest point?

Example 2.7: A ball is thrown upward with an initial velocity of 15.0 m/s from the top of a 60.0 m tall building.

(a) With what velocity will the ball strike the ground?

(b) How long does it take the ball to strike the ground?

Solution: The y in the equations stands for displacement from the launch point, not distance. When the ball strikes the ground, it will displace −60.0 m, or 60 m below the launch point.

Given: $y = -60.0$ m, $v_o = +15.0$ m/s.

Find: (a) t (b) v.

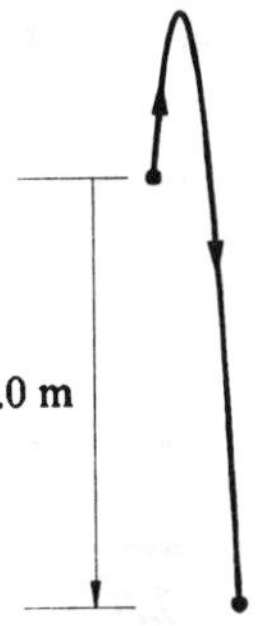

(a) $v^2 = v_o^2 - 2\,g\,y = (+15.0 \text{ m/s})^2 - 2\,(9.80 \text{ m/s}^2)(-60.0 \text{ m}) = 1.40\times10^3 \text{ m}^2/\text{s}^2$.

So $v = \sqrt{1.40\times10^3 \text{ m}^2/\text{s}^2} = \pm 37.4$ m/s.

The positive answer is discarded since the ball is falling when it lands (moving downward). Therefore $v = -37.4$ m/s.

(b) From $v = v_o - g\,t$, we have

$$t = \frac{v_o - v}{g} = \frac{(+15.0 \text{ m/s}) - (-37.4 \text{ m/s})}{9.80 \text{ m/s}^2} = \frac{52.4 \text{ m/s}}{9.80 \text{ m/s}^2} = 5.35 \text{ s}.$$

Try to solve this problem without using the overall displacement concept. You could break it into two phases. First, you would have to find out how high the ball goes, then secondly determine the velocity when it strikes the ground, and the total time it is in the air.

6. Graphical Interpretation

Graphical analysis is often helpful in understanding motion and its related quantities. On a position versus time graph we can extract the average velocity by finding the slope of a line connecting two points. Instantaneous velocity is equal to the slope of a straight line tangent to the curve at a specific point. For a velocity versus time graph, the average acceleration is the slope of a straight line connecting two points and instantaneous acceleration is the slope of a straight line tangent to the curve at a specific point. The area under the curve in a v vs. t gives displacement and the area under the curve in a vs. t graph yields velocity.

Example 2.8: The graph represents the position of a particle as a function of time.

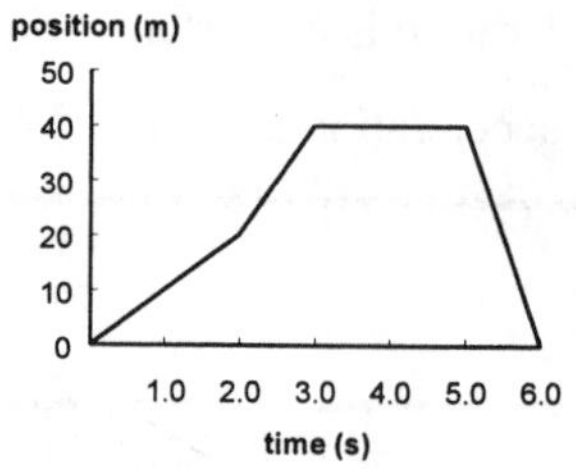

(a) What is the velocity at 1.0 s?

(b) What is the velocity at 2.5 s?

(c) What is the velocity at 4.0 s?

(d) What is the average velocity from 0 to 4.0 s?

(e) What is the average velocity for the 6.0 s interval?

Solution: (a) Velocity is the slope of the line. $v = \dfrac{\Delta x}{\Delta t} = \dfrac{20\text{ m} - 0\text{ m}}{2.0\text{ s} - 0\text{ s}} = 10\text{ m/s}.$

(b) Velocity is the slope of the line. $v = \dfrac{40\text{ m} - 20\text{ m}}{3.0\text{ s} - 2.0\text{ s}} = 20\text{ m/s}.$

(c) The slope of the line is zero and $v = 0$.

(d) $\bar{v} = \dfrac{\Delta x}{\Delta t} = \dfrac{40\text{ m} - 0\text{ m}}{4.0\text{ s} - 0\text{ s}} = 10\text{ m/s}.$

(e) $\bar{v} = \dfrac{0\text{ m} - 0\text{ m}}{6.0\text{ s} - 0\text{ s}} = 0.$

Example 2.9: The graph represents the velocity of a particle as a function of time.

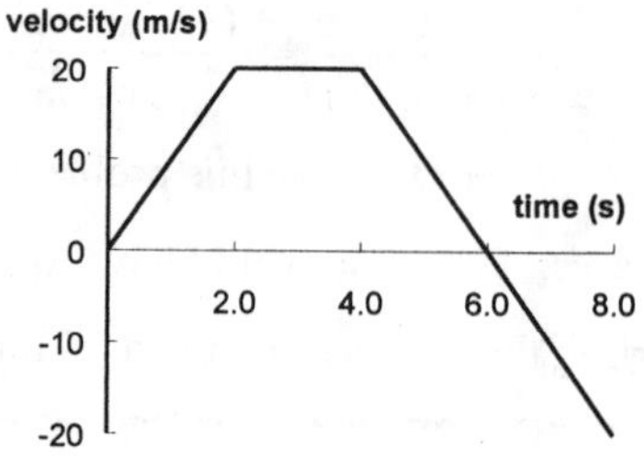

(a) What is the acceleration at 1.0 s?

(b) What is the acceleration at 3.0 s?

(c) What is the average acceleration between 0 and 5.0 s?

(d) What is the average acceleration for the 8.0 s interval?

(e) What is the displacement for the 8.0 s interval?

Solution: (a) Acceleration is the slope of the line. $a = \frac{\Delta v}{\Delta t} = \frac{20 \text{ m/s} - 0 \text{ m/s}}{2.0 \text{ s} - 0 \text{ s}} = 10 \text{ m/s}^2$.

(b) The slope of the line is zero and $a = 0$.

(c) $\bar{a} = \frac{\Delta v}{\Delta t} = \frac{10 \text{ m/s} - 0 \text{ m/s}}{5.0 \text{ s} - 0 \text{ s}} = 2.0 \text{ m/s}^2$.

(d) $a = \frac{\Delta v}{\Delta t} = \frac{-20 \text{ m/s} - 0 \text{ m/s}}{8.0 \text{ s} - 0 \text{ s}} = -2.5 \text{ m/s}^2$.

(e) The net area equals the displacement.

The area of a rectangle is length × width and the area of a triangle is $\frac{1}{2}$× base × height.

$\Delta x_{0\text{-}2} = \frac{1}{2}(2.0 \text{ s} - 0 \text{ s})(20 \text{ m/s}) = 20 \text{ m}$; $\Delta x_{2\text{-}4} = (4.0 \text{ s} - 2.0 \text{ s})(20 \text{ m/s}) = 40 \text{ m}$;

$\Delta x_{4\text{-}6} = \frac{1}{2}(6.0 \text{ s} - 4.0 \text{ s})(20 \text{ m/s}) = 20 \text{ m}$; $\Delta x_{6\text{-}8} = \frac{1}{2}(8.0 \text{ s} - 6.0 \text{ s})(-20 \text{ m/s}) = -20 \text{ m}$.

So $\Delta x = 20 \text{ m} + 40 \text{ m} + 20 \text{ m} + (-20 \text{ m}) = 60 \text{ m}$.

IV. Mathematical Summary

Average speed	$\bar{s} = \frac{\Delta d}{\Delta t}$	(2.2)	Defines average speed
Average velocity	$\bar{v} = \frac{\Delta x}{\Delta t}$	(2.3)	defines average velocity
Kinematic equation 1	$x = \bar{v}\, t$	(2.4)	Relates displacement with average velocity and time
Kinematic equation 2	$\bar{v} = \frac{v + v_o}{2}$	(2.8)	Defines average velocity for motion with constant acceleration
Kinematic equation 3	$v = v_o + a\, t$	(2.7)	Relates final velocity with initial velocity, acceleration, and time (constant acceleration only)
Kinematic equation 4	$x = v_o\, t + \frac{1}{2} a\, t^2$	(2.9)	Relates displacement with initial velocity, acceleration, and time (constant acceleration only)
Kinematic equation 5	$v^2 = v_o^{\,2} + 2\, a\, x$ $v = \sqrt{v_o^2 + 2ax}$	(2.10)	Relates final velocity with initial velocity, acceleration, and displacement (constant acceleration only)
Equation 1 in free fall	$y = \bar{v}\, t$	(2.4')	Relates displacement with average velocity and time
Equation 2 in free fall	$\bar{v} = \frac{v + v_o}{2}$	(2.8')	Defines average velocity
Equation 3 in free fall	$v = v_o - g\, t$	(2.7')	Relates final velocity with initial velocity, acceleration, and time

Equation 4 in free fall	$y = v_o t - \frac{1}{2} g t^2$ (2.9')	Relates displacement with initial velocity, acceleration, and time
Equation 5 in free fall	$v^2 = v_o^2 - 2gy$ (2.10')	Relates final velocity with initial velocity, acceleration, and displacement

V. Solutions of Selected Exercises and Paired Exercises

12. (a) $\bar{s} = \dfrac{\Delta d}{\Delta t} = \dfrac{150 \text{ km}}{2.5 \text{ h}} = 60 \text{ km/h}.$ Return trip $\bar{s} = \dfrac{150 \text{ km}}{2.0 \text{ h}} = 75 \text{ km/h}.$

(b) Total trip $\bar{s} = \dfrac{150 \text{ km} + 150 \text{ km}}{2.5 \text{ h} + 2.0 \text{ h}} = 67 \text{ km/h}.$

18. (a) $\bar{v} = \dfrac{\Delta x}{\Delta t}$, So $\bar{v}_{AB} = \dfrac{1.0 \text{ m} - 1.0 \text{ m}}{1.0 \text{ s} - 0} = 0;$

$\bar{v}_{BC} = \dfrac{7.0 \text{ m} - 1.0 \text{ m}}{3.0 \text{ s} - 1.0 \text{ s}} = 3.0 \text{ m/s};$ $\bar{v}_{CD} = \dfrac{9.0 \text{ m} - 7.0 \text{ m}}{4.5 \text{ s} - 3.0 \text{ s}} = 1.3 \text{ m/s};$

$\bar{v}_{DE} = \dfrac{7.0 \text{ m} - 9.0 \text{ m}}{6.0 \text{ s} - 4.5 \text{ s}} = -1.3 \text{ m/s};$ $\bar{v}_{EF} = \dfrac{2.0 \text{ m} - 7.0 \text{ m}}{9.0 \text{ s} - 6.0 \text{ s}} = -1.7 \text{ m/s}^2;$

$\bar{v}_{FG} = \dfrac{2.0 \text{ m} - 2.0 \text{ m}}{11.0 \text{ s} - 9.0 \text{ s}} = 0;$ $\bar{v}_{BG} = \dfrac{2.0 \text{ m} - 1.0 \text{ m}}{11.0 \text{ s} - 1.0 \text{ s}} = 0.10 \text{ m/s}.$

(b) The motion of BC, CD, and DE are not uniform since they are not straight lines.

(c) zero.

24. The minimum speed is $\bar{s} = \dfrac{\Delta d}{\Delta t} = \dfrac{675 \text{ km}}{7.00 \text{ h}} = (96 \text{ km/h})(1 \text{ mi}/1.609 \text{ km}) = 60.0 \text{ mi/h} < 65 \text{ mi/h}.$

No, she does not have to exceed the speed limit.

#8 34. The average velocity is $\bar{v} = \dfrac{\Delta x}{\Delta t} = \dfrac{100 \text{ m}}{4.5 \text{ s}} = 22.2 \text{ m/s}.$ Also $\bar{v} = \dfrac{v_o + v}{2} = \dfrac{v}{2}.$

So the final velocity must be $v = 2\,\bar{v} = 2(22.2 \text{ m/s}) = 44.4 \text{ m/s}.$

From $\bar{a} = \dfrac{\Delta v}{\Delta t}$, we have $\Delta t = \dfrac{\Delta v}{\bar{a}} = \dfrac{44.4 \text{ m/s} - 0}{9.0 \text{ m/s}^2} = 4.9 \text{ s} > 4.5 \text{ s}.$ So no, the drive did not do it.

The acceleration must be $\dfrac{44.4 \text{ m/s} - 0}{4.5 \text{ s}} = 9.9 \text{ m/s}^2.$

42. Given: $v_o = 70 \text{ km/h} = 19.4 \text{ m/s}$, $v = 35 \text{ km/h} = 9.72 \text{ m/s}$, $x = 50 \text{ m}$. Find: a.

From $v^2 = v_o^2 + 2ax$, we have $a = \dfrac{v^2 - v_o^2}{2x} = \dfrac{(9.72 \text{ m/s})^2 - (19.4 \text{ m/s})^2}{2(50 \text{ m})} = -2.8 \text{ m/s}^2.$

The negative sign indicates deceleration.

46. Given: $v_o = 0$, $v = 560$ km/h = 155.6 m/s, $x = 400$ m. Find: t.

From $x = \bar{v}\,t = \dfrac{v_o + v}{2}\,t$, we have $t = \dfrac{2\,x}{v_o + v} = \dfrac{2(400\text{ m})}{0 + 155.6\text{ m/s}} = 5.14$ s.

52. 40 km/h = 11.11 m/s. During reaction, the car travels $d = s\,t = (11.11\text{ m/s})(0.25\text{ s}) = 2.78$ m.

So the car really has 17 m – 2.78 m = 14.2 m to decelerate.

Let's calculate the stopping distance of the car.

Given: $v_o = 11.1$ m/s, $v = 0$, $a = -8.0\text{ m/s}^2$. Find: x.

From $v^2 = v_o^2 + 2ax$, we have $x = \dfrac{v^2 - v_o^2}{2\,a} = \dfrac{0 - (11.1\text{ m/s})^2}{2(-8.0\text{ m/s}^2)} = 7.70$ m.

So it takes the car only 2.78 m + 7.70 m = 10.5 m < 17 m to stop. The answer is yes.

54. Given: $v_o = 40$ m/s, $a = -3.5\text{ m/s}^2$, $x = 0$ ("returns to the origin"). Find: t.

From $x = v_o t + \frac{1}{2}at^2$, we have $0 = (40\text{ m/s})t + \frac{1}{2}(-3.5\text{ m/s}^2)t^2$.

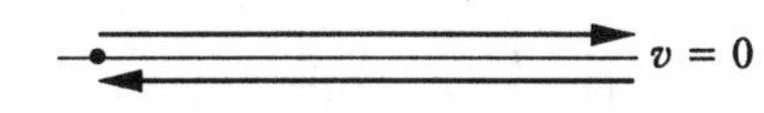

Reduce to a quadratic equation: $1.75t^2 - 40t = 0$,

or $t(1.75t - 40) = 0$,

Solve for $t =$ 0 or 23 s.

The $t = 0$ answer is corresponding to the initial time. So the answer is $t = 23$ s.

56. For constant acceleration, the v vs. t plot is a straight line.

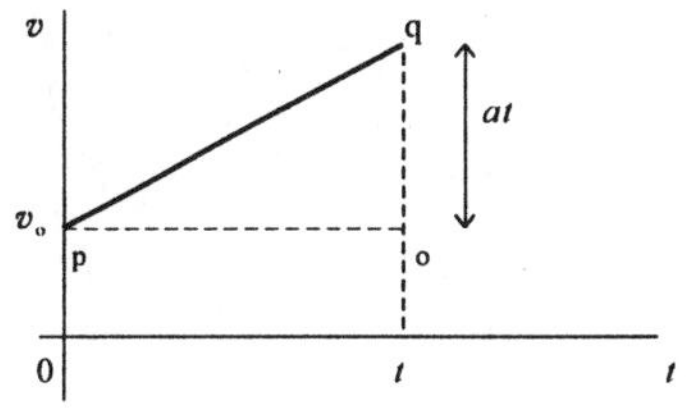

Point p has coordinates of $(0, v_o)$ and point q has coordinates of $(t, v_o + at)$. The distance from point q to point o is therefore at.

The area under the curve is the area of the triangle $\frac{1}{2}(at)t$ plus the area of the rectangle $v_o t$. So $A = v_o t + \frac{1}{2}at^2 = x$.

(a) $A = v_o t + 0 = v_o t$.

(b) $A = 0 + \frac{1}{2}at^2 = \frac{1}{2}a\,t^2$.

(c) $A = v_o t + \frac{1}{2}\,a\,t^2$.

60. Given: $x = 65$ m – 5.0 m = 60 m, $v_o = 10$ m/s, $t = 2.5$ s. Find: a.

From $x = v_o t + \frac{1}{2}at^2$, we have $a = \dfrac{x - v_o t}{\frac{1}{2}t^2} = \dfrac{60\text{ m} - (10\text{ m/s})(2.5\text{ s})}{\frac{1}{2}(2.5\text{ s})^2} = 11\text{ m/s}^2$.

80. (a) First calculate the time it takes for her to fall from the maximum height.

Given: $v_o = 0$ ("maximum height"), $y = -0.50$ m. Find: t.

From $y = v_o t - \frac{1}{2}gt^2 = -\frac{1}{2}gt^2$, we have $t = \sqrt{-\dfrac{2\,y}{g}} = \sqrt{-\dfrac{2(-0.50\text{ m})}{9.80\text{ m/s}^2}} = 0.319$ s.

So the total time in the air is 2(0.319 s) = 0.64 s.

(b) $v = v_o - gt = -(9.80\text{ m/s}^2)(0.319\text{ s}) = -3.1$ m/s. The negative sign indicates downward.

86. (a) Given: $v_o = 12.50$ m/s, $y = -60.0$ m. Find: t.

From $y = v_o t - \frac{1}{2}gt^2$, we have $-60.0 \text{ m} = (12.50 \text{ m/s})t - (4.90 \text{ m/s}^2)t^2$.

Reduce to quadratic equation: $4.90t^2 - 12.50t - 60.0 = 0$.

Solve for $t = \dfrac{-b \pm \sqrt{b^2 - 4ac}}{2a} = \dfrac{-(-12.50) \pm \sqrt{(-12.50)^2 - 4(4.90)(-60.0)}}{2(4.90)} = 5.00 \text{ s or } -2.45 \text{ s}$.

The negative root is physically meaningless.

(b) $v = v_o - gt = 12.50 \text{ m/s} - (9.80 \text{ m/s}^2)(5.00 \text{ s}) = -36.5 \text{ m/s}$. The negative sign indicates downward.

88. First find the time it takes for the ball to reach the level of the prof's head.

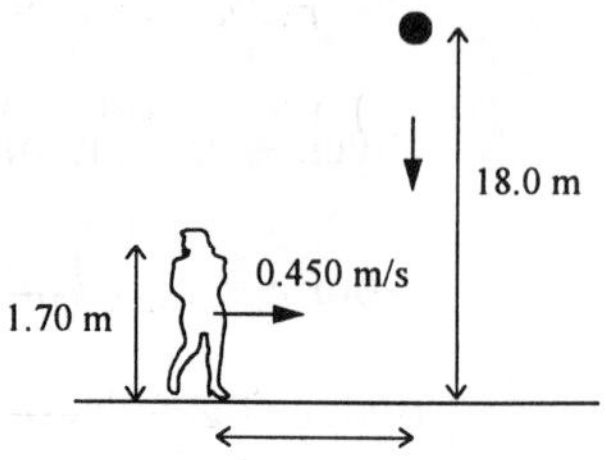

Given: $y = -(18.0 \text{ m} - 1.70 \text{ m}) = -16.3$ m, $v_o = 0$. Find: t.

From $y = v_o t - \frac{1}{2}gt^2 = -\frac{1}{2}gt^2$,

we have $t = \sqrt{-\dfrac{2y}{g}} = \sqrt{-\dfrac{2(-16.3 \text{ m})}{9.80 \text{ m/s}^2}} = 1.824$ s.

During this time, the Prof advances a distance equal to

$(0.450 \text{ m/s})(1.824 \text{ s}) = 0.821 \text{ m} < 1.00$ m. No, it does not hit his head.

Now calculate the time it takes for the ball to hit the ground level. $t = \sqrt{-\dfrac{2(-18.0 \text{ m})}{9.80 \text{ m/s}^2}} = 1.917$ s.

During this time, the prof advances $(0.450 \text{ m/s})(1.917 \text{ s}) = 0.862 \text{ m} < 1.00$ m.

So the ball hits $1.00 \text{ m} - 0.862 \text{ m} = 0.14 \text{ m} = 14$ cm in front of the prof.

Therefore, generally speaking, it does not hit the prof at all.

89. The key to this Exercise is to find the velocity of the object when it reaches the top of the window (it is not zero). This velocity is the initial velocity for Motion 1 and the final velocity for Motion 2.

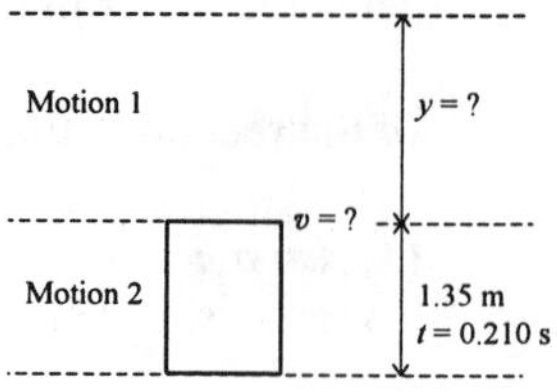

Consider Motion 2 first.

Given: $y = -1.35$ m, $t = 0.210$ s. Find: v_o.

From $y = v_o t - \frac{1}{2}gt^2$,

we have $v_o = \dfrac{x}{t} + \frac{1}{2}gt = \dfrac{-1.35 \text{ m}}{0.210 \text{ s}} + (4.90 \text{ m/s}^2)(0.210 \text{ s}) = -5.40$ m/s.

Now consider Motion 1. Given: $v_o = 0$, $v = -5.40$ m. Find: y.

From $v^2 = v_o^2 - 2gy$, we have $y = \dfrac{v_o^2 - v^2}{2g} = \dfrac{0 - (-5.40 \text{ m/s})^2}{2(9.80 \text{ m/s}^2)} = 1.49$ m.

VI. Practice Quiz

1. If you run half a lap around a circle of radius 25 m in 100 s, the magnitude of your average velocity is

   (a) zero. (b) 0.20 m/s. (c) 0.50 m/s. (d) 1.0 m/s. (e) 3.14 m/s.

2. An object moving in the $+x$ axis experiences an acceleration of 5.0 m/s^2. This means the object is

   (a) traveling 5.0 m in every second.

   (b) traveling at 5.0 m/s in every second.

   (c) changing its velocity by 5.0 m/s.

   (d) increasing its velocity by 5.0 m/s in every second.

   (e) moving with constant velocity.

3. A cart with an initial velocity of 5.0 m/s and experiences a constant acceleration of 2.0 m/s^2. What is the cart's displacement during the first 6.0 s of its motion?

   (a) zero (b) 10 m (c) 55 m (d) 66 m (e) 80 m

4. An object is thrown straight up. When it is at the highest point in the air

   (a) both its velocity and acceleration are zero.

   (b) neither its velocity nor its acceleration is zero.

   (c) its velocity is zero and its acceleration is not zero.

   (d) its velocity is not zero and its acceleration is zero.

   (e) none of the above.

5. Human reaction time is usually greater than 0.10 s. If your lab partner holds a ruler between your fingers and releases it without warning, how far can you expect the ruler to fall before you catch it?

   (a) at least 3.0 cm (b) at least 4.9 cm (c) at least 6.8 cm (d) at least 9.8 cm (e) at least 11.0 cm

6. Which one of the following quantity is an example of a scalar?

   (a) velocity (b) acceleration (c) force (d) mass (e) displacement

7. A ball is thrown vertically upward with a speed v. An identical second ball is thrown upward with a speed $2v$ (twice as fast). What is the ratio of the height of the second ball to that of the first ball? (how many times higher does the second ball go than the first ball?)

   (a) 4:1 (b) 2:1 (c) 1.7:1 (d) 1.4:1 (e) 1:1

8. A car starts from rest and accelerates for 4.0 m/s^2 for 5.0 s, then maintains that velocity for 10 s and then decelerates at the rate of 2.0 m/s^2 for 4.0 s. What is the final speed of the car?

(a) 20 m/s (b) 16 m/s (c) 12 m/s (d) 10 m/s (e) 8.0 m/s

9. An object moves 5.0 m north and then 3.0 m south. Find both the distance traveled and the magnitude of the displacement?

(a) 2.0 m; 8.0 m (b) 8.0 m; 2.0 m (c) 8.0 m; 8.0 m (d) 2.0 m; 2.0 m (e) 8.0 m, 15 m

10. An airplane travels at 300 mi/h south for 2.00 h and then at 250 mi/h north for 750 miles. What is the average speed for the trip?

(a) 260 mi/h (b) 270 mi/h (c) 275 mi/h (d) 280 mi/h (e) 285 mi/h

Answers to Practice Quiz:

1. c 2. d 3. d 4. c 5. b 6. d 7. a 8. c 9. b 10. b

CHAPTER 3

Motion in Two Dimensions

I. Chapter Objectives

Upon completion of this chapter, you should be able to:

1. analyze motion in terms of its components and apply the kinematic equations for components of motion.
2. add and subtract vectors (a) graphically and (b) analytically.
3. determine relative velocity through vector addition or subtraction.
4. analyze projectile motion to find position, time of flight, and range.

II. Key Terms

Upon completion of this chapter, you should be able to define and/or explain the following key terms:

components of motion	**unit vector**
vector addition (subtraction)	**component form**
triangle method	**analytical component method**
parallelogram method	**relative velocity**
polygon method	**projectile motion**
component method	**parabola**
magnitude-angle form	**range**

The definitions and/or explanations of the most important key terms can be found in the following section: **III. Chapter Summary and Discussion.**

III. Chapter Summary and Discussion

1. Components of Motion

Motion in two dimensions, or curvilinear motion, is motion in which an object moves in a plane which can be described by a rectangular coordinate system. To analyze such motion, quantities are usually resolved into rectangular **components**. The displacement vector is resolved into $x = d\cos\theta$ and $y = d\sin\theta$, the rectangular coordinates; velocity and acceleration vectors, **v** and **a**, are resolved into $v_x = v \cos\theta$, $v_y = v \sin\theta$, and $a_x = a\cos\theta$, $a_y = a \sin\theta$, their horizontal and vertical components. Note: the angle θ is the angle relative to the x axis.

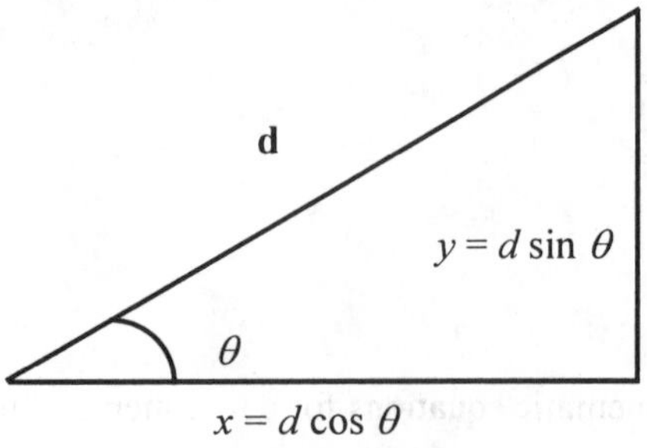

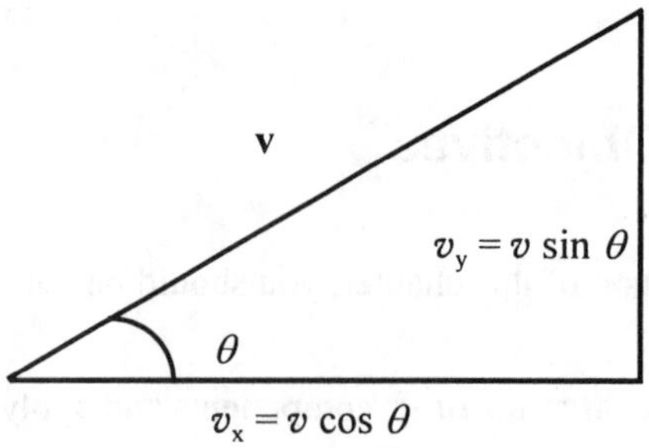

Once the displacement, velocity, and acceleration vectors are resolved into their respective components, we can apply the kinematic equations from Chapter 2 to the motion in the x and y directions. For example:

$$v_x = v_{xo} + a_x t, \quad v_y = v_{yo} + a_y t, \quad x = v_{xo}t + \tfrac{1}{2}a_x t^2, \quad y = v_{yo}t + \tfrac{1}{2}a_y t^2, \quad \text{etc.}$$

The key to success in solving two dimensional motion is to resolve the motion in components. Remember to treat the components as independent, i.e., a_x has nothing to do with a_y, etc. However, the time in all the equations is the same, providing a common link. Always think about resolving vectors into components when working problems in two dimensions.

Example 3.1: An airplane is moving at 300 mi/h in a direction of 37° N of E. Find the components of the plane's velocity in the eastward and northward directions.

Solution: Given: **v** = 300 mi/h in a direction of 37° N of E,

or, v = 300 mi/h and θ = 37°.

Find: v_x and v_y.

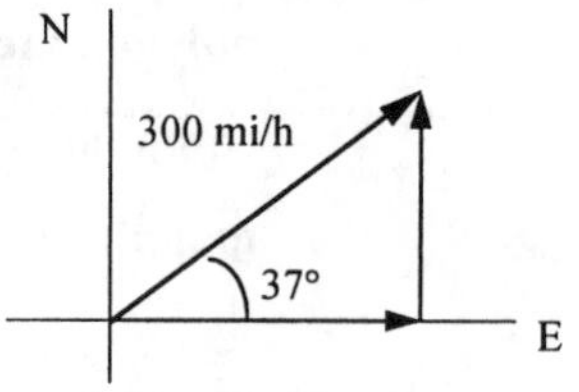

$v_x = v\cos\theta$ = (300 mi/h) cos 37° = 240 mi/h,

$v_y = v\sin\theta$ = (300 mi/h) sin 37° = 181 mi/h.

Example 3.2: A boat travels with a speed of 5.0 m/s in a straight path on a still lake. Suddenly, a strong, steady wind pushes the boat perpendicularly to its straight line path with a speed of 3.0 m/s for 5.0 s. Relative to its position just when the wind started to blow, where is the boat located at the end of this time?

Solution: Both motions are motion with constant velocity.

Choose the straight path of the boat as x axis and the direction of wind as the y axis.

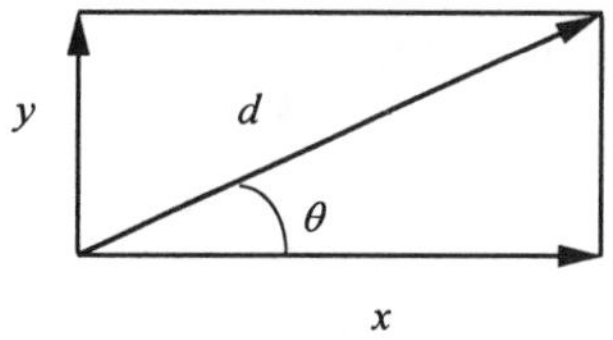

Given: $v_{xo} = 5.0$ m/s, $a_x = 0$, $v_{yo} = 3.0$ m/s, $a_y = 0$, $t = 5.0$ s.

Find: x and y.

$$x = v_{xo}\,t + \tfrac{1}{2}a_x\,t^2 = (5.0\text{ m/s})(5.0\text{ s}) + 0 = 25\text{ m},$$

$$y = v_{yo}\,t + \tfrac{1}{2}a_y\,t^2 = (3.0\text{ m/s})(5.0\text{ s}) + 0 = 15\text{ m}.$$

Or $d = \sqrt{x^2 + y^2} = \sqrt{(25\text{ m})^2 + (15\text{ m})^2} = 29$ m, and $\theta = \tan^{-1}\dfrac{y}{x} = \dfrac{15\text{ m}}{25\text{ m}} = 31°$.

2. Vector Addition and Subtraction

Vector addition can be done graphically with the triangle method and the parallelogram method for two vectors and the polygon method for more than two vectors. **Vector subtraction** is a special case of vector addition: **A** − **B** = **A** + (−**B**), and a negative vector is defined as a vector having the same magnitude but opposite in direction to the positive vector. For example, the negative vector of a velocity vector at 45 m/s north is simply 45 m/s south. When adding or subtracting vectors graphically, a convenient scale must be used.

Example 3.3: Two vectors **A** and **B** are given. Show

(a) **A** + **B** with the triangle method.

(b) **A** + **B** with the parallelogram method.

(c) **A** − **B** with the triangle method.

(d) **A** − **B** with the parallelogram method.

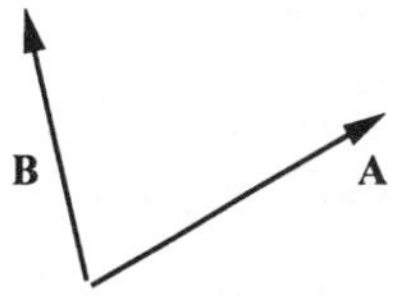

Solution: (a)

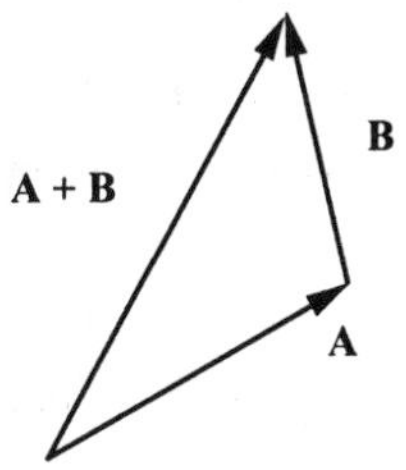

(b)

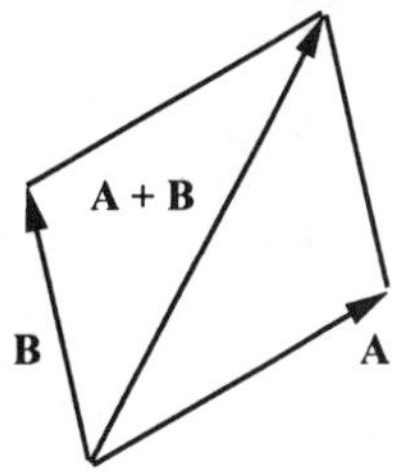

(c)

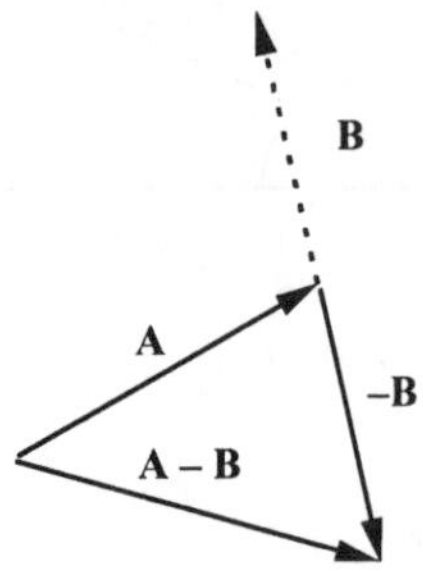

(d)

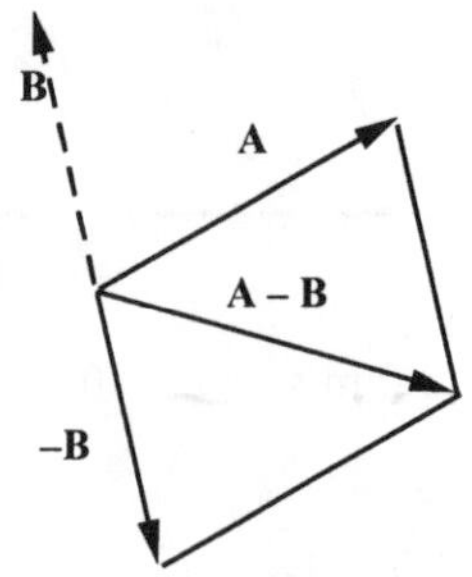

Vector addition is conveniently done by the **analytical component method**. The recommended procedure is as follows:

(1) Resolve the vectors to be added into their x and y components. Include directional plus or minus signs in the components.

(2) Add, algebraically, all the x components together and all the y components together to get the x and y components of the resultant vector.

(3) Express the resultant vector using (a) the component form, e.g., $\mathbf{A} = A_x\,\mathbf{x} + A_y\,\mathbf{y}$, or

(b) in magnitude-angle form, e.g., $A = \sqrt{A_x^2 + A_y^2}$,

$\theta = \tan^{-1}\dfrac{A_y}{A_x}$ (relative to x axis).

A detailed treatment of this procedure can be found on page 73 to 74 in the text.

Example 3.4: Find the resultant velocity of the sum of

(i) $\mathbf{v}_1 = 35$ m/s 30° N of E; (ii) $\mathbf{v}_2 = 55$ m/s 45° N of W.

Solution: **(1)** Resolve the vectors to be added into their x and y components.

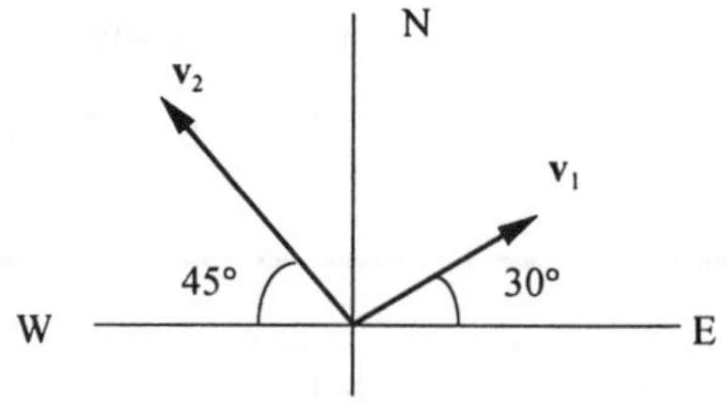

$v_{1x} = v_1 \cos 30° = (35 \text{ m/s}) \cos 30° = 30.3$ m/s,

$v_{1y} = (35 \text{ m/s}) \sin 30° = 17.5$ m/s.

$v_{2x} = -\,v_2 \cos 45° = (55 \text{ m/s}) \cos 45° = -38.9$ m/s, ($-x$ direction)

$v_{2y} = (55 \text{ m/s}) \sin 45° = 38.9$ m/s.

(2) Add components.

$v_x = v_{1x} + v_{2x} = 30.3 \text{ m/s} + (-38.9 \text{ m/s}) = -8.6$ m/s,

$v_y = v_{1y} + v_{2y} = 17.5 \text{ m/s} + 38.9 \text{ m/s} = 56.4$ m/s.

(3) Express the resultant vector

in component form: $\mathbf{v} = \mathbf{v}_1 + \mathbf{v}_2 = -8.6 \text{ m/s}\,\mathbf{x} + 56 \text{ m/s}\,\mathbf{y}$, and

in magnitude-angle form:

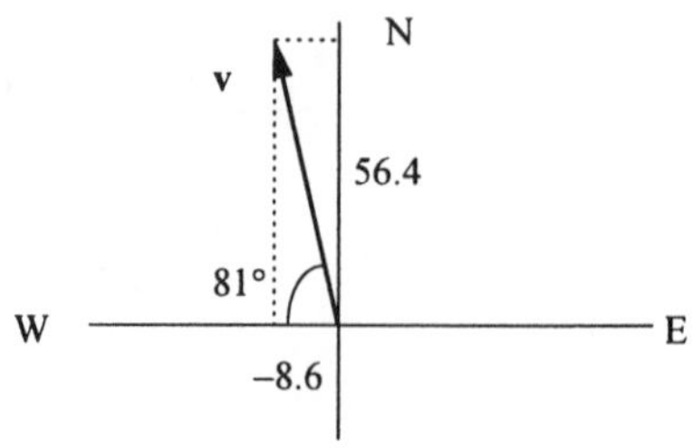

$$v = \sqrt{v_x^2 + v_y^2} = \sqrt{(-8.6 \text{ m/s})^2 + (56.4 \text{ m/s})^2} = 57 \text{ m/s},$$

$$\text{and } \theta = \tan^{-1}\frac{v_y}{v_x} = \tan^{-1}\frac{56.4 \text{ m/s}}{-8.6 \text{ m/s}} = 81° \text{ N of W}.$$

Note: Once you have the components for the resultant, draw a diagram like the one above to determine the angle and in which quadrant the vector is.

3. Relative Velocity

Physical phenomena can be observed from different **frames of reference**. The velocity of a ball tossed by a passenger on a moving train will be measured differently by the passenger than by an observer on the Earth. As a matter of fact, any velocity we measure is **relative**. The velocity of a moving car is measured relative to the ground, and the revolving motion of our Earth around the Sun is relative to the Sun. Relative velocity can be determined with vector addition or subtraction. The symbols used in relative velocity such as $\mathbf{v}_{cg}$ (where c stands for car and g stands for ground) means the velocity of a car relative to the ground.

Example 3.5: A river flows with a velocity of 1.0 m/s south. A boat, whose speed in still water is 5.0 m/s, is directed east across the 150 m wide river.

(a) How long does it take the boat to reach the opposite shore?

(b) How far downstream will the boat land?

(c) What is the velocity of the boat relative to the shore?

Solution: We use the following subscripts: r = river, b = boat, s = shore.

Given: $v_{rs} = 1.0$ m/s, $v_{br} = 5.0$ m/s.

Find: (a) t (b) y (c) v.

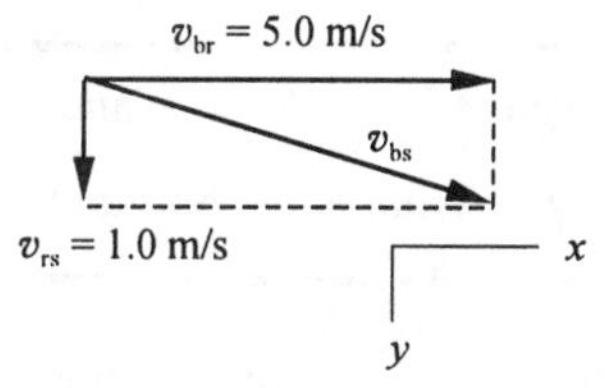

(a) From the concept of components of motion, the time it takes the boat to reach the opposite shore is simply $t = \dfrac{x}{v_{br}} = \dfrac{150 \text{ m}}{5.0 \text{ m/s}} = 30$ s.

(b) $y = v_{rs}\, t = (1.0 \text{ m/s})(30 \text{ s}) = 30$ m.

(c) $v = \sqrt{v_{br}^2 + v_{rs}^2} = \sqrt{(5.0 \text{ m/s})^2 + (1.0 \text{ m/s})^2} = 5.1$ m/s.

Example 3.6: The person in the boat in Example 3.5 wants to travel directly across the river.

(a) What angle upstream must the boat be directed?

(b) With what speed will the boat cross the river?

(c) How long will it take the boat to reach the opposite shore?

Solution: To travel directly across the river, the relative velocity of the boat to the shore must be directly across the river.

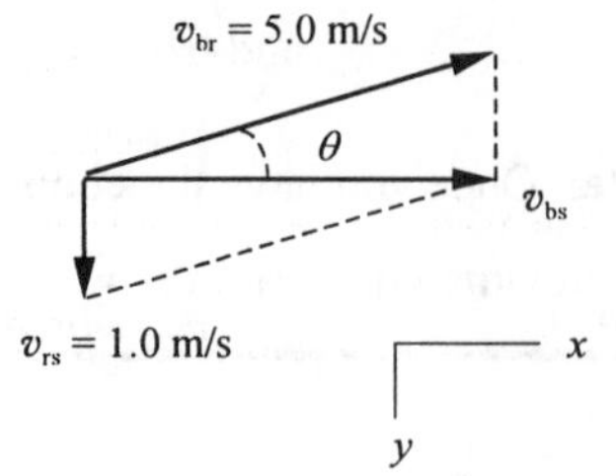

(a) $\theta = \sin^{-1}\dfrac{v_{rs}}{v_{br}} = \sin^{-1}\dfrac{1.0\text{ m/s}}{5.0\text{ m/s}} = 12°$ N of E.

(b) From the triangle in the diagram, we have $v_{bs}^2 + v_{rs}^2 = v_{br}^2$,

so $v_{bs} = \sqrt{v_{br}^2 - v_{rs}^2} = \sqrt{(5.0\text{ m/s})^2 - (1.0\text{ m/s})^2} = 4.9$ m/s.

(c) The time is then $t = \dfrac{x}{v_{bs}} = \dfrac{150\text{ m}}{4.9\text{ m/s}} = 31$ s.

4. Projectile Motion

Projectile motion is motion in two dimensions under the action of gravity only (downward), horizontal and vertical. Therefore the horizontal motion has zero acceleration if air resistance is ignored. The vertical motion is free fall and so the acceleration is the acceleration due to gravity, $a_y = -g$. The motion is analyzed using components, with time of flight being the common factor.

Applying the general kinematic equations in component form to projectile motion ($a_x = 0$ and $a_y = -g$ if the upward direction is chosen positive), we have $v_x = v_{xo}$, $v_y = v_{yo} - g\,t$, $x = v_{xo}\,t$, $y = v_{yo}\,t - \frac{1}{2}g\,t^2$, etc. Again, the key to success in solving projectile motion is to resolve the motion in components. Treat the components as independent. However, the time for both components is the same. Always think about resolving vectors into components. Usually, the time of flight is something you have to find first.

Example 3.7: An airplane traveling with a constant horizontal speed of 250 m/s at a altitude of 500 m releases a package. What is the horizontal distance the package travels before hitting the ground (range)?

Solution: Given: horizontal motion (taken in the x direction) $v_{xo} = 250$ m/s; vertical motion $v_{yo} = 0$, $y = -500$ m.

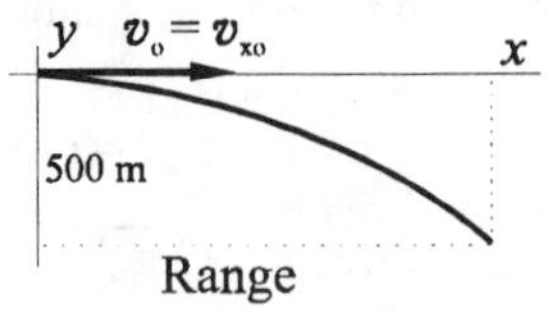

Find: x (range).

Since the range is given by $x = v_{xo}\,t$, we have to find the time of flight t first.

From the vertical motion, we use $y = v_{yo}\,t - \frac{1}{2}g\,t^2$.

$-500\text{ m} = 0 - \frac{1}{2}(9.80\text{ m/s}^2)\,t^2$, so $t = 10.1$ s.

Therefore $x = (250\text{ m/s})(10.1\text{ s}) = 2500$ m.

Example 3.8: A long jumper leaves the ground at 10 m/s at an angle of 35 degrees above the horizontal.

(a) How long is the jumper in the air?

(b) What is the distance of the jump?

Solution: Given: horizontal motion | vertical motion

$v_{xo} = v \cos\theta = (10 \text{ m/s}) \cos 35° = 8.19 \text{ m/s}$,

$v_{yo} = v \sin\theta = (10 \text{ m/s}) \sin 35° = 5.74 \text{ m/s}$.

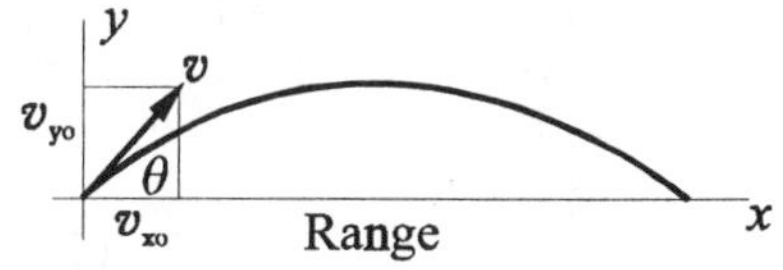

Find: (a) t (b) x.

(a) On landing, $y = 0$; and from $y = v_{yo}\, t - \frac{1}{2} g\, t^2$, we have $0 = (5.74 \text{ m/s})\, t - \frac{1}{2}(9.80 \text{ m/s}^2)\, t^2$.

Solving, $t = 0$ or 1.17 s. The $t = 0$ root corresponds to the position at the start ($x = 0$ and $y = 0$) and the $t = 1.17$ s corresponds to the landing position (x = range and $y = 0$). So the time of flight is 1.2 s, (to two significant figures).

(b) $x = v_{xo}\, t = (8.19 \text{ m/s})(1.17 \text{ s}) = 9.6 \text{ m}$.

Or, using the range equation, we have $x = R = \dfrac{v_o^2 \sin 2\theta}{g} = \dfrac{(10 \text{ m/s})^2 \sin 2(35°)}{9.80 \text{ m/s}^2} = 9.6 \text{ m}$

IV. Mathematical Summary

Components of initial velocity	$v_{xo} = v \cos\theta$ $v_{yo} = v \sin\theta$	(3.1a) (3.1b)	Relate the x and y components to the magnitude and the angle of the initial velocity. (θ is from x axis)
Components of displacement	$x = v_{xo}\, t + \frac{1}{2} a_x\, t^2$ $y = v_{yo}\, t + \frac{1}{2} a_y\, t^2$	(3.3a) (3.3b)	Relate the displacement components to initial velocity components, acceleration components and time.
Components of velocity	$v_x = v_{xo} + a_x\, t$ $v_y = v_{yo} + a_y\, t$	(3.3c) (3.3d)	Relate the velocity components to initial velocity components and acceleration components.
Component method of vector addition	$v_x = \Sigma v_{xi}$ $v_y = \Sigma v_{yi}$ $\mathbf{v} = v_x\, \mathbf{x} + v_y\, \mathbf{y}$ $v = \sqrt{v_x^2 + v_y^2}$, $\theta = \tan^{-1}\left\lvert \frac{v_y}{v_x} \right\rvert$	(3.7) (3.4a) (3.4b)	Outlines the analytical component method of vector addition.
Projectile motion	$R = \dfrac{v_o^2 \sin 2\theta}{g}$	(3.11)	Calculate the range of a projectile. ($y_{initial} = y_{final}$ only)

V. Solutions of Selected Exercises and Paired Exercises

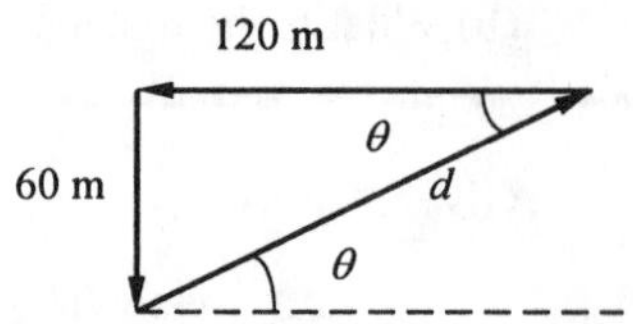

6. $d = \sqrt{x^2 + y^2} = \sqrt{(60\text{ m})^2 + (120\text{ m})^2} = 1.3 \times 10^2\text{ m}.$

$\theta = \tan^{-1}\left|\frac{v_y}{v_x}\right| = \tan^{-1}\left(\frac{60}{120}\right) = 27°\text{ N of E}.$

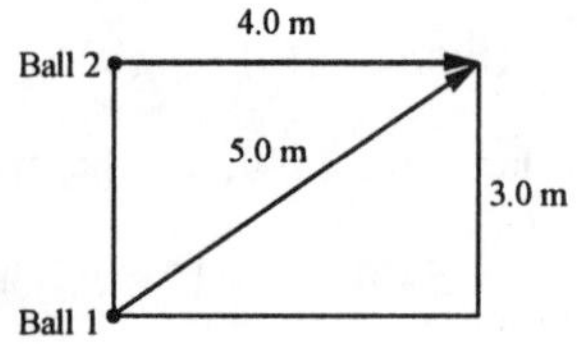

14. $d = \sqrt{x^2 + y^2} = \sqrt{(3.0\text{ m})^2 + (4.0\text{ m})^2} = 5.0\text{ m}.$

Use $x = v\,t$.

For ball 1, $t = \frac{x}{v} = \frac{5.0\text{ m}}{0.75\text{ m/s}} = 6.67\text{ s}.$

For ball 2, $v = \frac{x}{t} = \frac{4.0\text{ m}}{6.67\text{ s}} = 0.60\text{ m/s}.$

20. (a)

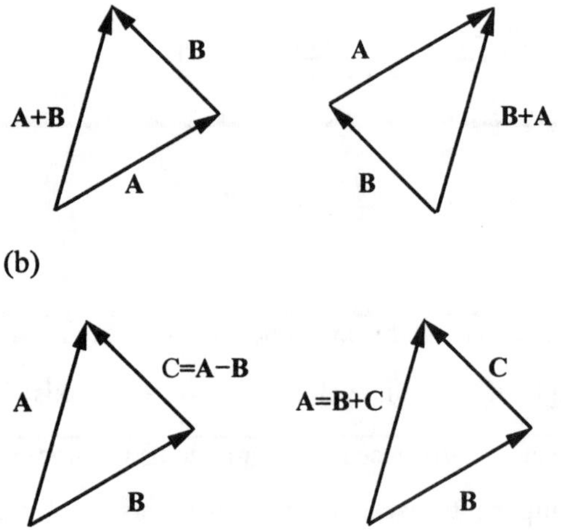

34. From $\mathbf{F}_1 + \mathbf{F}_2 + \mathbf{F}_3 = 0$, we have $\mathbf{F}_3 = -\mathbf{F}_1 - \mathbf{F}_2 = 3.0\text{ N }\mathbf{x} - 1.5\text{ N }\mathbf{y}$.

49. Use the following subscripts: b = boat, w = water, and g = ground.

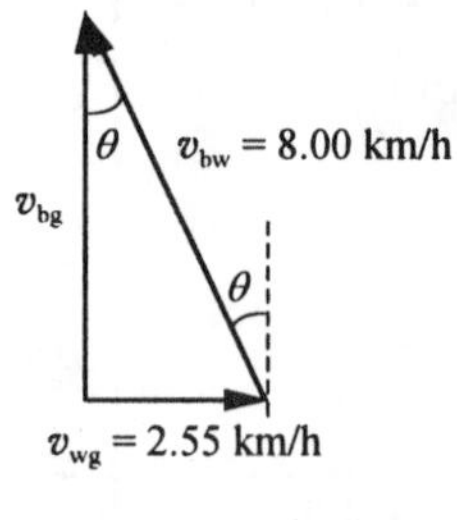

(a) $\theta = \sin^{-1}\left(\frac{2.55}{8.00}\right) = \boxed{18.6°\text{ upstream}}$.

(b) $v_{bg} = (8.00\text{ km/h})\cos\theta = 7.58\text{ km/h} = 2.11\text{ m/s}$,

so the time to cross is $t = \frac{500\text{ m}}{2.11\text{ m/s}} = 237\text{ s} = \boxed{3.96\text{ min}}$.

50. Use the following subscripts: b = boat, w = water, and g = ground

First use the distances to calculate some angles.

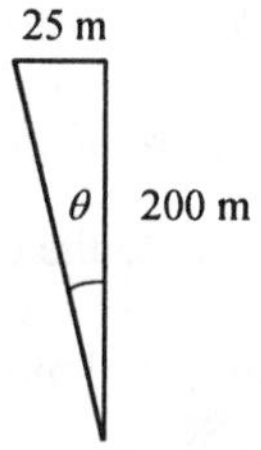

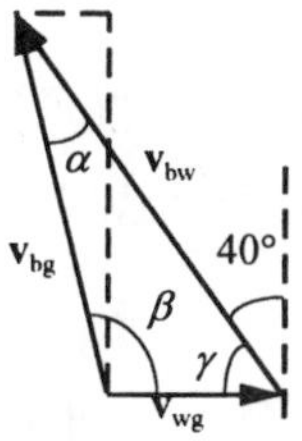

$\theta = \tan^{-1}\left(\dfrac{25}{200}\right) = 7.13°.$

So $\beta = 90° + 7.13° = 97.13°$, $\gamma = 50°$,

$\alpha = 180° - \beta - \gamma = 32.88°.$

Use sine law: $\dfrac{v_{bw}}{\sin 97.13°} = \dfrac{v_{wg}}{\sin 32.88°}$,

or $\dfrac{3.50 \text{ m/s}}{\sin 97.13°} = \dfrac{v_{wg}}{\sin 32.88°}$, so $v_{wg} = 1.91$ m/s.

56. The velocity of the passenger relative to the ground is 0.30 m/s for the first 25 m and (0.30 m/s + 0.50 m/s) = 0.80 m/s for the rest. So the total time is

$$\frac{25 \text{ m}}{0.30 \text{ m/s}} + \frac{50 \text{ m}}{0.80 \text{ m/s}} = 146 \text{ s} = 2.43 \text{ min}.$$

66. From $x = v_{xo}\, t$, we have $t = \dfrac{x}{v_{xo}} = \dfrac{18.4 \text{ m}}{23.5 \text{ m/s}} = 0.7830$ s.

$y = v_{yo}t - \frac{1}{2}gt^2 = 0 - (4.9 \text{ m/s}^2)(0.7830 \text{ s})^2 = -3.00$ m. So it drops by 3.00 m..

This is not realistic. 3.00 m is about 10 ft.

72. (a) At maximum height, $v_y = 0$. From $v_y^2 = v_{yo}^2 - 2gy$, we have $y_{max} = \dfrac{v_y^2}{2g} = \dfrac{v_1^2 \sin^2\theta}{2g}$.

$R = \dfrac{v_o^2 \sin 2\theta}{g} = 4\,\dfrac{v_o^2 \sin\theta \cos\theta}{2g}$. When $\theta = 45°$, $\sin\theta = \cos\theta$,

so $R = 4\, y_{max}$. Therefore the answer is yes.

(b) Since $y_{max} = R$, we have $\dfrac{v_o^2 \sin^2\theta}{2g} = 4\,\dfrac{v_o^2 \sin\theta \cos\theta}{2g}$.

So $\sin\theta = 4\cos\theta$, or $\tan\theta = 4$. Therefore $\theta = 76°$.

76. The range $R = 15$ m. So $R = \dfrac{v_o^2 \sin 2\theta}{g} = 15$ m.

or $R = \dfrac{(55 \text{ m/s})^2 \sin 2\theta}{9.80 \text{ m/s}^2} = 15$ m, so $\sin 2\theta = 0.0486$.

Therefore $2\theta = \sin^{-1}(0.0486) = 2.79°$, hence $\theta = 1.4°$.

83. (a) From $v_y^2 = v_{yo}^2 - 2gy$, we have $y = \dfrac{v_{yo}^2}{2g} = \dfrac{v_o^2 \sin^2\theta}{2g} = \dfrac{(8.0\text{ m/s})^2 \sin^2 30°}{2(9.80\text{ m/s}^2)} = 0.82\text{ m}.$

So the maximum height above the water is 4.0 m + 0.82 m = 4.8 m.

(b) First calculate the time of flight. $y = v_{yo}t - \frac{1}{2}gt^2$,

so $-4.0\text{ m} = (8.0\text{ m/s})\sin 30° t - (4.90\text{ m/s}^2)t^2$.

Reduce to a quadratic equation: $4.90\,t^2 - 4.0\,t - 4.0 = 0$.

Solve for $t = \dfrac{-b \pm \sqrt{b^2 - 4ac}}{2a} = \dfrac{4.0 \pm \sqrt{(-4.0)^2 - 4(4.90)(-4.0)}}{2(4.90)} = 1.40\text{ s or } -0.58\text{ s}.$

The −0.58 s means it would have taken 0.58 s for the diver to reach the height of the board had the diver started the "dive" from the water level. Therefore $R = x = v_{xo}\,t = (8.0\text{ m/s})(\cos 30°)(1.40\text{ s}) = 9.7\text{ m}.$

91. (a) $v_x = v_o \cos\theta = (25.0\text{ m/s})\cos 40° = 19.2\text{ m/s}$, $v_y = v_o \sin\theta = (25.0\text{ m/s})\sin 40° = 16.1\text{ m/s}.$

(b) $x = v_x\,t = (19.2\text{ m/s})(2.50\text{ s}) = 48.0\text{ m}$, $y = v_y\,t = (16.1\text{ m/s})(2.50\text{ s}) = 40.3\text{ m}.$

92. First find the time of flight.

From $y = v_{yo}t + \frac{1}{2}gt^2$, we have $-30\text{ m} = -(25\text{ m/s})\sin 40°\,t - (4.90\text{ m/s}^2)t^2$.

Reduce to a quadratic equation: $4.90t^2 + 16.1t - 30 = 0$.

Solve for $t = \dfrac{-b \pm \sqrt{b^2 - 4ac}}{2a} = \dfrac{-16.1 \pm \sqrt{(16.1)^2 - 4(4.90)(-30)}}{2(4.90)} = 1.33\text{ s or } -4.61\text{ s}.$

Discard the negative answer since it is physically unreasonable.

So $x = v_{xo}\,t = (25\text{ m/s})(\cos 40°)(1.33\text{ s}) = 25\text{ m}.$

102. $d_{\text{north}} = (150\text{ km/h})(2.00\text{ h}) = 300\text{ km}$, $d_{\text{west}} = (30\text{ km/h})(2.00\text{ h}) = 60\text{ km}.$

So the distance $d = \sqrt{(300\text{ km})^2 + (60\text{ km})^2} = 3.1 \times 10^2\text{ km}.$

VI. Practice Quiz

1. The resultant of two vectors is smallest when the angle between them is

(a) 0°. (b) 45°. (c) 60°. (d) 90°. (e) 180°.

2. If a ball is thrown with a velocity of 25 m/s at an angle of 37° above the horizontal, what is the horizontal component of the velocity?

(a) 25 m/s (b) 20 m/s (c) 18 m/s (d) 15 m/s (e) 10 m/s

3. An object is moving in the x–y plane. The acceleration in the x and y directions are 1.0 m/s^2 and 3.0 m/s^2, respectively. If the object starts from rest, what are the coordinates of the object when t = 4.0 s?

(a) (8.0 m, 24 m) (b) (24 m, 8.0 m) (c) (4.0 m, 12 m) (d) (12 m, 4.0 m) (e) (8.0 m, 12 m)

4. A boat, whose speed in still water is 8.0 m/s, is directed across a river with a current of 6.0 m/s. What is the speed of the boat as it crosses the river?

(a) 2.7 m/s (b) 5.3 m/s (c) 6.0 m/s (d) 8.0 m/s (e) 10 m/s

5. Find the resultant of the following vectors:

$\mathbf{v}_1$ = 2.0 m/s $\mathbf{x}$ + 3.0 m/s $\mathbf{y}$, $\mathbf{v}_2$ = −4.0 m/s $\mathbf{x}$ + 7.0 m/s $\mathbf{y}$, $\mathbf{v}_3$ = 10 m/s, 37° above the $-x$ axis.

(a) 19 m/s, 58° above $+x$ axis (b) 19 m/s, 58° above $-x$ axis (c) 19 m/s, 22° above $+x$ axis

(d) 19 m/s, 22° above $-x$ axis (e) none of the above

6. A track star in the broad jump goes into the jump with a speed of 12 m/s at an angle of 20° above the horizontal. How long is the jumper in the air before returning to Earth?

(a) 0.21 s (b) 0.42 s (c) 0.84 s (d) 1.25 s (e) 1.66 s

7. A stone is thrown horizontally with an initial speed of 10 m/s from the edge of a cliff. A stop watch measures the stone's trajectory time from top of cliff to bottom to be 4.3 s. What is the height of the cliff?

(a) 11 m (b) 22 m (c) 43 m (d) 77 m (e) 91 m

8. The resultant shown in the vector diagram is for

(a) **A** + **B**. (b) **A** − **B**. (c) **B** + **A**. (d) **B** − **A**. (e) **A** × **B**.

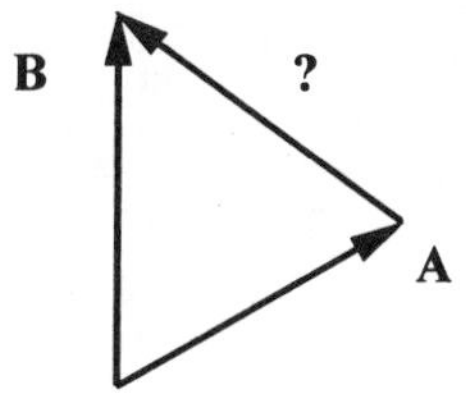

9. You are traveling at 55 mi/h relative to a straight, level road and pass a car traveling at 45 mi/h. The relative velocity of the other car to you is

(a) 10 mi/h. (b) −10 mi/h. (c) 65 mi/h. (d) 35 mi/h. (e) 100 mi/h.

10. If you walk 6.0 km in a straight line in a direction north of east and you end up 2.0 km east and several kilometers north. How many degrees north of east have you walked?

(a) 18° (b) 19° (c) 45° (d) 60° (e) 71°

Answers to Practice Quiz:

1. e 2. b 3. a 4. e 5. b 6. c 7. e 8. d 9. b 10. e

CHAPTER 4

Force and Motion

I. Chapter Objectives

Upon completion of this chapter, you should be able to:

1. relate force and motion and explain what is meant by a net or unbalanced force.
2. state and explain Newton's first law of motion, and describe inertia and its relationship to mass.
3. state and explain Newton's second law of motion and apply it to physical situations.
4. apply Newton's second law, including the component form, to various situations.
5. state and explain Newton's third law of motion and identify action–reaction pairs.
6. explain the causes of friction and how it is described using coefficients of friction.

II. Key Terms

Upon completion of this chapter, you should be able to define and/or explain the following key terms:

force
net (unbalanced) force
inertia
Newton's first law (law of inertia)
Newton's second law
newton (unit)
weight
free-body diagram
translational equilibrium
normal force
condition for translational equilibrium
Newton's third law
force of friction
static (kinetic) friction
kinetic friction
rolling friction
coefficient of static (kinetic) friction
air resistance
terminal velocity

The definitions and/or explanations of the most important key terms can be found in the following section: **III. Chapter Summary and Discussions.**

III. Chapter Summary and Discussion

1. Force and Net force

Force F is the cause of *acceleration*, and it is the technical term for what we commonly call a push, pull, kick, or shove. Force is a vector quantity (it has both magnitude and direction). There must be a net force acting on an object for the object to change its velocity (either magnitude or direction) or to accelerate.

Net Force ΣF is the vector sum, the resultant, or the unbalanced force acting on an object.

For example, in the diagram below two forces 10 N and 30 N each are acting on an object in opposite directions, as illustrated. There are two forces on the object. However, there is only one net force (the resultant of the sum of the two forces) of 20 N (to the right) acting on the object.

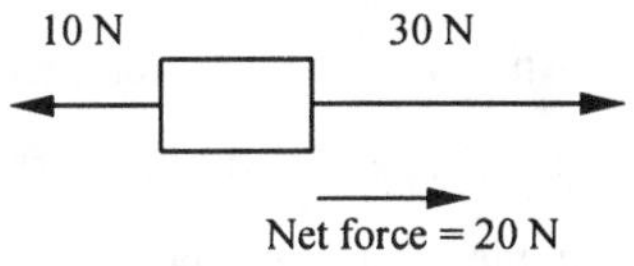

Note: The net force is *not* a separate force. It is simply the vector sum of the individual forces.

To analyze the forces acting on an object, you should draw a **free-body diagram**. This is done by

(1) Isolating the object of interest (this should be the only object in the diagram).

(2) Drawing all the forces acting on the object as vectors (including directions).

Example 4.1: Draw a free-body diagram of a book sitting on a desk.

Solution: First we isolate the book and then we analyze the forces acting on the book. There are two forces acting on the book, the gravitational force (or weight) and the supporting force on the book by the desk. The gravitational force (an action-at-a-distance force with no physical contact) is always present if we are dealing with objects on the earth. Whenever an object makes a physical contact with another object, a force results. Here the book makes a contact with the desk and so there is a supporting force. In most cases, these contact forces are perpendicular to the contact surface and therefore are called normal forces. Now try to draw a free-body diagram of an object sliding freely on a frictionless ramp.

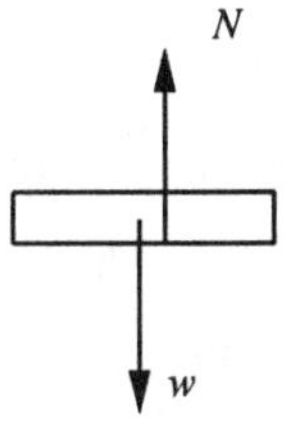

If there are *no forces* acting on an object, or if there are *equal and opposite forces* acting on it, the net force *is zero* and the object *will not* accelerate. This state of motion is called **translational equilibrium**.

Notes: The symbol Σ means "the sum of ". ΣF means the vector sum of forces. The unit of force in SI is a combination of the fundamental units of mass, length, and time and it is called **newton** (N). 1 N is the net force required to accelerate a 1–kg mass with an acceleration of 1 m/s^2 or, 1 N = 1 kg·m/s^2.

2. Newton's Laws of Motion

(1) **Newton's first law** describes the state of motion of an object when there is *no net force* (net force equals zero) acting on the object.

An object at rest will remain at rest and an object in motion will keep moving with constant velocity if the net force on the object is zero.

An object at rest has no change in velocity and so its acceleration is zero. An object moving with constant velocity (same speed and same direction) has no change in velocity either and its acceleration is also zero. Therefore, Newton's first law can be summarized in a very simple mathematical relation:

If $\Sigma \mathbf{F} = 0$, then $\mathbf{v}$ = constant (always zero is just a special case of velocity being constant), or $\mathbf{a} = 0$.

Note: Newton's first law is often called the law of inertia. **Inertia** is the property of an object that resists acceleration or change in motion and it is measured quantitatively by its mass.

A common misconception about Newton's first law is that a force is required to keep an object in motion. This is not so. Experiments done on air tracks (where there is negligible friction) show that no force is required to keep an object moving with constant velocity. We get this misconception because friction is always present in our everyday lives. To maintain constant velocity of a moving car, you have to push the accelerator. Does this contradict Newton's first law? No! It is seen clearly in the diagram below that the forward force on the car is equal in magnitude and opposite to the backward friction force (air friction plus ground friction) and so they cancel each other. Therefore, the net force is still zero and the car moves away with constant velocity.

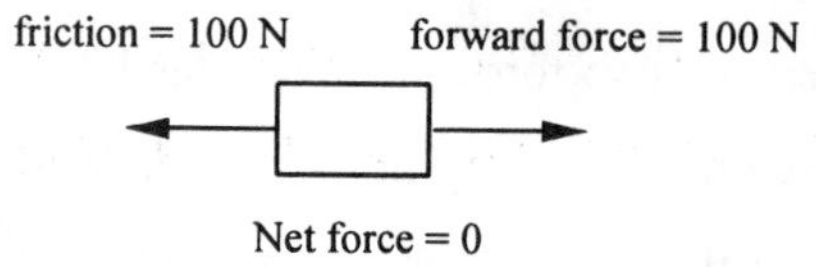

What happens if the *net force is not zero*? Newton's second law describes the relation among force, mass, and acceleration when there is a non-zero net force on an object.

(2) **Newton's second law** states that the acceleration depends on the net force ΣF and on the mass *m* of the object (note it is net force here, not just force). Mathematically, it is equivalent to

$$\mathbf{a} = \frac{\Sigma \mathbf{F}}{m} \quad \text{or} \quad \Sigma \mathbf{F} = m\,\mathbf{a}.$$

If you think of inertia as the qualitative term for the property of a body that resists acceleration, then **mass** (a scalar quantity) is the quantitative measure of inertia. If the mass is large, the acceleration produced by a given net force will be small.

Newton's second law is a vector equation. Note **a** is in the direction of Σ**F**, not **v**. You must include *all* the forces acting on an object to determine the net force, and then the acceleration. However, many times you will find that a force on an object is balanced by an equal and opposite force, such as weight balanced by a normal force. Since these pair of forces cancel out each other, you do not need to include them in your calculation.

Example 4.2: A 10-kg box sitting on a horizontal surface is pulled by a 5.0-N horizontal force. A 3.0-N friction force retards the motion. What is the acceleration of the object?

Solution: Given: $m = 10$ kg, F (pulling) = 5.0 N, f (friction) = 3.0 N.

Find: a.

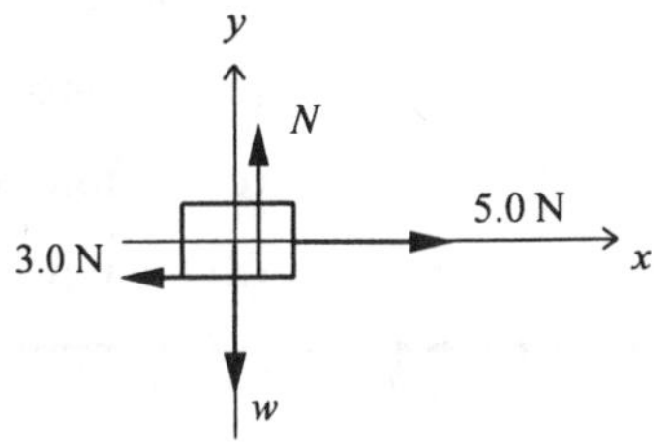

First we draw a free–body diagram of the box. The two vertical forces w, and N are equal and opposite (if they were not, the box would accelerate up or down). So they cancel out and we do not need to include them in our calculation, or $\Sigma F_y = m\, a_y = 0$. Thus we have only two forces, 5.0 N to the right and 3.0 N to the left, to deal with. The net force in this case is (using x direction)

$\Sigma \mathbf{F} = F - f = 5.0$ N $(+x) - 3.0$ N $(-x) = 2.0$ N (to the right or $+x$).

Hence the acceleration of the object is $\mathbf{a} = \dfrac{\Sigma \mathbf{F}}{m} = \dfrac{2.0 \text{ N}}{10 \text{ kg}} = 0.20 \text{ m/s}^2$ (to the right or $+x$).

If you apply Newton's second law to gravity near the Earth's surface, you get the relation among weight $\mathbf{w}$ (gravitational force), mass m, and gravitational acceleration $\mathbf{g}$: $\mathbf{w} = m\,\mathbf{g}$.

Example 4.3: Find the weight of a 3.50-kg object.

Solution: Given: $m = 3.50$ kg. Find: w.

We use the relation: $\mathbf{w} = m\,\mathbf{g} = (3.50 \text{ kg})(9.80 \text{ m/s}^2 \text{ downward}) = 34.3$ N downward.

---

(3) **Newton's Third Law** states that if object 1 exerts a force on object 2, then object 2 exerts an equal and opposite force on object 1. In mathematical terms, it can be written as $\mathbf{F}_{12} = -\,\mathbf{F}_{21}$.

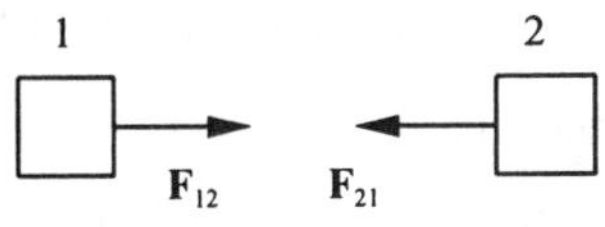

The force by object 1 is *on* object 2, and the force by object 2 is *on* object 1, and so the two forces are always acting on *two different* objects. Therefore, even though these two forces are equal and opposite, they *do not* cancel out each other because of this fact. If you analyze only object 1 in the diagram, there is only *one* force acting on it and object 1 will accelerate.

There is another misconception concerning the third law. The third law states that the two forces are *equal* no matter what. For example, an egg and a stone collide with each other, the egg breaks and the stone is in tact. Since the egg breaks, we often conclude that the force by the stone on the egg is greater than the force by the egg on the stone. This is not so. The forces are always equal. The egg breaks because it is simply easier to break. It takes a smaller force to break the egg than to break the stone.

Example 4.4: A brick hits a window and breaks the glass. Since the brick breaks the glass,

(a) the force on the brick is greater than the force on the glass,

(b) the force on the brick is equal to the force on the glass,

(c) the force on the brick is smaller than the force on the glass,

(d) the force on the brick is in the same direction as the force on the glass.

Solution: According to Newton's third law, the answer is (b). Why isn't answer (c) correct since the brick breaks the glass? It takes a considerably smaller force to break a glass than to break a brick. When the brick hits the glass, they exert equal but opposite forces on each other, say 150 N. It may take only 100 N to break the glass and 1000 N to break the brick. So the glass breaks and the brick remains basically undamaged even though the forces by the brick and by the glass are equal.

---

3. Force of Friction

Consider one object in contact with another, such as a book on a desk. Suppose that the desk is stationary and that the book is either *sliding* or *on the verge of sliding* on the desk. There is interaction between the book and the desk based on intermolecular attractions and repulsions. It is convenient, however, to consider the interaction as if it were two independent ones: one perpendicular to the contact surface (the *normal* force mentioned earlier) and another one parallel to the contact surface, which is called *force of friction* or *friction force.*

There are two kinds of friction forces: static and kinetic. Static friction force ($f_s \le \mu_s N$) is parallel to the contact surface when there is *no relative motion* between the contact surfaces and kinetic friction force ($f_k = \mu_k N$) is also parallel to the contact surface when there is *relative motion* between the contact surfaces. Here $\mu_{s\,(k)}$ is called the coefficient of static (kinetic) friction, which is a measure of the strength of the molecular interaction, and N is the normal force.

Kinetic friction force rarely causes much conceptual difficulty, but static friction force is often misunderstood and therefore incorrectly used. If a book is at rest on a desk and no forces with horizontal components are applied to the book, there is no static friction force. If there were, an unbalanced force would exist and the book would move by itself. Now suppose a small horizontal force is applied to the book but the book still remains at rest. Then we must conclude that an equal and opposite force acts on the book to prevent it from moving. This is the static friction force. If the horizontally applied force increases, the static friction force will increase by the same amount until its maximum value is reached ($f_s = \mu_s N$). So the static friction force is *not* a fixed value, but is always equal to the applied force, unless the maximum static friction force is exceeded.

Another misconception about static friction force is about its direction. Somehow we have the idea that friction force is always opposite to the direction of motion. This is true only if the friction force is kinetic. Static friction force can be in directions other than opposite to motion. Think about a machine part moving on a conveying belt in a factory and a car turning on a flat surface. What is moving the part and turning the car? It is static friction force. The static friction force is in the direction of motion on the part and perpendicular to the direction of motion on the car. Think carefully when you walk the next time. What enables you to walk? In which direction is that force? (Hint: why is it difficult to walk on ice? Is it possible to walk on perfectly smooth surface?)

Example 4.5: The coefficients of static and kinetic frictions between a 3.0-kg box and a desk are 0.40, and 0.30, respectively. What is the net force on the box when each of the following horizontal forces is applied to the box? (a) 5.0 N, (b) 10 N, (c) 15 N.

Solution: Given: $\mu_s = 0.40$, $\mu_k = 0.30$, $m = 3.0$ kg. $F = 5.0$ N (a), 10 N (b), 15 N (c).

Find: ΣF (net force) in (a), (b), and (c).

In this example, we know that if the applied force is greater than the maximum static friction force, the box will accelerate. Since both friction static and kinetic forces depend on the normal forces. We must find it first. Applying Newton's second law in the vertical direction, $\Sigma F_y = N - w = m\, a_y = 0$ so the normal force between the box and the desk is $N = w = mg = (3.0 \text{ kg})(9.80 \text{ m/s}^2) = 29.4$ N.

The maximum static friction force is then $f_{s\,max} = \mu_s N = (0.40)(29.4 \text{ N}) = 11.8$ N.

(a) Since the applied force of 5.0 N is smaller than 11.8 N, the maximum static friction force, the object remains at rest and therefore the net force is zero.

(b) The applied force is still smaller than the maximum static friction force and the net force is still zero.

(c) Now the applied force of 15 N is greater than the maximum static friction force, the object will accelerate and there is a relative motion. Therefore we need to replace the static friction force with the kinetic friction force.

$$f_k = \mu_k N = (0.30)(29.4 \text{ N}) = 8.82 \text{ N}.$$

Hence the net force is $\Sigma F = 15 \text{ N} - 8.82 \text{ N} = 6.18 \text{ N}$ and the box will accelerate at a rate of

$$a = \frac{\Sigma F}{m} = \frac{6.18 \text{ N}}{3.0 \text{ kg}} = 2.1 \text{ m/s}^2.$$

---

4. Application of Newton's Laws

When it comes to applying Newton's laws to various mechanical systems, there is no short cut to take. You *must* follow a certain set of procedures to analyze and solve the unknown(s). Here is a simplified version of the procedures outlined in the text:

(1) Draw a free-body diagram for each object involved in the analysis.

(2) Select a rectangular coordinate system. The solutions will be much easier if you select the $+x$ axis in the direction of acceleration and the $+y$ axis perpendicular to the x axis.

(3) Find the x and y components of all forces if they are not pointing in the x or y directions.

(4) Add all the x components and y components of the forces, respectively.

(5) Set $\Sigma F_x = m\,a$ and $\Sigma F_y = 0$ and solve for the unknown(s).
Since you have chosen the $+x$ axis in the direction of acceleration, the object will not accelerate in the y direction. So its acceleration in the y direction is zero and $\Sigma F_y = m\,a_y = 0$. If the system is in equilibrium, set $\Sigma F_x = 0$ also.

Example 4.6: A student pulls a box of books on a smooth horizontal floor with a force of 100 N in a direction of 37° above the horizontal surface. If the mass of the box and the books is 40.0 kg, what is the acceleration of the box and the normal force on the box by the floor?

Solution: Given: $F = 100$ N, $\theta = 37°$, $m = 40.0$ kg.

Find: a and N.

(1) Free-body diagram:

Since the box is on the Earth, it has weight $w = mg$ pointing toward the center of the earth (or perpendicular to the horizontal direction). The box is also making physical contact with the floor and the student. So there will be two contact forces. The contact force with the floor is a normal force and is directed straight upward perpendicular to the contact surface. The force by the student is directed at 37° above the horizontal direction (smooth floor implies negligible friction).

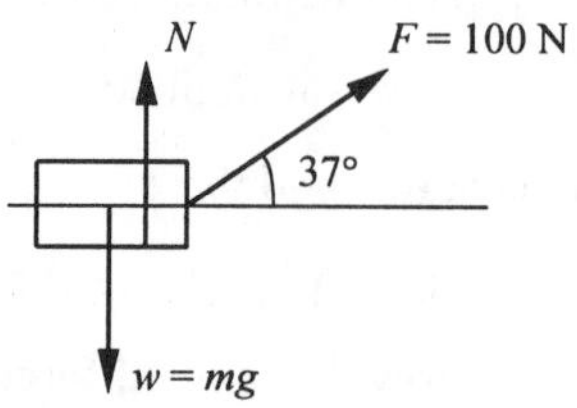

(2) Coordinate system:

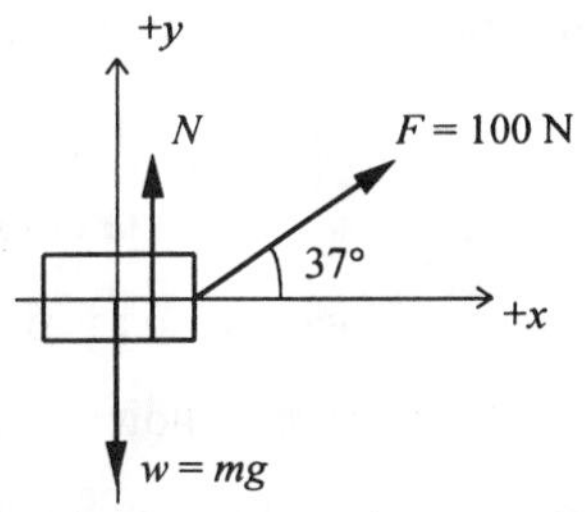

Even though the student is pulling at an angle above the horizontal direction, the box will move or accelerate along the horizontal direction. So we choose the horizontal direction (to the right) as the $+x$ axis and the vertical direction (upward) as the $+y$ axis.

(3) x and y components:

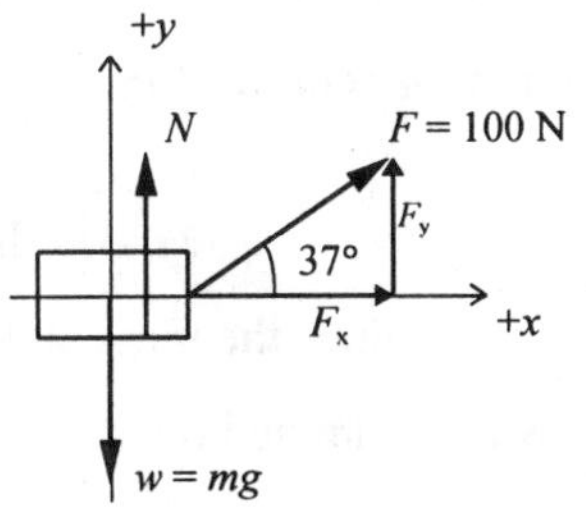

Among the three forces, only the student's pulling force is not completely along either the x or the y axis. So we need to find its x and y components.

From trigonometry, we can see that the x component is adjacent to the 37° angle and the y component is opposite to the 37° angle. Therefore we use cos 37° to calculate the x component and sin 37° to calculate the y component.

$F_x = F\cos 37° = (100\text{ N})\cos 37° = 80.0\text{ N},$

$F_y = F\sin 37° = (100\text{ N})\sin 37° = 60.0\text{ N}.$

(4) Adding x and y components, respectively:

In the x direction, there is only one force, the x component of the student's pulling force, 80 N. In the y direction, there are three forces: the upward normal force N, the upward y component of the student's pulling force (60 N), and the downward weight of the box and books of $w = mg = (40.0\text{ kg})(9.80\text{ m/s}^2) = 392\text{ N}$.

So $\quad \Sigma F_x = 80.0\text{ N}$ and $\Sigma F_y = N + 60.0\text{ N} - w = N + 60.0\text{ N} - 392\text{ N} = N - 332\text{ N}.$

(5) Setting $\Sigma F_x = m\,a$ and $\Sigma F_y = 0$ and solving for the unknown(s):

$\Sigma F_x = 60.0\text{ N} = m\,a = (40.0\text{ kg})a.$ Eq. (1) and

$\Sigma F_y = N - 332\text{ N} = 0.$ Eq. (2)

From Eq. (1), we have $a = \dfrac{80.0\text{ N}}{40.0\text{ kg}} = 2.00\text{ m/s}^2.$

From Eq. (2), we have $N = 332$ N.

Here the normal force (332 N) is not equal to the weight of the box and the books (392 N). Why?

Example 4.7: An object with a mass m slides down a rough 37° inclined plane where the coefficient of kinetic friction is 0.20.

(a) Find the acceleration of the mass.

(b) If the plane is 10 m long and the mass starts from rest, what will be its speed at the bottom of the plane?

Solution: Given: $\theta = 37°$, $\mu_k = 0.20$, $v_o = 0$, $x = 10$ m.

Find: (a) a (b) v.

(a) **(1)** Free-body diagram:

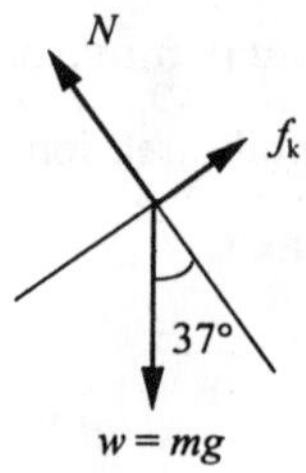

Since the box is on the Earth, it has weight $w = mg$ pointing toward the center of the Earth (or perpendicular to the horizontal direction). The box is also making physical contact with the inclined plane. So there will be a contact forces (normal force) perpendicular to the inclined plane. Furthermore, since the inclined plane is rough, there is a kinetic friction force parallel to the inclined plane in the direction opposite to the direction of motion or acceleration.

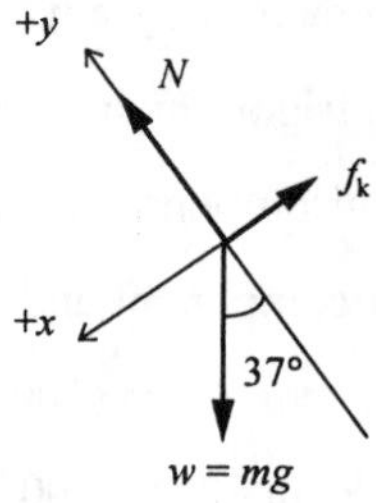

(2) Coordinate system:

Since the mass slides down the inclined plane, we choose the $+x$ axis down the inclined plane and the $+y$ axis perpendicular to the inclined plane.

(3) x and y components:

The weight of the mass $w = mg$ is not along either the x axis or the y axis. So we need to find its x and y components. From trigonometry, we can see that the x component is adjacent to the 37° angle and the y component is opposite to the 37° angle.

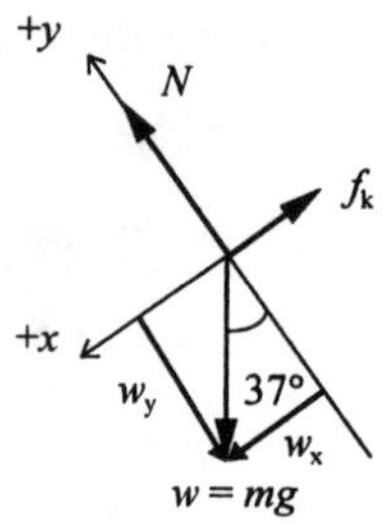

So $w_x = w \sin 37° = mg \sin 37°$ and $w_y = w \cos 37° = mg \cos 37°$.

(4) Adding x and y components, respectively:

In x direction, there are two forces, w_x and f_k.

So $\Sigma F_x = mg \sin 37° - f_k = mg \sin 37° - \mu_k N$.

In y direction, there are also two forces, N and w_y.

So $\Sigma F_y = N - mg \cos 37°$.

(5) Setting $\Sigma F_x = m\,a$ and $\Sigma F_y = 0$ and solving for the unknown(s):

$\Sigma F_x = mg \sin 37° - f_k = m\,a$. Eq. (1) $\Sigma F_y = N - mg \cos 37°$. Eq. (2)

From Eq. (2), we have $N = mg \cos 37°$.

Substituting this into Eq. (1), we have $mg \sin 37° - \mu_k (mg \cos 37°) = m\,a$.

Solve for $a = g (\sin 37° - \mu_k \cos 37°) = (9.80 \text{ m/s}^2)[\sin 37° - (0.20) \cos 37°] = 4.3 \text{ m/s}^2$.

(b) Since velocity is a kinematic quantity, we need to use one of the kinematic equations.

We use $v^2 = v_o^2 + 2\,a\,x = (0)^2 + 2(4.3 \text{ m/s}^2)(10 \text{ m}) = 86 \text{ m}^2/\text{s}^2$, so $v = 9.3$ m/s.

IV. Mathematical Summary

Newton's first law of motion	If $\Sigma\mathbf{F} = 0$, then $\mathbf{v}$ = constant or $\mathbf{a} = 0$ (4.4)	Describes the state of motion of an object when the net force on it is zero
Newton's second law of motion	$\Sigma\mathbf{F} = \mathrm{m}\,\mathbf{a}$ (4.1) (Note: it is a vector equation)	Relates acceleration with mass and net force when the net force on it is not zero
Newton's third law of motion	$\mathbf{F}_{12} = -\,\mathbf{F}_{21}$	States that if object 1 exerts a force on object 2, then object 2 exerts an equal and opposite force on object 1
Weight and mass	$\mathbf{w} = m\,\mathbf{g}$ (4.2)	Relates weight (gravitational force) with mass and gravitational acceleration
Force of static friction	$f_s \le \mu_s N$ (4.6) $f_{smax} = \mu_s N$ (4.7) (Note: it is not a fixed value)	Relates the static friction force with the coefficient of static friction and normal force when there is no relative motion between the two surfaces
Force of kinetic friction	$f_k = \mu_k N$ (4.8) (Note: it is a fixed value)	Relates the kinetic friction force with the coefficient of kinetic friction and normal force when there is relative motion between the two surfaces

V. Solutions of Selected Exercises and Paired Exercises

10. $\mathbf{F}_1 = (5.0\text{ N})\cos 30°\,\mathbf{x} + (5.0\text{ N})\sin 30°\,\mathbf{y} = 4.33\text{ N}\,\mathbf{x} + 2.50\text{ N}\,\mathbf{y}$,

$\mathbf{F}_2 = -(3.0\text{ N})\cos 37°\,\mathbf{x} - (3.0\text{ N})\sin 37°\,\mathbf{y} = -2.40\text{ N}\,\mathbf{x} - 1.81\text{ N}\,\mathbf{y}$.

$\Sigma\mathbf{F} = \mathbf{F}_1 + \mathbf{F}_2 + \mathbf{F}_3 = 0$,

so $\mathbf{F}_3 = -(\mathbf{F}_1 + \mathbf{F}_2) = -1.9\text{ N}\,\mathbf{x} - 0.70\text{ N}\,\mathbf{y}$.

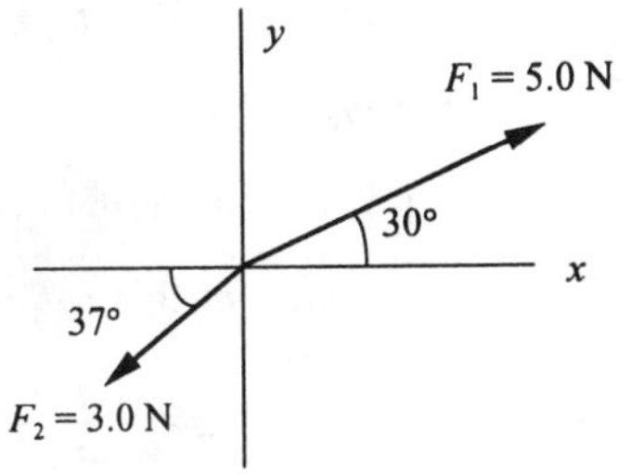

22. (a) The total force = 18(415 N) = 7470 N = weight,

From $w = mg$, we have $m = \dfrac{w}{g} = \dfrac{7470\text{ N}}{9.80\text{ m/s}^2} = 762\text{ kg}$.

(b) $w = (7470\text{ N}) \times \dfrac{1\text{ lb}}{4.45\text{ N}} = 1.68 \times 10^3\text{ lb}$.

24. (a) From $w = mg$, we have $m = \dfrac{w}{g} = \dfrac{98.0\text{ N}}{(9.80\text{ m/s}^2)/6} = 60.0\text{ kg}$.

From Newton's second law $a = \dfrac{F_{net}}{m} = \dfrac{12.0\text{ N}}{60.0\text{ kg}} = 0.200\text{ m/s}^2$.

(b) $m = 60.0$ kg is the same. So the acceleration is the same.

30. First find acceleration from kinematics.

Given: $v_o = 90$ km/h = 25 m/s, $v = 0$, $t = 4.5$ s. Find: a.

From $v = v_o + a\,t$, we have $a = \dfrac{v - v_o}{t} = \dfrac{0 - 25\text{ m/s}}{4.5\text{ s}} = -5.56\text{ m/s}^2$.

From Newton's second law $F_{\text{net}} = ma = (60\text{ kg})(-5.56\text{ m/s}^2) = -3.3\times10^2$ N.

The − sign indicates that the force is opposite to velocity.

35. (a) There are two strings. So the tension on each string is

$T = \dfrac{w}{2} = \dfrac{mg}{2} = \dfrac{(75\text{ kg})(9.80\text{ m/s}^2)}{2} = 3.7\times10^2$ N.

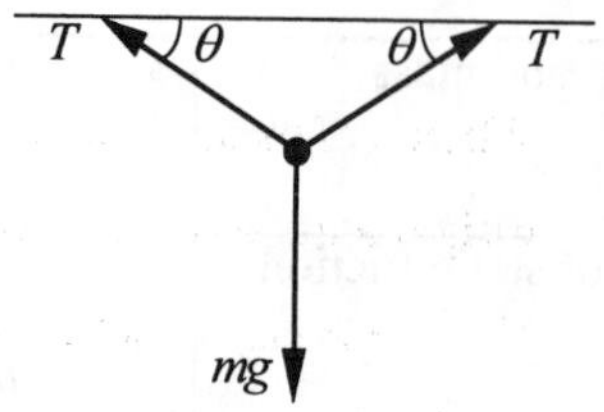

(b) In the vertical direction: $\Sigma F_y = T\sin\theta + T\sin\theta - mg = 0$.

Solve for $T = \dfrac{mg}{2\sin\theta} = \dfrac{(75\text{ kg})(9.80\text{ m/s}^2)}{2\sin 45°} = 5.2\times10^2$ N.

(c) Since $T \propto \dfrac{1}{\sin\theta}$, T becomes large as θ becomes small.

36. From Example 4.6, $a = \dfrac{(m_2 - m_1)g}{m_1 + m_2} = \dfrac{(0.20\text{ kg} - 0.15\text{ kg})(9.80\text{ m/s}^2)}{0.20\text{ kg} + 0.15\text{ kg}} = 1.4\text{ m/s}^2$, up,

46. The free-body diagrams of the three objects.

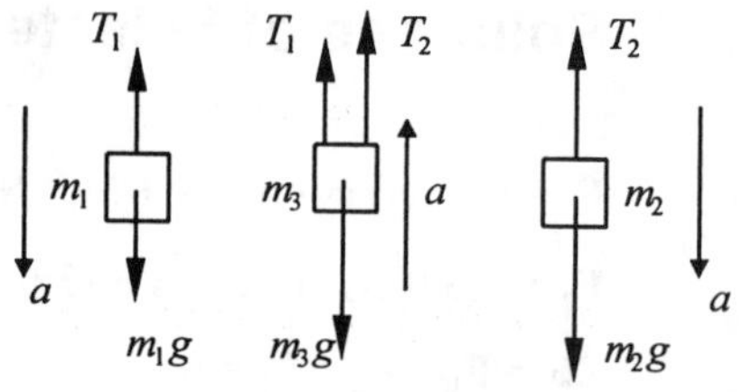

For m_1: $m_1 g - T_1 = m_1 a$, Eq. (1)

For m_3: $T_2 + T_1 - m_3 g = m_3 a$, Eq. (2)

For m_2: $m_2 g - T_2 = m_2 a$, Eq. (3)

Eq. (1) + Eq. (2) + Eq. (3) gives

$(m_2 + m_1 - m_3)g = (m_1 + m_2 + m_3)a$.

Solve for $a = \dfrac{(m_1 + m_2 - m_3)g}{m_1 + m_2 + m_3}$.

(a) $a = \dfrac{(3.0\text{ kg} + 3.0\text{ kg} - 4.0\text{ kg})(9.80\text{ m/s}^2)}{3.0\text{ kg} + 3.0\text{ kg} + 4.0\text{ kg}} = 2.0\text{ m/s}^2$.

(b) Since $m_1 = m_2$, $T_1 = T_2$. From Eq. (3) $T_2 = T_1 = m_2(g - a) = (3.0\text{ kg})(9.80\text{ m/s}^2 - 1.96\text{ m/s}^2) = 24$ N.

48. The scale reading is equal in magnitude to the normal force on the person by the scale.

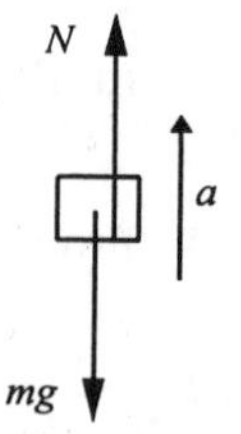

In the vertical direction: $F_{\text{net}} = N - mg = ma$, so $N = m(g + a)$.

(a) $a = 0$, $N = mg = (70.0\text{ kg})(9.80\text{ m/s}^2) = 686$ N.

(b) $a = 0.525\text{ m/s}^2$, $N = (70.0\text{ kg})(9.80\text{ m/s}^2 + 0.525\text{ m/s}^2) = 723$ N.

(c) $a = -0.525\text{ m/s}^2$, $N = (70.0\text{ kg})(9.80\text{ m/s}^2 - 0.525\text{ m/s}^2) = 649$ N.

58. (a) For A: $T = ma = (50\text{ kg})(0.92\text{ m/s}^2) = 46\text{ N}$.

From Newton's third law, the force on B is also 46 N, so $a_B = \dfrac{46\text{ N}}{60\text{ kg}} = 0.77\text{ m/s}^2$.

(b) From kinematic equations, we have $x_B = \frac{1}{2} a_B t^2$, $x_A = \frac{1}{2} a_A t^2$. So $\dfrac{x_A}{x_B} = \dfrac{a_A}{a_B} = \dfrac{0.92}{0.767} = 1.2$.

Also $x_B + x_A = 10\text{ m}$, or $x_B + 1.2\, x_B = 10\text{ m}$,

solve for $x_B = 4.5\text{ m}$ from B's original position.

Since mass can never be negative, this means the friction is too large for constant velocity.

63. When it is pushed, the normal force equals the sum of the weight and the vertical component of the pushing force, i.e., $N = w + F \sin\theta$. When it is pulled, $N = w - F\sin\theta$. Since friction force is directly proportional to normal force it is easier to pull than to push.

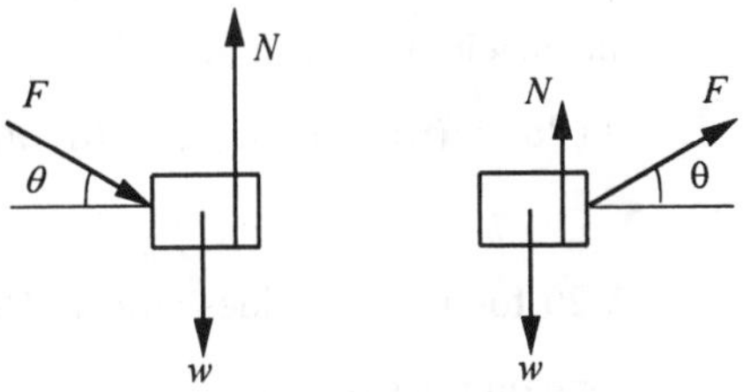

76. First find the acceleration from kinematics.

Given: $v_o = 80\text{ km/h} = 22.2\text{ m/s}$, $v = 0$, $x = 22\text{ m}$. Find: a.

From $v^2 = v_o^2 + 2ax$, we have $a = \dfrac{v^2 - v_o^2}{2x} = \dfrac{0 - (22.2\text{ m/s})^2}{2(22\text{ m})} = -11.2\text{ m/s}^2$.

$F_x = -f_s = -\mu_s mg = ma$, so $\mu_s = -\dfrac{a}{g} = -\dfrac{-11.2\text{ m/s}^2}{9.80\text{ m/s}^2} = 1.1$.

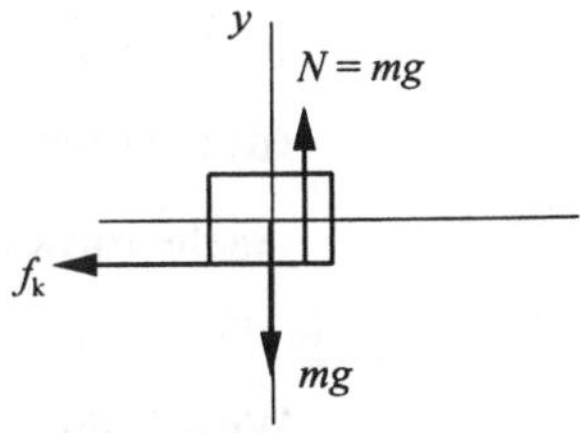

82. First find the acceleration from dynamics.

$F_x = mg\sin\theta = ma$, so $a = g\sin\theta = (9.80\text{ m/s}^2)\sin 20° = 3.35\text{ m/s}^2$.

Now find the acceleration from kinematics.

Given: $v_o = 0$, $v = 24\text{ m/s}$, $x = 135\text{ m}$. Find: a.

From $v^2 = v_o^2 + 2ax$,

we have $a = \dfrac{v^2 - v_o^2}{2x} = \dfrac{(24\text{ m/s})^2 - 0}{2(135\text{ m})} = 2.13\text{ m/s}^2 < 3.35\text{ m/s}^2$.

Therefore the answer is no.

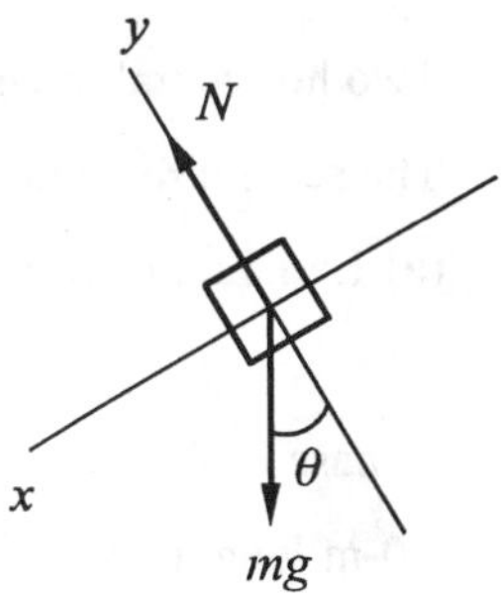

94. $F_y = N - mg\cos\theta = 0$, so $N = mg\cos\theta$.

$F_x = F - mg\sin\theta - f_k = 0$,

or $F - mg\sin\theta - \mu_s N = F - mg\sin\theta - \mu_k(mg\cos\theta) = 0$.

Solve for $F = mg(\sin\theta + \mu_k\cos\theta)$

$= (9.80\times10^3\text{ N})[\sin 37° + (0.750)\cos 37°] = 1.18\times10^4\text{ N}$.

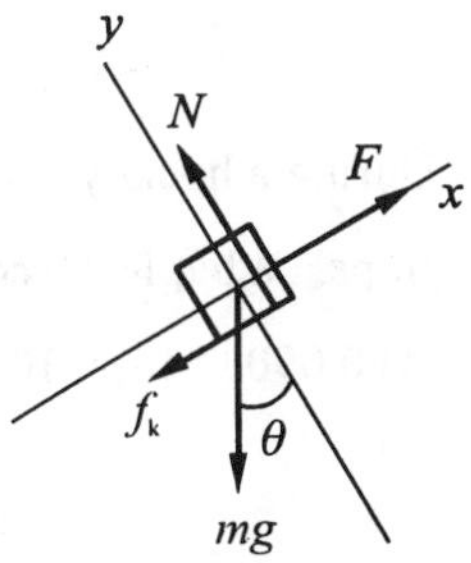

VI. Practice Quiz

1. Which of Newton's laws of motion *best* explains why motorists should buckle-up?

   (a) the first law (b) the second law (c) the third law (d) the law of gravitation

2. A net force F accelerates a mass m with an acceleration a. If the same net force is applied to mass $2m$, then the acceleration will be

   (a) $4a$. (b) $2a$. (c) a. (d) $a/2$. (e) $a/4$.

3. A 20-ton truck collides with a 1500-lb car and causes a lot of damage to the car. Since a lot of damage is done on the car

   (a) the force on the truck is greater than the force on the car.

   (b) the force on the truck is equal to the force on the car.

   (c) the force on the truck is smaller than the force on the car.

   (d) the truck did not slow down during the collision.

   (e) the truck did not gave any damage.

4. What is the mass of an object that weighs 100 N on the surface of Earth?

   (a) 100 kg (b) 9.8 kg (c) 10.2 kg (d) 9,800 kg (e) none of the above

5. Two horizontal forces act on a 5.0-kg mass. One force has a magnitude of 8.0 N and is directed due north. The second force toward the east has a magnitude of 6.0 N. What is the acceleration of the mass?

   (a) 1.6 m/s^2 due north (b) 1.2 m/s^2 due east (c) 2.0 m/s^2 at 53° N of E (d) 2.0 m/s^2 at 53° E of N

6. A mass is placed on a smooth inclined plane with an angle of 37° to the horizontal. If the inclined plane is 5.0-m long, how long does it take for the mass to reach the bottom of the inclined plane after it is released from rest?

   (a) 1.3 s (b) 1.2 s (c) 1.1 s (d) 1.0 s (e) 0.90 s

7. During a hockey game, a hockey puck is given an initial speed of 10 m/s. It slides 50 m on the ice before it stops. What is the coefficient of kinetic friction between the puck and ice?

   (a) 0.090 (b) 0.10 (c) 0.11 (d) 0.12 (e) 0.13

8. A person on a scale rides in an elevator. If the mass of the person is 60 kg and the elevator accelerate downward with an acceleration of 4.9 m/s^2, what is the reading on the scale in newtons?

(a) 147 N (b) 294 N (c) 588 N (d) 882 N (e) 1176 N

9. A traffic light of weight 100 N is supported by two ropes as shown in the diagram. What are the tensions in the ropes?

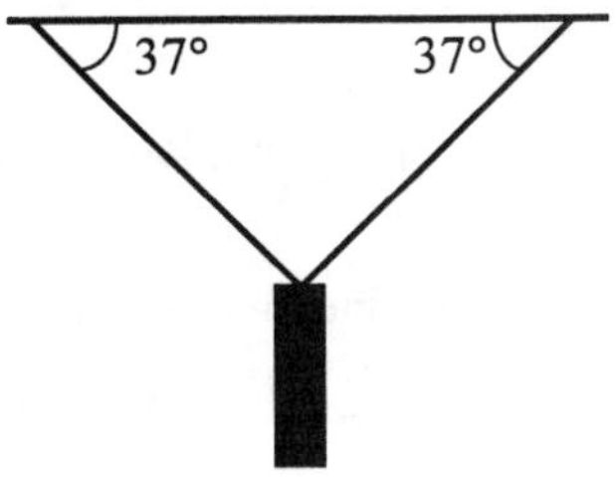

(a) 50 N

(b) 63 N

(c) 66 N

(d) 83 N

(e) 100 N

10. Find the magnitudes of the acceleration of the system and the tension in the connecting string. (Neglect friction and mass of the pulley)

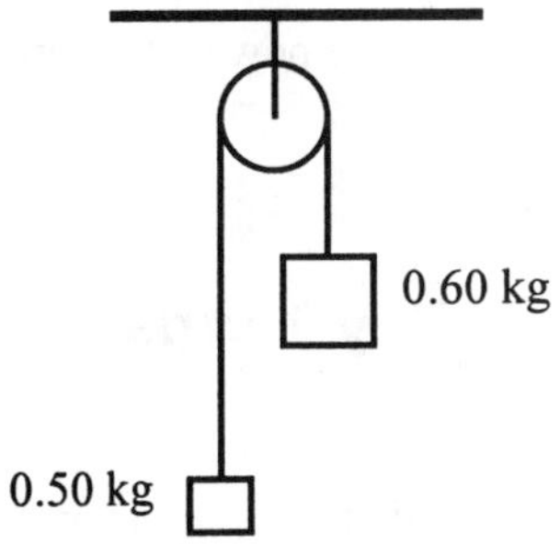

(a) 5.3 m/s^2, 7.5 N

(b) 4.5 m/s^2, 7.1 N

(c) 0.89 m/s^2, 5.3 N

(d) 0.090 m/s^2, 4.9 N

(e) 1.2 m/s^2, 4.3 N.

Answers to Practice Quiz:

1. a 2. d 3. b 4. c 5. c 6. a 7. b 8. b 9. d 10. c

CHAPTER 5

Work and Energy

I. Chapter Objectives

Upon completion of this chapter, you should be able to:

1. define mechanical work and compute the work done in various situations.
2. differentiate work done by constant and variable force and compute the work done by a spring force.
3. show how the work–energy theorem is used in defining kinetic energy.
4. explain how potential energy depends on position and compute values of gravitational potential energy.
5. distinguish between conservative and nonconservative forces, and explain their effects on the conservation of energy.
6. define power and describe mechanical efficiency.

II. Key Terms

Upon completion of this chapter, you should be able to define and/or explain the following key terms:

work
joule (unit)
foot-pound (unit)
spring (force) constant
kinetic energy
work-energy theorem
potential energy
gravitational potential energy
law of conservation of total energy
conservative force
nonconservative force
total mechanical energy
conservative system
law of conservation of mechanical energy
power
watt (unit)
horsepower (unit)
efficiency

The definitions and/or explanations of the most important key terms can be found in the following section: **III. Chapter Summary and Discussion.**

III. Chapter Summary and Discussion

1. Work Done by a Constant Force

The **work** done by a constant force is defined as $W = F\,d\cos\theta$, where F and d are the magnitudes of the force vector and displacement vector, respectively, and θ is the angle between these vectors. Although force and displacement are vector quantities, work is a scalar quantity. The unit of work is N·m, which is called joule (J).

Note: The angle in the work definition is the *angle between the force vector and the displacement vector*, which is not always the angle from the horizontal.

Graphically, the work done by a force is equal to the area under the curve in a force versus position graph.

Example 5.1 A 500-kg elevator is pulled upward with a constant force of 5500 N for a distance of 50.0 m.

(a) Find the work done by the upward force.

(b) Find the work done by the gravitational force.

(c) Find the work done by the net force (the net work done on the elevator).

Solution: Given: $F_{up} = 5500$ N, $w = mg = (500\text{ kg})(9.80\text{ m/s}^2) = 4900$ N, $d = 50$ m.

Find: (a) W_{up}, (b) W_{grav}, and (c) W_{net}.

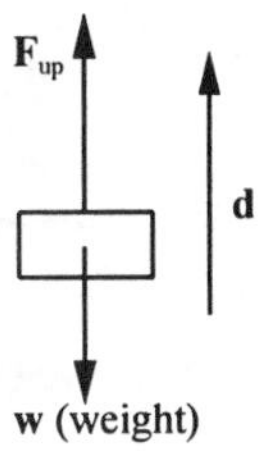

(a) The displacement is upward and the upward force is (of course) upward. So the angle between them is zero.

So $W_{up} = Fd\cos\theta = F_{up}\,d\cos 0° = (5500\text{ N})(50.0\text{ m})\cos 0° = 2.75\times10^5$ J.

(b) The angle between displacement and the gravitational force (weight vector) is 180°.

So $W_{grav} = Fd\cos\theta = wd\cos 180° = (4900\text{ N})(50.0\text{ m})\cos 180° = -2.45\times10^5$ J.

(c) The work done by the net force is equal to the net work done on the elevator.

$W_{net} = W_{up} + W_{grav} = 2.75\times10^5\text{ J} + (-2.45\times10^5\text{ J}) = 3.0\times10^4$ J.

Note: W_{net} is also equal to $W_{net} = F_{net}\,d\cos\theta$, where $F_{net} = F - w$.

Example 5.2 A force moves an object in the direction of the force. The graph on the right shows the force versus the object's position. Find the net work done when the object moves

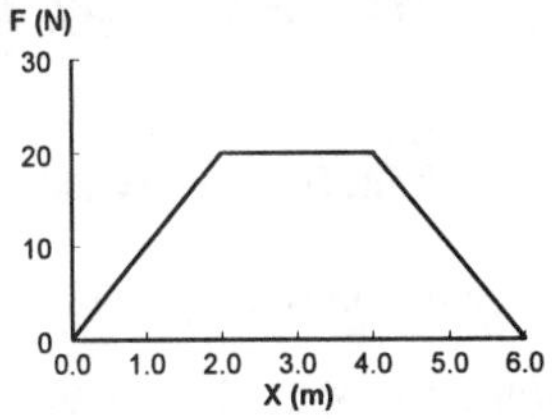

(a) from 0 to 2.0 m.

(b) from 2.0 to 4.0 m.

(c) from 4.0 to 6.0 m.

(d) from 0 to 6.0 m.

Solution: Work done is equal to the area under the curve.

(a) The area under the curve from 0 to 2.0 m is the left triangle. The area of a triangle is $\frac{1}{2}$ (base × height).

So $W_{0\text{-}2} = \frac{1}{2}(2.0\text{ m} - 0)(20\text{ N}) = 20\text{ J}.$

(b) The area under the curve from 2.0 m to 4.0 m is the rectangle.

So $W_{2\text{-}4} = (4.0\text{ m} - 2.0\text{ m})(20\text{ N}) = 40\text{ J}.$

(c) The area under the curve from 4.0 m to 6.0 m is the right triangle.

So $W_{4\text{-}6} = \frac{1}{2}(6.0\text{ m} - 4.0\text{ m})(20\text{ N}) = 20\text{ J}.$

(d) The area under the curve from 0 to 6.0 m is the sum of the areas of the two triangles and the rectangle.

So $W_{0\text{-}6} = W_{0\text{-}2} + W_{2\text{-}4} + W_{4\text{-}6} = 20\text{ J} + 40\text{ J} + 20\text{ J} = 80\text{ J}.$

2. Work Done by a Variable Force

Our discussion of variable force in this section is restricted to spring force, $F = -kx$ (with $x_o = 0$), where k is a constant called spring constant which measures the stiffness of a spring. The work done by an external force in stretching or compressing a spring (overcoming the spring force) is $W = \frac{1}{2}kx^2$, where x is the stretch or compression distance.

Example 5.3 Calculate the work required to compress an initially uncompressed spring with a spring constant of 20 N/m by 0.15 m.

Solution: Given: $k = 20\text{ N/m}$ and $x = -0.15\text{ m}.$

Find: $W.$

$W = \frac{1}{2}kx^2 = \frac{1}{2}(20\text{ N/m})(-0.15\text{ m})^2 = 0.23\text{ J}.$

3. Work-Energy Theorem: Kinetic Energy

Kinetic energy is the energy of motion and it is defined as $K = \frac{1}{2}mv^2$, where m is the mass and v is the velocity of the object. Although velocity is a vector quantity, kinetic energy is a scalar quantity because it depends on velocity squared. The unit of kinetic energy is joule (J). By combining the definition of work and kinematic equations, we can derive the **work–energy theorem**, which states that the net work done on an object is equal to its change in kinetic energy, i.e., $W_{net} = K - K_o = \Delta K$. In general, work is a measure of energy transfer or energy is the capacity of doing work.

Example 5.4 An object hits a wall and bounces back with half of its original speed. What is the ratio of the final kinetic energy to the initial kinetic energy?

Solution: Given: $v = \frac{v_o}{2}$. Find: $\frac{K}{K_o}$.

$K_o = \frac{1}{2} m v_o^2$ and $K = \frac{1}{2} m v^2 = \frac{1}{2} m \left(\frac{v_o}{2}\right)^2 = \frac{1}{4} (\frac{1}{2} m v_o^2)$. So $\frac{K}{K_o} = \frac{1}{4}$.

Why isn't the result $\frac{1}{2}$?

Example 5.5 The kinetic friction force between a 60.0-kg object and a horizontal surface is 50.0 N. If the initial speed of the object is 25.0 m/s, what distance will it slide before coming to a stop?

Solution: Given: $m = 60.0$ kg, $v_o = 25.0$ m/s, $v = 0$, $f_k = 50.0$ N.

Find: d.

The angle between the friction force and the displacement is 180°. With $W_{net} = K - K_o$, since f_k is the unbalanced force, we have

$W_{net} = f_k\, d \cos 180° = -(50\text{ N})\, d = K - K_o = \frac{1}{2} m v^2 - \frac{1}{2} m v_o^2 = \frac{1}{2}(60.0\text{ kg})(0)^2 - \frac{1}{2}(60.0\text{ kg})(25.0\text{ m/s})^2$,

solving for d, we have $d = 3.75 \times 10^2$ m.

Note: This problem can also be solved with dynamics and kinematics. We can find the acceleration of the object from Newton's second law and then use kinematic equations to find the distance.

4. Potential Energy

Potential energy is the energy of position. Two forms of potential energy are considered here, **gravitational potential energy** and **elastic (spring) potential energy**. A reference point/level is always required in measuring position and therefore potential energy. This choice of reference point/level is arbitrary, i.e., you can choose anywhere as your reference point/level in your calculation. For example, with the zero potential as a reference point/level,

Gravitational potential energy: $U = mgh$ (with $h_o = 0$)

Elastic potential energy: $U = \frac{1}{2} kx^2$ (with $x_o = 0$)

Example 5.6 An 5.00-kg object is moved from a height of 3.00 m above a floor to a height of 7.00 m above the floor. What is the change in gravitational potential energy?

Solution: We choose the floor level as our reference level ($h_o = 0$).

Given: $m = 5.00$ kg, $h_1 = 3.00$ m, $h_2 = 7.00$ m. Find: ΔU.

$\Delta U = U_2 - U_1 = mgh_2 - mgh_1 = mg(h_2 - h_1) = (5.00 \text{ kg})(9.80 \text{ m/s}^2)(7.00 \text{ m} - 3.00 \text{ m}) = 1.96\times 10^2$ J.

As expected, ΔU is positive, since positive external work is required to move the object.

If we choose the reference level at 3.00 m above the floor (height is zero at that level), then $h_1 = 3.00 \text{ m} - 3.00 \text{ m} = 0$, and $h_2 = 7.00 \text{ m} - 3.00 \text{ m} = 4.00$ m.

So $\Delta U = mg(h_2 - h_1) = (5.00 \text{ kg})(9.80 \text{ m/s}^2)(4.00 \text{ m} - 0) = 1.96\times 10^2$ J, independent of reference choice.

Example 5.7 A spring with a spring constant of 15 N/m is initially compressed by 3.0 cm. How much work is required to compress the spring an additional 4.0 cm?

Solution: The work required goes into the change in elastic potential energy. $W = \Delta K + \Delta U = \Delta U$ ($\Delta K = 0$). We choose the uncompressed position as $x_o = 0$.

Given: $k = 15$ N/m, $x_1 = 0.030$ m, $x_2 = x_1 + \Delta x = 0.030 \text{ m} + 0.040 \text{ m} = 0.070$ m.

Find: W.

$W = \Delta U = U_2 - U_1 = \frac{1}{2}k x_2^2 - \frac{1}{2}k x_1^2 = \frac{1}{2}k (x_2^2 - x_1^2) = \frac{1}{2}(15 \text{ N/m})[(0.070 \text{ m})^2 - (0.030 \text{ m})^2] = 0.030$ J.

Note: Why $W \neq \frac{1}{2}k (x_2 - x_1)^2 = \frac{1}{2}(14 \text{ m/s})(0.070 \text{ m} - 0.030 \text{ m})^2 = 0.012$ J?

5. The Conservation of Energy

A force is **conservative** if the work done by or against the force is independent of the path. Gravitational force is an example of a conservative force. A force is **nonconservative** if the work done by or against it depends on the path. Force of friction is an example of a nonconservative force.

The **total mechanical energy**, E, of a system is defined as the sum of the kinetic energy and potential energy, i.e, $E = K + U$. If the working forces (the forces which are doing work) in a system are conservative, the total mechanical energy of the system is conserved. This is **the conservation of total mechanical energy**.

$$K_o + U_o = K + U \quad \text{or} \quad \tfrac{1}{2}m v^2 + U = \tfrac{1}{2}m v_o^2 + U_o.$$

Note: When applying the conservation of mechanical energy, you should choose a reference point/level for position for determining U values. The choice is arbitrary, that is, you can choose the reference point/level to be anywhere.

Example 5.8 A 60-kg skier starts from rest from the top of a 50-m high slope. What is the speed of the skier on reaching the bottom of the slope? (Neglect friction.)

Solution: Choose the bottom of the slope as the reference level ($h = 0$).

Given: $v_o = 0$, $h_o = 50$ m, $h = 0$.

Find: v.

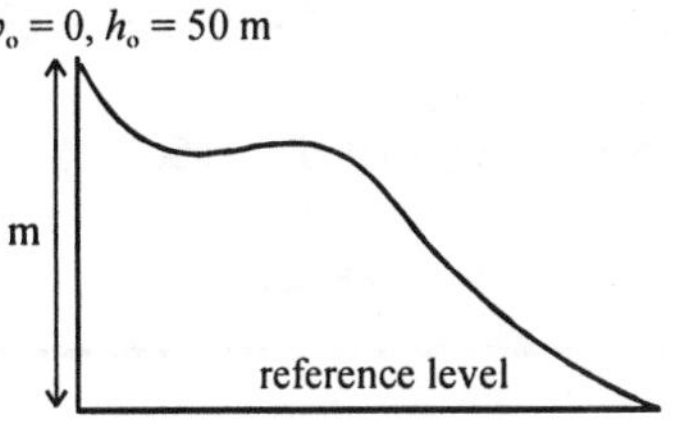

From $\frac{1}{2}m v^2 + U = \frac{1}{2}m v_o^2 + U_o$,

we have $\frac{1}{2}m v^2 + mgh = \frac{1}{2}m v_o^2 + mgh_o$.

So $\frac{1}{2}m v^2 + mg(0) = \frac{1}{2}m (0)^2 + mgh_o$,

therefore $v = \sqrt{2gh_o} = \sqrt{2(9.80 \text{ m/s}^2)(50 \text{ m})} = 31$ m/s.

Note: The mass of the skier cancels out in the equation because it is canceled. Note the example does not even give the mass of the skier!

Example 5.9 A 1500-kg car moving at 25 m/s hits an initially uncompressed horizontal spring with spring constant of 2.0×10^6 N/m. What is the maximum compression of the spring? (Neglect the mass of the spring)

Solution: Choose the uncompressed position of the spring as the reference point ($x_o = 0$).

Before the car hits the spring, it is moving with an initial speed of 25 m/s and the compression of the spring is zero. When the spring is maximally compressed, the car stops (transfer all its kinetic energy to elastic potential energy in the spring) and the spring is compressed by a distance of x.

Given: $m = 1500$ kg, $k = 2.0\times10^6$ N/m, $v_o = 25$ m/s, $v = 0$, $x_o = 0$.

Find: x.

From $\frac{1}{2}m v^2 + U = \frac{1}{2}m v_o^2 + U_o$, we have $\frac{1}{2}m v^2 + \frac{1}{2}k x^2 = \frac{1}{2}m v_o^2 + \frac{1}{2}k x_o^2$.

So $\frac{1}{2}m (0) + \frac{1}{2}k x^2 = \frac{1}{2}m v_o^2 + \frac{1}{2}k (0)^2$,

therefore $x = \sqrt{\dfrac{m v_o^2}{k}} = \sqrt{\dfrac{m}{k}}\, v_o = \sqrt{\dfrac{1500 \text{ kg}}{2.0\times10^6 \text{ N/m}}}(25 \text{ m/s}) = 0.68$ m.

If there is a nonconservative force doing work in a system, the total mechanical energy of the system is not conserved. However, the total energy (not mechanical!) of the system is still conserved. Some of the total energy is used to overcome the work done by the nonconservative force. $W_{nc} = E_o - E = -\Delta E$.

Example 5.10 In Example 5.8, if the work done by the friction force is -6.0×10^3 J (the work done by friction force is negative because the angle between the friction force and the displacement is 180°). What is the speed of the skier at the bottom of the slope?

Solution: $W_{nc} = E - E_o = (\frac{1}{2}m\,v^2 + mgh) - (\frac{1}{2}m\,v_o^{\,2} + mgh_o)$.

So $-6.0\times10^3\text{ J} = \frac{1}{2}(60\text{ kg})v^2 + mg(0) - \frac{1}{2}m\,(0)^2 - (60\text{ kg})(9.80\text{ m/s}^2)(50\text{ m})$.

Solve for $v = 28$ m/s.

---

6. Power

Average **power** is the average rate of doing work (work done divided by time), $\overline{P} = \frac{W}{t}$. The SI unit of power is watt (W). A common British unit of power is horsepower (hp) and 1 hp = 746 W. If the work is done by a constant force in the direction of motion ($\theta = 0°$) and the object is moved through a distance of d, $\overline{P} = \frac{Fd}{t} = F\,\overline{v}$, where $\frac{d}{t} = \overline{v}$ is the average speed.

Example 5.11 A 1500-kg car accelerates from 0 to 25 m/s in 7.0 s. What is the average power delivered by the of the engine?

Solution: Given: $m = 1500$ kg, $v_o = 0$, $v = 25$ m/s, $t = 7.0$ s. Find: $\overline{P}$.

From work-energy theorem, $W = \Delta K = \frac{1}{2}m\,v^2 - \frac{1}{2}m\,v_o^{\,2} = \frac{1}{2}(1500\text{ kg})[(25\text{ m/s})^2 - (0)^2] = 4.69\times10^5$ J.

So $\overline{P} = \frac{W}{t} = \frac{4.69\times10^5\text{ J}}{7.0\text{ s}} = 6.7\times10^4$ W (= 90 hp).

Alternate method: The average velocity of the car is $\overline{v} = \frac{v + v_o}{2} = \frac{25\text{ m/s} + 0}{2} = 12.5$ m/s.

$F = ma = m\frac{v - v_o}{t} = (1500\text{ kg})\frac{25\text{ m/s} - 0}{7.0\text{ s}} = 5.36\times10^3$ N.

So $\overline{P} = F\,\overline{v} = (5.36\times10^3\text{ N})(12.5\text{ m/s}) = 6.7\times10^4\text{ W} = 90$ hp.

IV. Mathematical Summary

Work	$W = F\,d\cos\theta$ (5.2)	Defines work.
Hooke's Law (spring force)	$F = -k(x - x_o)$ (5.3) $F = -k\,x$ (with $x_o = 0$)	Relates spring force with spring constant and change in length.
Work (or Potential Energy Change) in a Spring	$W = \frac{1}{2}k(x^2 - x_o^2)$ (5.4) $W = \frac{1}{2}k\,x^2$ (with $x_o = 0$)	Relates work done by external force (or potential energy change) in stretching or compressing a spring.
Kinetic Energy	$K = \frac{1}{2}m\,v^2$ (5.5)	Defines kinetic energy in terms of mass and velocity.
Work-Energy Theorem	$W_{net} = K - K_o = \Delta K$ (5.6)	Relates net work to change in kinetic energy.
Elastic (spring) Potential Energy	$U = \frac{1}{2}k\,x^2$ (with $x_o = 0$) (5.7)	Defines elastic potential energy.

V. Solutions of Selected Exercises and Paired Exercises

10. Gravity does work. $W = F_{||}\,d = mgd\sin\theta = (3.00\text{ kg})(9.80\text{ m/s}^2)(1.50\text{ m})\cos 20° = 15.1\text{ J}.$

16. $F_y = N + F\sin\theta - mg = 0$, so $N = mg - F\sin\theta.$

$F_x = F\cos\theta - f_k = 0$, or $F\cos\theta = \mu_k N = \mu_k(mg - F\sin\theta) = 0.$

Solve for $F = \dfrac{\mu_k\,mg}{\cos\theta + \mu_k\sin\theta} = \dfrac{(0.25)(35\text{ kg})(9.80\text{ m/s}^2)}{\cos 30° + (0.25)\sin 30°} = 86.5\text{ N}.$

Therefore $W = F\,d\cos\theta = (86.5\text{ N})(10\text{ m})\cos 30° = 7.5\times10^2\text{ J}.$

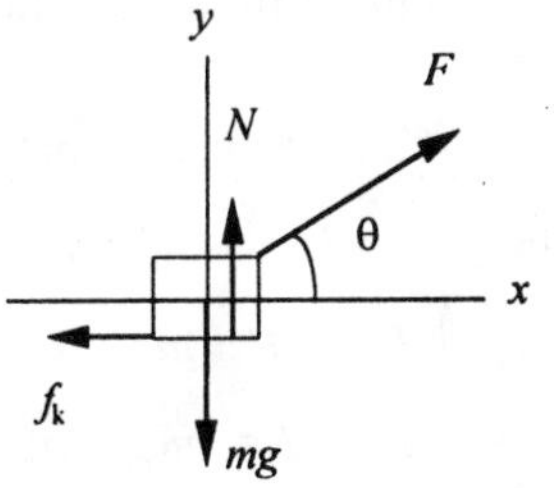

24. (a) $W = \frac{1}{2}k\,x^2 = \frac{1}{2}(2.5\times10^3\text{ N/m})(0.060\text{ m})^2 = 4.5\text{ J}.$

(b) $W = \frac{1}{2}k(x_2^2 - x_1^2) = \frac{1}{2}(2.5\times10^3\text{ N/m})[(0.080\text{ m})^2 - (0.060\text{ m})^2] = 3.5\text{ J}.$

28. $W = \text{area} = \frac{1}{2}(6.0\text{ N})(2.0\text{ m}) + \frac{1}{2}(-6.0\text{ N})(5.0\text{ m} - 2.0\text{ m}) = -3.0\text{ J}.$

38. $W_{net} = \frac{1}{2}m\,v^2 - \frac{1}{2}m\,v_o^2 = 0 - \frac{1}{2}m\,v_o^2.$ Also $W_{net} = -f\,d.$ So $d = \dfrac{v_o^2}{2f} \propto v_o^2.$

49. Since $U = mgh$, $\Delta U = mg\Delta h$. So $\Delta h = \dfrac{\Delta U}{mg} = \dfrac{-10\ \text{J}}{(2.0\ \text{kg})(9.80\ \text{m/s}^2)} = -0.51\ \text{m}$.

Therefore it is lowered by 0.51 m.

50. (a) $U_b = mgh = (1.5\ \text{kg})(9.80\ \text{m/s}^2)(-3.0\ \text{m}) = -44\ \text{J}$,

$U_a = (1.5\ \text{kg})(9.80\ \text{m/s}^2)(4.5\ \text{m}) = 66\ \text{J}$.

(b) $\Delta U_{\text{a to b}} = mg\Delta h = (1.5\ \text{kg})(9.80\ \text{m/s}^2)(-4.5\ \text{m}) = -66\ \text{J}$,

$\Delta U_{\text{b to a}} = (1.5\ \text{kg})9.80\ \text{m/s}^2)[4.5\ \text{m} - (-3.0\ \text{m})] = 110\ \text{J}$.

64. (a) $L - h = L\cos 25°$, so $h = L(1 - \cos 25°)$.

(b) From conservation of mechanical energy, $K + U = K_o + U_o = 0 + U_o$, we have

$K = U_o - U = mg\,(h_o - h) = mg\,L[(1 - \cos 25°) - (1 - \cos 9.0°)]$

$= (0.15\ \text{kg})(9.80\ \text{m/s}^2)(0.75\ \text{m})(\cos 9.0° - \cos 25°) = 9.0\times10^{-2}\ \text{J}$.

(c) $K = (0.15\ \text{kg})(9.80\ \text{m/s}^2)(0.75\ \text{m})(\cos 0° - \cos 25°) = 0.103\ \text{J} = \frac{1}{2}m\,v^2$,

so $v = \sqrt{\dfrac{2(0.103\ \text{J})}{0.15\ \text{kg}}} = 1.2\ \text{m/s}$.

78. 90 km/h = 25 m/s. $W = \Delta K = \frac{1}{2}m\,v^2 - 0$.

So $P = \dfrac{W}{t} = \dfrac{m\,v^2}{2\,t} = \dfrac{(1500\ \text{kg})(25\ \text{m/s})^2}{2(5.0\ \text{s})} = 9.4\times10^4\ \text{W}$.

80. From the kinematic equation $v^2 = v_o^2 - 2gy = 0 - 2(9.80\ \text{m/s}^2)(-5.0\ \text{m})$, we have $v = 9.90\ \text{m/s}$.

So the average power is $\overline{P} = F\bar{v} = mg\,\dfrac{v_o + v}{2} = (1.0\ \text{kg})(9.80\ \text{m/s}^2) \times \dfrac{0 + 9.90\ \text{m/s}}{2} = 49\ \text{W}$.

82. $F_{\text{net}} = F - mg = ma$,

so $F = m(g + a) = (1500\ \text{kg})(9.80\ \text{m/s}^2 + 0.50\ \text{m/s}^2) = 1.545\times10^4\ \text{N}$.

$\bar{v} = \dfrac{v_o + v}{2} = \dfrac{0.25\ \text{m/s} + 0.75\ \text{m/s}}{2} = 0.50\ \text{m/s}$.

So $\overline{P} = F\bar{v} = (1.545\times10^4\ \text{N})(0.50\ \text{m/s}) = 7.7\times10^3\ \text{W}$.

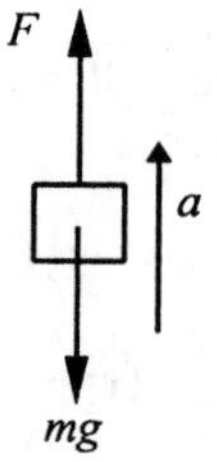

92. (a) 90 km/h = 25 m/s. From conservation of mechanical energy $\frac{1}{2}mv^2 + U = \frac{1}{2}mv_o^2 + U_o$,

we have $0 + mgh = \frac{1}{2}m\,v_o^2 + 0$, so $h = \dfrac{v_o^2}{2\,g} = \dfrac{(25\ \text{m/s})^2}{2(9.80\ \text{m/s}^2)} = 32\ \text{m}$.

(b) $h' = 0.70\,h = 22\ \text{m}$.

99. $\frac{1}{2}mv^2 + U = \frac{1}{2}mv_o^2 + U_o$

so $\frac{1}{2}mv^2 + mg(30\text{ m}) = 0 + mg(45\text{ m})$,

$v = \sqrt{2g(45\text{ m} - 30\text{ m})} = \sqrt{2(9.80\text{ m/s}^2)(15\text{ m})} = \boxed{17\text{ m/s}}$.

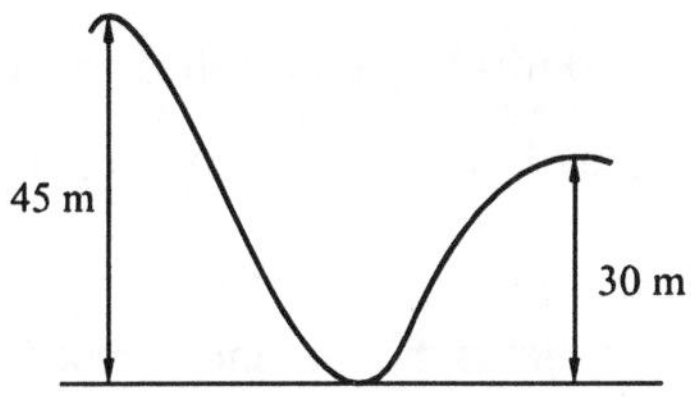

101. $E_{top} = 0 + mg(25\text{ m}) = 25mg\text{ J} = 245m\text{ J}$. $E_{bottom} = \frac{1}{2}m(20\text{ m/s})^2 + 0 = 200m\text{ J}$.

So because $E_{top} > E_{bottom}$, it is not conserved.

VI. Practice Quiz

1. Which one of the following is the correct unit of work expressed in SI base units?
(a) kg·m/s (b) kg·m/s^2 (c) kg·m^2/s (d) kg·m^2/s^2 (e) kg^2·m/s^2

2. A 40-N crate is pulled 5.0 m up along a 37° inclined plane. What is the magnitude of the work done by the weight (gravitational force) of the crate? (Hint: draw a diagram.)
(a) 6.0 J (b) 12 J (c) 1.2×10^2 J (d) 2.0×10^2 J (e) 1.2×10^3 J.

3. What work is required to stretch a spring of spring constant 40 N/m from $x = 0.20$ m to 0.25 m? (The unstretched position is at $x = 0$.)
(a) 0.45 J (b) 0.80 J (c) 1.3 J (d) 0.050 J (e) 0.90 J

4. A horizontal force of 200 N is applied to move a 55-kg cart (initially at rest) across a 10 m level surface. What is the final kinetic energy of the cart?
(a) 1.0×10^3 J (b) 2.0×10^3 J (c) 2.7×10^3 J (d) 4.0×10^3 J (e) 5.4×10^3 J

5. A bobsled makes a run down an ice track from a vertical distance of 150 m up the hill. If there is no friction and the bobsled starts from rest, what is its speed at the bottom of the hill?
(a) 14 m/s (b) 27 m/s (c) 36 m/s (d) 45 m/s (e) 54 m/s

6. A 10-N force is needed to move an object with a constant velocity of 5.0 m/s. What power must be delivered to the object by the force?
(a) 0.50 W (b) 1.0 W (c) 2.0 W (d) 50 W (d) 100 W

7. If it takes 50 m to stop a car initially moving at 25 m/s, what distance is required to stop a car moving at 50 m/s? (Assume the same braking force.)

(a) 25 m (b) 50 m (c) 100 m (d) 150 m (e) 200 m

8. What is the minimum speed of the ball at the bottom of its swing (point B) in order for it to reach point A, which is 1.0-m above the bottom of the swing.

(a) 1.1 m/s (b) 2.2 m/s (c) 3.1 m/s (d) 4.4 m/s (e) 4.9 m/s

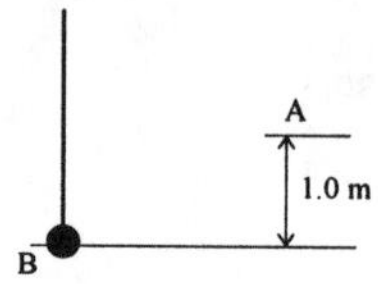

9. A force of 10 N is applied horizontally to a 2.0-kg mass on a level surface. The coefficient of kinetic friction between the mass and the surface is 0.20. If the mass is moved a distance of 10 m, what is the change in its kinetic energy?

(a) 100 J (b) 61 J (c) 46 J (d) 39 J (e) 20 J

10. A 10-kg mass is moving with a speed of 4.0 m/s. How much work is required to stop the mass?

(a) 20 J (b) 40 J (c) 60 J (d) 80 J (e) 100 J

Answers to Practice Quiz:

1.d 2.c 3.a 4.b 5.e 6.d 7.e 8.d 9.b 10.d

CHAPTER 6

Momentum and Collision

I. Chapter Objectives

Upon completion of this chapter, you should be able to:

1. compute linear momentum and the components of momentum.
2. explain the condition for the conservation of linear momentum and apply it to physical situations.
3. relate impulse and momentum, and kinetic energy and momentum.
4. describe the conditions on kinetic energy and momentum in elastic and inelastic collisions.
5. explain the concept of the center of mass, compute its location for simple systems, and describe how the center of mass and center of gravity are related.
6. apply the conservation of momentum in the explanation of jet propulsion and the operation of rockets.

II. Key Terms

Upon completion of this chapter, you should be able to define and/or explain the following key terms:

linear momentum	**inelastic collision**
total linear momentum	**completely inelastic collision**
conservation of linear momentum	**center of mass (CM)**
impulse	**center of gravity (CG)**
impulse-momentum theorem	**jet propulsion**
elastic collision	**reverse thrust**

The definitions and/or explanations of the most important key terms can be found in the following section: **III. Chapter Summary and Discussion**.

III. Chapter Summary and Discussion

1. Linear Momentum

The **linear momentum** of an object is defined as the product of its mass and velocity, $\mathbf{p} = m\,\mathbf{v}$. Since velocity is a vector, so is momentum. The SI unit of momentum is kg·m/s.

The **total linear momentum** of a system is the vector sum of the momenta of the individual particles.

$$\mathbf{P} = \mathbf{p}_1 + \mathbf{p}_2 + \mathbf{p}_3 + \ldots = \sum_i \mathbf{p}_i$$

Example 6.1 Which has more linear momentum:

(a) a 75-kg running back moving at 9.5 m/s or

(b) a 100-kg lineman moving at 6.8 m/s?

Solution: Given: mass, m, and velocity, v. Find: magnitude of momentum, p.

(a) $p = m\,v = (75 \text{ kg})(9.5 \text{ m/s}) = 7.1\times10^2 \text{ kg·m/s}$.

(b) $p = (100 \text{ kg})(6.8 \text{ m/s}) = 6.8\times10^2 \text{ kg·m/s}$.

The much lighter running back has more momentum than the lineman.

Example 6.2 Two identical 1500-kg cars are moving perpendicular to each other. One moves with a speed of 25 m/s due north and the other moves at 15 m/s due east. What is the total linear momentum of the system?

Solution: Given: $m_1 = m_2 = 1500$ kg, $v_1 = 25$ m/s, $v_2 = 15$ m/s.

Find: **P**.

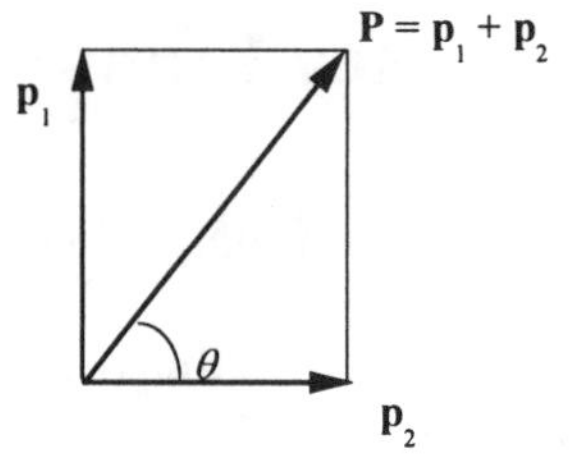

$p_1 = m_1\,v_1 = (1500 \text{ kg})(25 \text{ m/s}) = 3.75\times10^4 \text{ kg·m/s}$.

$p_2 = m_2\,v_2 = (1500 \text{ kg})(15 \text{ m/s}) = 2.25\times10^4 \text{ kg·m/s}$.

So $P = \sqrt{p_1^2 + p_2^2} = \sqrt{(3.75\times10^4 \text{ kg·m/s})^2 + (2.25\times10^4 \text{ kg·m/s})^2}$

$= 4.4\times10^4$ kg·m/s.

$$\theta = \tan^{-1}\frac{p_1}{p_2} = \frac{3.75\times10^4 \text{ kg·m/s}}{2.25\times10^4 \text{ kg·m/s}} = 59° \text{ N of E.}$$

Note: Momentum is a vector quantity. You must use vector addition when adding momenta.

Newton's second law can also be expressed in terms of momentum: $\overline{\mathbf{F}} = \dfrac{\Delta \mathbf{p}}{\Delta t}$. It is equivalent to $\mathbf{F} = m\mathbf{a}$ if mass is a constant. For system in which mass is not a constant, such rocket propulsion, $\overline{\mathbf{F}} = \dfrac{\Delta \mathbf{p}}{\Delta t}$ should be used. The impulse-momentum theorem discussed later is a variation of this form of Newton's second law.

2. The Conservation of Linear Momentum

The total linear momentum of a system is conserved if the net external force on the system is zero.

$$\mathbf{P} = \mathbf{P}_o, \quad \text{or} \quad \mathbf{p}_1 + \mathbf{p}_2 + \ldots = \mathbf{p}_{1o} + \mathbf{p}_{2o} + \ldots.$$

Note: Momentum is a vector quantity. You must use vector addition when applying the conservation of linear momentum.

Example 6.3 A 50-kg pitching machine (excluding the baseball) is placed on a frozen pond. The machine fires a 0.40-kg baseball with a speed of 35 m/s in the horizontal direction. What is the recoil velocity of the pitching machine? (Assume negligible friction.)

Solution: Before the baseball is fired, neither the machine nor the baseball is moving ($P_o = 0$). After the ball is fired, the ball moves in one direction and the machine recoils in the opposite direction so that $P = 0$.

Given: $m_1 = 50$ kg (machine) $m_2 = 0.40$ kg (ball)

$v_{1o} = 0$ $v_{2o} = 0$,

$v_2 = 35$ m/s

Find: v_1.

From momentum conservation, we have $\mathbf{P} = \mathbf{P}_o$, or $\mathbf{p}_1 + \mathbf{p}_2 = \mathbf{p}_{1o} + \mathbf{p}_{2o}$.

So $m_1 v_1 + m_2 v_2 = m_1 v_{1o} + m_2 v_{2o} = 0$,

therefore $v_1 = -\dfrac{m_2 v_2}{m_1} = -\dfrac{(0.40 \text{ kg})(35 \text{ m/s})}{50 \text{ kg}} = -0.28$ m/s.

The negative sign here indicates the machine is moving in exactly the opposite direction to the ball.

Example 6.4 A 10-g bullet moving at 300 m/s is fired into a 1.0-kg block. The bullet emerges (the bullet does not stay embedded in the block) with half of its original speed. What is the velocity of the block right after the collision?

Solution: Right before the bullet hits the block, the bullet is moving at 300 m/s and the block is at rest. Right after the bullet hits the block, it emerges with (300 m/s)/2 = 150 m/s and the block should be moving in the same direction as the bullet. (Why?)

Given: $m_1 = 0.010$ kg (bullet) $m_2 = 1.0$ kg (block)

$v_{1o} = 300$ m/s $v_{2o} = 0$

$v_1 = 150$ m/s

Find: v_2.

From momentum conservation, we have $\mathbf{P} = \mathbf{P}_o$, or $\mathbf{p}_1 + \mathbf{p}_2 = \mathbf{p}_{1o} + \mathbf{p}_{2o}$.

So $m_1 v_1 + m_2 v_2 = m_1 v_{1o} + m_2 v_{2o}$,

or $(0.010 \text{ kg})(150 \text{ m/s}) + (1.0 \text{ kg})v_2 = (0.010 \text{ kg})(300 \text{ m/s}) + (1.0 \text{ kg})(0)$.

Solve for $v_2 = +1.5$ m/s, it is positive because the block is moving in the same direction as the bullet.

---

3. Impulse

From Newton's second law, $\overline{\mathbf{F}} = \frac{\Delta \mathbf{p}}{\Delta t}$, we derived the **impulse-momentum theorem**, *impulse is equal to the change in momentum*, or $\overline{\mathbf{F}}\,\Delta t = \Delta \mathbf{p} = \mathbf{p} - \mathbf{p}_o$, where $\overline{\mathbf{F}}\,\Delta t$ is called impulse.

Impulse-momentum is very useful in explaining some everyday phenomena. For example, why do tennis players following through? Why do football players wear pads? Why can some not-so-strong martial artists break objects like bricks with their bare fists?

Note: Again, momentum is a vector quantity and it must be treated as a vector.

Example 6.5 A 0.10-kg ball is dropped onto a table top. The speeds of the ball right before hitting the table top and right after hitting the table top are 5.0 m/s and 4.0 m/s, respectively. If the collision between the ball and the tabletop lasts 0.15 s, what is the average force exerted on the ball by the table top?

Solution: The velocity of the ball before the collision is downward and the velocity of the ball after the collision is upward. Since these two velocities are opposite, we have to assign signs (+ or −) to them to find $\Delta \mathbf{v}$ properly. Conventionally, we chose the upward direction as positive so the downward direction is negative.

Given: $m = 0.10$ kg, $v_o = -5.0$ m/s, $v = +4.0$ m/s, $\Delta t = 0.15$ s.

Find: $\overline{\mathbf{F}}$.

From impulse-momentum theorem, we have $\overline{\mathbf{F}}\,\Delta t = \Delta \mathbf{p} = \mathbf{p} - \mathbf{p}_o$.

So $$\overline{\mathbf{F}} = \frac{\mathbf{p} - \mathbf{p}_o}{\Delta t} = \frac{m\,v - m\,v_o}{\Delta t} = \frac{(0.10 \text{ kg})(4.0 \text{ m/s}) - (0.10 \text{ kg})(-5.0 \text{ m/s})}{0.15 \text{ s}} = +6.0 \text{ N}.$$

The positive sign indicates that the force on the ball by the tabletop is upward, which makes sense.

---

4. Elastic and Inelastic Collision

Momentum is always conserved in a collision, as long as the net external force is zero on the system, which is approximately true at least during the small time interval Δt of the collision. However, in an **elastic collision**, kinetic energy is also conserved while in an **inelastic collision**, kinetic energy is not conserved. (Note: when kinetic energy is not conserved as is the usual case, this does not mean the total energy is not conserved. Some of the kinetic energy is converted to heat, sound, etc. during a collision. Total energy is always conserved as we shall see!). If the objects in a system stick together after collision, the collision is called **completely inelastic**. In an inelastic collision, the kinetic energy of the system before the collision is always greater than the kinetic energy of the system after the collision. Why?

Example 6.6 A 1000-kg automobile traveling at 25 m/s runs into the rear of a stopped car that has a mass of 1500 kg and they stick together.

(a) What is the speed of the composite system (two cars) immediately after collision?

(b) How much kinetic energy is lost during the collision?

Solution: Since the two cars stick together after the collision, they must have the same final velocity.

Given: $m_1 = 1000$ kg $\qquad m_2 = 1500$ kg

$v_{1o} = 25$ m/s $\qquad v_{2o} = 0$

Find: $v_1 \qquad v_2$.

$v_1 = v_2 = v$.

(a) From momentum conservation during the small Δt of the collision (ignore friction), we have

$\mathbf{P} = \mathbf{P}_o$, or $\mathbf{p}_1 + \mathbf{p}_2 = \mathbf{p}_{1o} + \mathbf{p}_{2o}$.

So $m_1 v_1 + m_2 v_2 = m_1 v + m_2 v = (m_1 + m_2)v = m_1 v_{1o} + m_2 v_{2o}$,

therefore $v = \dfrac{m_1 v_{1o} + m_2 v_{2o}}{m_1 + m_2} = \dfrac{(1000 \text{ kg})(25 \text{ m/s}) + (1500 \text{ kg})(0)}{1000 \text{ kg} + 1500 \text{ kg}} = 10$ m/s.

(b) The initial kinetic energy $K_o = \frac{1}{2} m_1 v_{1o}^2 + \frac{1}{2} m_2 v_{2o}^2 = \frac{1}{2}(1000 \text{ kg})(25 \text{ m/s})^2 + \frac{1}{2} m_2 (0)^2 = 3.13\times10^5$ J.

The final kinetic energy $K = \frac{1}{2} m_1 v_1^2 + \frac{1}{2} m_2 v_2^2 = \frac{1}{2}(1000 \text{ kg})(10 \text{ m/s})^2 + \frac{1}{2}(1500 \text{ kg})(10 \text{ m/s})^2 = 1.25\times10^5$ J.

So the kinetic energy lost in the collision is $|\Delta K| = K_o - K = 3.13\times10^5 \text{ J} - 1.25\times10^5 \text{ J} = 1.9\times10^5$ J.

The percentage of kinetic energy loss is $\dfrac{K - K_o}{K_o} = \dfrac{-1.88\times10^5 \text{ J}}{3.13\times10^5 \text{ J}} = -0.60 = -60\%$.

Example 6.7 A rubber ball with a speed of 5.0 m/s collides head on elastically with an identical ball at rest. Find the velocity of each object after the collision.

Solution: Given: $m_1 = m \qquad m_2 = m$ (identical ball)

$v_{1o} = 5.0$ m/s $\qquad v_{20} = 0$

Find: $v_1 \qquad v_2$.

We use Equations (6.15) and (6.16).

$$v_1 = \left(\frac{m_1 - m_2}{m_1 + m_2}\right) v_{1o} = \frac{m - m}{m + m}(5.0 \text{ m/s}) = 0. \qquad v_2 = \left(\frac{2m_1}{m_1 + m_2}\right) v_{1o} = \frac{2\,m}{m + m}(5.0 \text{ m/s}) = 5.0 \text{ m/s}.$$

The two balls exchange velocities. This is always the case if $m_1 = m_2$ in such collisions.

---

5. Center of Mass

The **center of mass** of a system is the point at which all the mass of the system may be considered to be concentrated. If the acceleration due to gravity, g, is a constant, then the **center of gravity**, or the point at which all the weight of the system may be considered to be concentrated, is at the center of mass. For a system of particles, the coordinates of the center of mass are calculated by $X_{CM} = \frac{\sum_i m_i x_i}{M}$ and $Y_{CM} = \frac{\sum_i m_i y_i}{M}$, where x_i and y_i are the coordinates of the particle m_i, and M is the total mass of the system.

The motion of the center of mass also obeys Newton's second law: $\mathbf{F}_{\text{net external}} = M\,\mathbf{A}_{CM}$, where $\mathbf{F}_{\text{net external}}$ is the net external force, M is the total mass of the system, and $\mathbf{A}_{CM}$ is the acceleration of the center of mass.

Example 6.8 Find the location of the center of mass of the three mass system.

mass	location
$m_1 = 1.0$ kg	(0, 0)
$m_2 = 2.0$ kg	(1.0 m, 1.0 m)
$m_3 = 3.0$ kg	(2.0 m, −2.0 m)

Solution:

$$X_{CM} = \frac{\sum_i m_i x_i}{M} = \frac{(1.0\text{ kg})(0) + (2.0\text{ kg})(1.0\text{ m}) + (3.0\text{ kg})(2.0\text{ m})}{1.0\text{ kg} + 2.0\text{ kg} + 3.0\text{ kg}} = 1.3\text{ m}.$$

$$Y_{CM} = \frac{\sum_i m_i y_i}{M} = \frac{(1.0\text{ kg})(0) + (2.0\text{ kg})(1.0\text{ m}) + (3.0\text{ kg})(-2.0\text{ m})}{1.0\text{ kg} + 2.0\text{ kg} + 3.0\text{ kg}} = -0.67\text{ m}.$$

So the location of the center of mass is at $(X_{CM}, Y_{CM}) = (1.3\text{ m}, -0.67\text{ m})$.

IV. Mathematical Summary

Linear Momentum	$\mathbf{p} = m\,\mathbf{v}$ (6.1)	Defines linear momentum.
Total Linear Momentum	$\mathbf{P} = \mathbf{p}_1 + \mathbf{p}_2 + \mathbf{p}_3 + \ldots = \sum_i \mathbf{p}_i$ (6.2)	Computes total linear momentum of a system.
Newton's Second Law	$\overline{\mathbf{F}} = \frac{\Delta \mathbf{p}}{\Delta t}$ (6.3)	Rewrites Newton's second law in terms of momentum

Impulse-momentum theorem	Impulse $= \overline{\mathbf{F}}\,\Delta t = \Delta \mathbf{p}$ $= m\,\mathbf{v} - m\,\mathbf{v}_o$ (6.6)	Relates impulse with momentum
Conditions for Elastic Collision	$\mathbf{P}_f = \mathbf{P}_i, \quad K_f = K_i$ (6.8)	Define the conditions for an elastic collision
Conditions for Inelastic Collision (6.9)	$\mathbf{P}_f = \mathbf{P}_i, \quad K_f < K_i$ (6.9)	Define the conditions for an inelastic collision
Final Velocities in Head-On Elastic Collision ($v_{2_o} = 0$)	$v_1 = \left(\dfrac{m_1 - m_2}{m_1 + m_2}\right) v_{1_o}$ (6.15) $v_2 = \left(\dfrac{2m_1}{m_1 + m_2}\right) v_{1_o}$ (6.16)	Express the final velocities in head-on elastic collisions
Coordinate of center of mass	$X_{CM} = \dfrac{\sum_i m_i x_i}{M}$ (6.19)	Calculates the coordinate the center of mass of a system (using signs for directions)

V. Solutions of Selected Exercises and Paired Exercises

14. $v_o = 3.0$ km/h $= 0.833$ m/s.

$$F = \frac{\Delta p}{\Delta t} = \frac{mv - mv_o}{\Delta t} = \frac{(5.0\times10^3\text{ kg})(0 - 0.833\text{ m/s})}{0.64\text{ s}} = -6.5\times10^3\text{ N}.$$

So the magnitude is 6.5×10^3 N.

18. $\Delta p_x = m\Delta v_x = (0.20\text{ kg})[(15\text{ m/s})\sin 60° - (15\text{ m/s})\sin 60°] = 0,$

$\Delta p_y = m\Delta v_y = (0.20\text{ kg})[-(15\text{ m/s})\cos 60° - (15\text{ m/s})\cos 60°] = -3.0\text{ kg·m/s}.$

So $\Delta\mathbf{p} = -3.0\text{ kg·m/s }\mathbf{y}$.

24. The recoil velocity of the gun firing a blank cartridge will be smaller because the blank cartridge has less mass, therefore less momentum.

30. Apply momentum conservation $\mathbf{P}_o = \mathbf{P}$

in x axis: $(2.0\text{ kg})(0) = (0.50\text{ kg})(-2.8\text{ m/s}) + (1.3\text{ kg})(0) + (1.2\text{ kg})v_x$, so $v_x = 1.17$ m/s;

in y axis: $(3.0\text{ kg})(0) = (0.50\text{ kg})(0) + (1.3\text{ kg})(-1.5\text{ m/s}) + (1.2\text{ kg})v_y$, so $v_y = 1.63$ m/s.

Therefore $v = \sqrt{(1.17\text{ m/s})^2 + (1.63\text{ m/s})^2} = 2.0$ m/s, and

$\theta = \tan^{-1}\left(\dfrac{1.63}{1.17}\right) = 54°$ above $+x$ axis.

46. Consider the horizontal motion.

$$\bar{F}\Delta t = mv - mv_o = mv, \quad \text{so} \quad v = \frac{\bar{F}\Delta t}{m} = \frac{4.0\ \text{N·s}}{0.20\ \text{kg}} = 20\ \text{m/s}.$$

54. For elastic collision:

$$v_1 = \frac{m_1 - m_2}{m_1 + m_2} v_{1o} = \frac{4.0\ \text{kg} - 2.0\ \text{kg}}{4.0\ \text{kg} + 2.0\ \text{kg}} (4.0\ \text{m/s}) = 1.3\ \text{m/s}.$$

$$v_2 = \frac{2m_1}{m_1 + m_2} v_{1o} = \frac{2(4.0\ \text{kg})}{4.0\ \text{kg} + 2.0\ \text{kg}} (4.0\ \text{m/s}) = 5.3\ \text{m/s}.$$

64. 90.0 km/h = 25.0 m/s, 60.0 km/h = 16.67 m/s.

Using result from Exercise 6.63, we have

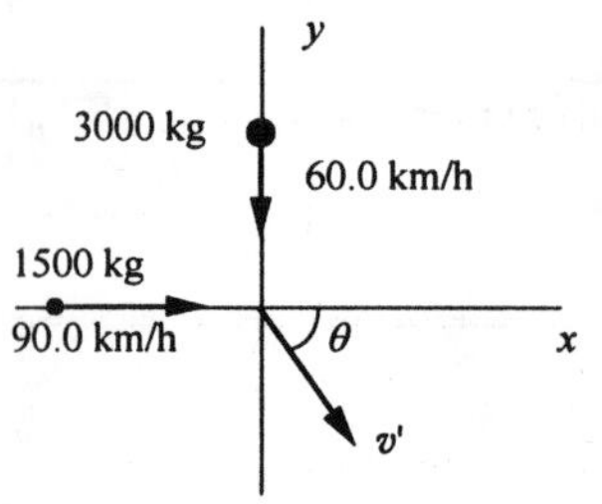

$$v' = \frac{\sqrt{m^2 v^2 + M^2 V^2}}{m + M}$$

$$= \frac{\sqrt{(1500\ \text{kg})^2 (25.0\ \text{m/s})^2 + (3000\ \text{kg})^2 (16.67\ \text{m/s})^2}}{1500\ \text{kg} + 3000\ \text{kg}}$$

$$= 13.9\ \text{m/s}. \quad \theta = \tan^{-1}\left[\frac{(3000\ \text{kg})(16.67\ \text{m/s})}{(1500\ \text{kg})(25.0\ \text{m/s}}\right] = 53.1\ \text{S of E}.$$

71. From momentum conservation $\mathbf{P}_o = \mathbf{P}$: $m_1 v_{1o} + m_2 (0) = (m_1 + m_2)v$, so $v = \frac{m_1 v_{1o}}{m_1 + m_2}$.

$$K_o = \tfrac{1}{2} m_1 v_{1o}^2 \quad \text{and} \quad K = \tfrac{1}{2}(m_1 + m_2)v^2 = \tfrac{1}{2}(m_1 + m_2)\left(\frac{m_1 v_{1o}}{m_1 + m_2}\right)^2 = \frac{1}{2}\frac{(m_1 v_{1o})^2}{m_1 + m_2}.$$

So the fraction of kinetic energy lost is

$$\frac{|\Delta K|}{K_o} = \frac{K_o - K}{K_o} = 1 - \frac{K}{K_o} = 1 - \frac{\frac{1}{2}\frac{(m_1 v_{1o})^2}{m_1 + m_2}}{\frac{1}{2} m_1 v_{1o}^2} = 1 - \frac{m_1}{m_1 + m_2} = \frac{m_2}{m_1 + m_2}.$$

82. $$X_{CM} = \frac{\Sigma_i m_i x_i}{M} = \frac{(3.0\ \text{kg})(-6.0\ \text{m/s}) + (2.0\ \text{kg})(1.0\ \text{m}) + (4.0\ \text{kg})(3.0\ \text{m})}{3.0\ \text{kg} + 2.0\ \text{kg} + 4.0\ \text{kg}} = -0.44\ \text{m},$$

$Y_{CM} = 0$. So the CM is at (−0.44 m, 0 m).

90. Due to the lack of external force, the CM is stationary where they meet.

$$X_{CM} = \frac{\Sigma_i m_i x_i}{M} = \frac{(4000\ \text{kg})(0) + (100\ \text{kg})(5.0\ \text{m})}{4000\ \text{kg} + 100\ \text{kg}}$$

$= 0.12$ m from capsule's original position.

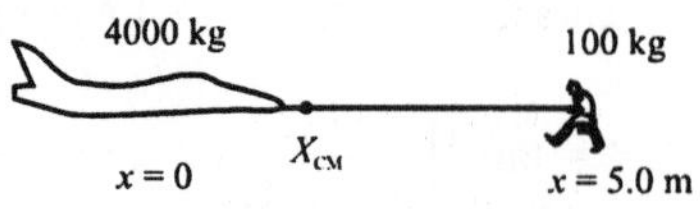

103. First find the recoil velocity of the astronaut from momentum conservation $\mathbf{P}_o = \mathbf{P}$.

$m_1 v_{1o} + m_2 v_{2o} = m_1 v_1 + m_2 v_2$, so $(0.50 \text{ kg})(0) + (90 \text{ kg})(0) = (0.50 \text{ kg})(4.0 \text{ m/s}) + (90 \text{ kg})v$,

therefore $v = -0.0222$ m/s. Hence it takes him $\dfrac{12 \text{ m}}{0.0222 \text{ m/s}} = 5.4 \times 10^2$ s.

VI. Practice Quiz

1. The SI unit of momentum is which one of the following?

(a) N·m (b) N/s (c) N·s (d) N/m (e) kg·m/s^2

2. A 10 N force acts on a 5.0-kg object, initially at rest, for 2.5 s. What is the final speed of the object?

(a) 1.0 m/s (b) 2.0 m/s (c) 3.0 m/s (d) 4.0 m/s (e) 5.0 m/s

3. A 70-kg astronaut is space walking outside the space capsule and is stationary when the tether line breaks. As a means of returning to the capsule he throws his 2.0-kg space wrench at a speed of 14 m/s away from the capsule. At what speed does the astronaut move toward the capsule?

(a) 0.40 m/s (b) 1.5 m/s (c) 3.5 m/s (d) 5.0 m/s (e) 7.0 m/s

4. A railroad freight car, mass 15,000 kg, is allowed to coast along a level track at a speed of 2.0 m/s. It collides and couples with a 50,000-kg second car, initially at rest and with brakes released. What is the speed of the two cars after coupling?

(a) 0.46 m/s (b) 0.60 m/s (c) 1.2 m/s (d) 1.8 m/s (e) 86 m/s

5. A 0.060-kg tennis ball, initially moving at a speed of 12 m/s, is struck by a racket causing it to rebound in the opposite direction at a speed of 18 m/s. What is the change in momentum of the ball?

(a) 0.18 kg·m/s (b) 0.36 kg·m/s (c) 0.72 kg·m/s (d) 1.1 kg·m/s (e) 1.8 kg·m/s.

6. The center of mass of a two-particle system is at the origin. One particle is located at (3.0 m, 0 m) and has a mass of 2.0 kg. What is the location of the second mass of 3.0 kg?

(a) −3.0 m (b) −2.0 m (c) 0 m (d) 2.0 m (e) 3.0 m

7. In an inelastic collision, if the momentum of the system is conserved, then which of the following statements is true about kinetic energy?

(a) kinetic energy is also conserved. (b) kinetic energy is gained.

(c) kinetic energy is lost. (d) kinetic energy is halved.

(e) none of the above.

8. A 0.10-kg object with a velocity of 0.20 m/s in the $+x$ direction makes a head-on elastic collision with a 0.15 kg object initially at rest. What is the final velocity of the 0.10-kg object after collision?

(a) −0.16 m/s (b) +0.16 m/s (c) 0 (d) −0.040 m/s (e) +0.040 m/s

9. If the momentum of an object is doubled, by what factor will its kinetic energy change?

(a) 2 (b) 4 (c) 0 (d) 1/2 (e) 1/4

10. A model rocket sits on the launch pad and its fuel is ignited, blasting the rocket upward. What happens to the center of mass of the rocket-fuel system?

(a) it goes up (b) it is stationary (c) it follows the path of the rocket

(d) it follows the path of the fuel (e) not enough information given

Answers to Practice Quiz:

1.c 2.e 3.a 4.a 5.e 6.b 7.c 8.d 9.b 10.b

CHAPTER 7

Circular Motion and Gravitation

I. Chapter Objectives

Upon completion of this chapter, you should be able to:

1. define units of angular measure and show how angular measure is related to circular arc length.
2. describe and compute angular speed and velocity and explain their relationship to tangential speed.
3. explain why there is a centripetal acceleration in constant or uniform circular motion and compute centripetal acceleration.
4. define angular acceleration and analyze rotational kinematics.
5. describe Newton's law of gravitation and how it relates to the acceleration due to gravity and apply the general formulation of gravitational potential energy.
6. state and explain Kepler's laws of planetary motion and describe the orbits and motions of satellites.

II. Key Terms

Upon completion of this chapter, you should be able to define and/or explain the following key terms:

angular distance
radian (rad)
average angular speed
angular velocity
- **average**
- **instantaneous**

tangential speed
period (*T*)
frequency (*f*)
hertz (Hz)
uniform circular motion
centripetal acceleration
centripetal force
average angular acceleration
***g*'s of force**
gravitational potential energy
Kepler's first law (the law of orbits)
Kepler's second law (the law of area)
Kepler's third law (the law of periods)
escape speed
tangential acceleration
universal law of gravitation
universal gravitational constant, *G*

The definitions and/or explanations of the most important key terms can be found in the following section: **III. Chapter Summary and Discussion**.

III. Chapter Summary and Discussion

1. Angular Measure

Circular motion is conveniently described using the polar coordinates (r, θ) because r is a constant and only θ varies. The relationships between rectangular coordinates and polar coordinates are as follows:

$x = r \cos \theta$ and $y = r \sin \theta$, where θ is measured ccw from the $+x$ axis

Angular distance ($\Delta\theta = \theta - \theta_o = \theta$ if $\theta_o = 0$) is measured in degrees or radians. A **radian** (rad) is defined as the angle that subtends an arc length that is equal to the radius: 1 rad = 57.3° or 2π rad = 360°.

$s = r\,\theta$, where s is the arc length, r is the radius, and θ is the angle in radians.

Example 7.1 What arc length does the Earth travel in a three month period in its nearly circular orbit about the Sun with a radius of 1.5×10^{11} m?

Solution: The Earth revolve the Sun once a year (2π rad/year). In a three month period, it sweeps through

$$\theta = \frac{2\pi \text{ rad}}{1 \text{ year}} \times \frac{1 \text{ year}}{12 \text{ months}} \times (3 \text{ months}) = \frac{\pi}{2} \text{ rad}.$$

Given: $\theta = \frac{\pi}{2}$ rad, $r = 1.5\times10^{11}$ m.

Find: s.

$$s = r\,\theta = (1.5\times10^{11} \text{ m}) \times (\frac{\pi}{2} \text{ rad}) = 2{,}4\times10^{11} \text{ m}.$$

---

2. Angular Speed and Velocity

Angular speed and **angular velocity** are defined analogously to their linear motion counterparts. The direction of angular velocity is determined by the right-hand rule: when the fingers of the right hand are curled in the direction of the rotation, the extended thumb points in the direction of the velocity vector. Therefore, there are only two possible directions for angular velocity, perpendicularly in or out of a plane, for a given observer. For convenience, they may be expressed with plus (ccw) and minus (cw) signs. They both have the units rad/s.

Average angular speed is defined by $\bar{\omega} = \frac{\Delta\theta}{\Delta t} = \frac{\theta - \theta_o}{t - t_o}$.

Like their linear counterparts, angular speed and angular velocity could both be average or instantaneous. Angular speed is the magnitude of angular velocity.

Sometimes angular speed and velocity are given in units of rpm (revolutions per minute). You should first convert them to rad/s before trying to solve the problems. For example:

$$(33\ \text{rmp}) = \frac{33\ \text{rev}}{\text{min}} \times \frac{2\pi\ \text{rad}}{1\ \text{rev}} \times \frac{1\ \text{min}}{60\ \text{s}} = 3.5\ \text{rad/s}.$$

Tangential (linear) speed and angular speed are related to each other through $v = r\,\omega$, where r is the radius. (Generally, linear quantity = radius × angular quantity, e.g., $s = r\,\theta$.)

In circular motion, the time it takes for an object to go though one revolution is called the **period** (T) and the number of revolutions in one second is called **frequency** (f). They are related through $f = \frac{1}{T}$. The SI unit of frequency is 1/s or hertz (Hz).

Example 7.2 A bicycle wheel rotates uniformly through 2.0 revolutions in 4.0 s.

(a) What is the average angular speed of the wheel?

(b) What is the tangential speed of a point 0.10 m from the center of the wheel?

(c) What is the period?

(d) What is the frequency?

Solution: Given: $\theta = 2.0\ \text{rev} = (2.0\ \text{rev}) \times \frac{2\pi\ \text{rad}}{1\ \text{rev}} = 4\pi\ \text{rad}$, $t = 4.0$ s, $r = 0.10$ m.

Find: (a) $\overline{\omega}$ (b) v (c) T (d) f.

(a) $\overline{\omega} = \frac{\Delta\theta}{\Delta t} = \frac{4\pi\ \text{rad}}{4.0\ \text{s}} = \pi\ \text{rad/s} = 3.1\ \text{rad/s}.$

(b) Since the rotation is uniform, $\omega = \overline{\omega}$. So $v = r\omega = (0.10\ \text{m})(3.1\ \text{rad/s}) = 0.31\ \text{m/s}.$

(c) Period is the time for one revolution. So $T = \frac{4.0\ \text{s}}{2.0} = 2.0\ \text{s}.$

(d) $f = \frac{1}{T} = \frac{1}{2.0\ \text{s}} = 0.50\ \text{Hz}.$

---

3. Uniform Circular Motion and Centripetal Acceleration

For **uniform circular motion**, the speed (tangential and angular) is a constant, but there is an acceleration. The linear tangential velocity vector changes direction as the object moves along the circle. Since velocity has both magnitude and direction, a change in direction results in a change in velocity, therefore, an acceleration (see the discussion in Chapter 2). This acceleration is called **centripetal acceleration** (center-seeking) because it is always directed toward the center of the circle. The magnitude of centripetal acceleration is given by $a_c = \dfrac{v^2}{r} = r\omega^2$.

From Newton's second law, net force = mass × acceleration, we conclude that there is a net force associated with centripetal acceleration. In the case of uniform circular motion, this force is called **centripetal force** and directed toward the center of the circle since we know the net force on an object is in the same direction as acceleration. Its magnitude is $F_c = ma_c = \dfrac{m v^2}{r} = mr\omega^2$.

When analyzing circular motions, always think about centripetal acceleration and centripetal force. (Note: centripetal force is not a separate force. It is a net force toward the center of the circle.) A centripetal force is required for objects to stay in circular path. Without the necessary centripetal force, an object will not stay in a circular orbit and instead fly out along a tangent line due to its inertia. Can you identify the centripetal force when a car makes a turn on a flat road and when cloth moves around in a dryer?

Example 7.3 A car is negotiating a flat circular curve of radius 50 m with a speed of 20 m/s. The centripetal force (provided by static friction) is 1.2×10^4 N.

(a) What is the magnitude of the centripetal acceleration of the car?

(b) What is the mass of the car?

Solution: Given: $r = 50$ m, $v = 20$ m/s, $F_c = 1.2\times10^4$ N.

Find: (a) a_c (b) m.

(a) $a_c = \dfrac{v^2}{r} = \dfrac{(20 \text{ m/s})^2}{50 \text{ m}} = 8.0 \text{ m/s}^2$.

(b) From $F_c = ma_c$, we have $m = \dfrac{F_c}{a_c} = \dfrac{1.2\times10^4 \text{ N}}{8.0 \text{ m/s}^2} = 1.5\times10^3$ kg.

---

4. Angular Acceleration

Average **angular acceleration** is defined analogously as its linear counterpart.

$\bar{\alpha} = \dfrac{\Delta\omega}{\Delta t} = \dfrac{\omega - \omega_o}{t - t_o}$. The unit of angular acceleration is rad/s^2.

An object having angular acceleration has also tangential acceleration, i.e., its tangential speed changes. They are related by $a_t = r\alpha$ (again, linear quantity = radius × angular quantity, as in $s = r\theta$ and $v = r\omega$).

If an object is in nonuniform circular motion, it will have both centripetal acceleration (due to circular motion) and tangential acceleration (due to the change in tangential speed). The total acceleration is given by $\mathbf{a} = a_t\,\mathbf{t} + a_c\,\mathbf{r}$, where **t** is the unit vector in the tangential direction and **r** is the unit vector pointing toward the center of the circle.

As you can see, circular motion and linear motion are analogous (at least mathematically). As a matter of fact, we can arrive at all the equations for angular motion from the kinematic equations in Chapter 2 if we just use the following substitutions: $\theta \to x$, $\omega \to v$, and $\alpha \to a$ (or add the word angular to every linear quantity except time). The angular kinematic equations are (their linear counterparts are in parenthesis):

$$\theta = \bar{\omega}\, t \qquad (x = \bar{v}\, t),$$

$$\bar{\omega} = \frac{\omega + \omega_o}{2} \qquad (\bar{v} = \frac{v + v_o}{2}),$$

$$\omega = \omega_o + \alpha\, t \qquad (v = v_o + a\, t),$$

$$\theta = \omega_o\, t + \tfrac{1}{2}\alpha\, t^2 \qquad (x = v_o\, t + \tfrac{1}{2} a\, t^2),$$

$$\omega^2 = \omega_o^2 + 2\alpha\theta \qquad (v^2 = v_o^2 + 2\, a\, x).$$

Example 7.4 A wheel is rotating with a constant angular acceleration of 3.5 rad/s^2. If the initial angular velocity is 2.0 rad/s and is speeding up, find

(a) the angle the wheel rotate through in 2.0 s,

(b) the angular speed at $t = 2.0$ s.

Solution: Given: $\alpha = 3.5$ rad/s^2, $\omega_o = 0$, $t = 2.0$ s.

Find: (a) θ (b) ω.

(a) $\theta = \omega_o t + \frac{1}{2}\alpha t^2 = (2.0 \text{ rad/s})(2.0 \text{ s}) + \frac{1}{2}(3.5 \text{ rad/s}^2)(2.0 \text{ s})^2 = 11$ rad.

(b) $\omega = \omega_o + \alpha\, t = 2.0 \text{ rad/s} + (3.5 \text{ rad/s})(2.0 \text{ s}) = 9.0$ rad/s.

Example 7.5 A Ferris wheel rotating at 20 rad/s decelerates with a constant angular acceleration of -5.0 rad/s^2. How many revolutions does it rotate before coming to rest?

Solution: The unknown (how many turns) is related to the angle, θ, since 2π rad is one revolution or turn.

Given: $\omega_o = 20$ rad/s, $\omega = 0$ (coming to rest), $\alpha = -5.0$ rad/s^2.

Find: θ.

From $\omega^2 = \omega_o^2 + 2\alpha\theta$, $\theta = \dfrac{\omega^2 - \omega_o^2}{2\,\alpha} = \dfrac{0 - (20 \text{ rad/s})^2}{2(-5.0 \text{ rad/s}^2)} = (40 \text{ rad}) \times \dfrac{1 \text{ rev}}{2\pi \text{ rad}} = 6.4$ rev (turns).

---

5. Newton's Law of Gravitation

Newton's **universal law of gravitation** states that the mutual gravitational attraction between any two point masses is directly proportional to the product of their masses and inversely proportional to the square of the distance between them. The magnitude is given by $F = \frac{G\,m_1\,m_2}{r^2}$, where $G = 6.67\times10^{-11}\ \text{N·m}^2/\text{kg}^2$, is the **universal gravitational constant.**

Note: The magnitude of the force depends on $\frac{1}{r^2}$ (inverse square). For example, if the distance between the masses is doubled, the force between them would decrease to $\frac{1}{2^2} = \frac{1}{4}$ (not $\frac{1}{2}$).

According to Newton's second law, $F = ma$, the acceleration due to gravity by a mass M acting on another object m is then equal to $a_g = \frac{F}{m} = \frac{G\,M}{r^2}$.

At the Earth's surface: $a_g = g = \frac{G\,M_E}{R_E^2}$, where M_E and R_E are the mass of radius of the Earth.

At an altitude h above the Earth: $a_g = \frac{G\,M_E}{(R_E + h)^2}$.

The **gravitational potential energy** between two masses is given by $U = -\frac{G\,m_1\,m_2}{r}$, **note:** it is just r (not r^2) in the denominator. The minus sign arises from the choice of zero reference point (point where $U = 0$), which is $r = \infty$.

Example 7.6 The hydrogen atom consists of a proton of mass 1.67×10^{-27} kg and an orbiting electron of mass 9.11×10^{-31} kg. In one of its orbits, the electron is 5.3×10^{-11} m from the proton and in another orbit, it is 10.6×10^{-11} m from the proton.

(a) What are the mutual attractive forces when the electron is in these orbits, respectively?

(b) If the electron jumps from the large orbit to the small one, what is the change in potential energy?

Solution: Given: $m_1 = 1.67\times10^{-27}$ kg, $m_2 = 9.11\times10^{-31}$ kg, $r_1 = 5.3\times10^{-11}$ m, $r_2 = 10.6\times10^{-11}$ m.

Find: (a) F_1 and F_2 (b) ΔU

$$\text{(a) } F_1 = \frac{G\,m_1\,m_2}{r_1^2} = \frac{(6.67\times10^{-11}\ \text{N·m}^2/\text{kg}^2)(1.67\times10^{-27}\ \text{kg})(9.11\times10^{-31}\ \text{kg})}{(5.3\times10^{-11}\ \text{m})^2} = 3.6\times10^{-47}\ \text{N}.$$

Since $r_2 = 2\,r_1$ and $F \propto \frac{1}{r^2}$, $\frac{F_2}{F_1} = \frac{r_1^2}{r_2^2} = \frac{1}{2^2} = \frac{1}{4}$.

So $F_2 = \frac{1}{4}F_1 = \frac{1}{4}(3.6\times10^{-47}\ \text{N}) = 9.0\times10^{-48}$ N.

(c) $\Delta U = U_1 - U_2 = -\frac{G\,m_1\,m_2}{r_1} - (-\frac{G\,m_1\,m_2}{r_2}) = G\,m_1\,m_2\,(\frac{1}{r_2} - \frac{1}{r_1})$

$= (6.67\times10^{-11}\ \text{N·m}^2/\text{kg}^2)(1.67\times10^{-27}\ \text{kg})(9.11\times10^{-31}\ \text{kg})\ (\frac{1}{10.6\times10^{-11}\ \text{m}} - \frac{1}{5.3\times10^{-11}\ \text{m}})$

$= -9.6\times10^{-58}$ J. So the atom loses potential energy.

Example 7.7 Calculate the acceleration due to gravity at the surface of the Moon.

Solution: Given: $R_M = 1750\ \text{km} = 1.75\times10^6\ \text{m}$, $M_M = 7.4\times10^{22}\ \text{kg}$ (from back inside cover).

Find: a_g.

$$a_g = \frac{G\,M}{r^2} = \frac{G\,M_M}{R_M^2} = \frac{(6.67\times10^{-11}\ \text{N·m}^2/\text{kg}^2)(7.4\times10^{22}\ \text{kg})}{(1.75\times10^6\ \text{m})^2} = 1.6\ \text{m/s}^2.$$

This is about $\frac{1}{6}$ of the value at the surface of the Earth, $g = 9.80\ \text{m/s}^2$.

---

6. Kepler's Laws and Earth Satellites

(1) **The law of orbits:** Planets move in elliptical orbits with the Sun at one of the focal points.

(2) **The law of areas:** A line from the Sun to a planet sweeps out equal area in equal time interval.

(3) **The law of periods:** The square of the period of a planet is directly proportional to the cube of the average distance of the planet from the Sun, that is, $T^2 \propto r^3$. Or $T^2 = K\,r^3$, where $K = 2.97\times10^{-19}\ \text{s}^2/\text{m}^3$ for the Sun.

The **escape speed** is the initial speed needed to escape from the surface of a planet or moon. For Earth satellites, this speed is $v_{esc} = \sqrt{\frac{2\,G\,M_E}{R_E}} = \sqrt{2\,g\,R_E}$, since $g = \frac{G\,M_E}{R_E}$.

Example 7.8 The planet Jupiter is 7.78×10^{11} m from the Sun. How long does it take for Jupiter to orbit once about the Sun?

Solution: Given: $r = 7.78\times10^{11}$ m.

Find: T.

From Kepler's law of periods, $T^2 = (2.97\times10^{-19}\ \text{s}^2/\text{m}^3)r^3$, we have

$$T = \sqrt{(2.97\times10^{-19}\ \text{s}^2/\text{m}^3)r^3} = \sqrt{(2.97\times10^{-19}\ \text{s}^2/\text{m}^3)(7.78\times10^{11}\ \text{m})^3} = 3.75\times10^8\ \text{s} \approx 12\ \text{years}.$$

Example 7.9 If a satellite were launched from the surface of the Moon, at what initial speed would it need to begin in order for it to escape the gravitational attraction of the Moon?

Solution: With the known data:

$$v_{esc} = \sqrt{\frac{2\,G\,M_M}{R_M}} = \sqrt{\frac{2(6.67\times10^{-11}\ \text{N·m}^2/\text{kg}^2)(7.4\times10^{22}\ \text{kg})}{1.75\times10^6\ \text{m}}} = 2.4\times10^3\ \text{m/s}.$$

IV. Mathematical Summary

Arc Length	$s = r\,\theta$ (7.3)	Relates arc length to radius and angle (in radians)
Angular Kinematic Equation 1	$\theta = \bar{\omega}\,t$ (7.5)	Relates angular displacement with average angular velocity and time
Angular Kinematic Equation 2	$\bar{\omega} = \frac{\omega + \omega_o}{2}$ (2, Table 7.2)	Defines average angular velocity for circular motion with constant angular acceleration
Angular Kinematic Equation 3	$\omega = \omega_o + \alpha\,t$ (7.12)	Relates final angular velocity with initial angular velocity, angular acceleration, and time (constant angular acceleration only)
Angular Kinematic Equation 4	$\theta = \omega_o\,t + \frac{1}{2}\alpha\,t^2$ (4, Table 7.2)	Relates angular displacement with initial angular velocity, angular acceleration, and time (constant angular acceleration only)
Angular Kinematic Equation 5	$\omega^2 = \omega_o^2 + 2\alpha\theta$ (5, Table 7.2)	Relates final angular velocity with initial angular velocity, angular acceleration, and angular displacement (constant angular acceleration only)
Tangential and Angular Speeds	$v = r\,\omega$ (7.6)	Relates tangential speed with radius and angular speed
Angular Speed	$\omega = \frac{2\pi}{T} = 2\pi\,f$ (7.8)	Defines angular speed for uniform circular motion
Frequency and Period	$f = \frac{1}{T}$ (7.7)	Relates frequency with period.
Centripetal Acceleration	$a_c = \frac{v^2}{r} = r\,\omega^2$ (7.10)	Defines centripetal acceleration.
Tangential and Angular Accelerations	$a_t = r\,\alpha$ (7.13)	Relates tangential acceleration with radius and angular acceleration.

Centripetal Force and Acceleration	$F_c = m\,a_c = \dfrac{m\,v^2}{r}$ (7.11)	Defines centripetal force.
Newton's Law of Gravitation	$F = \dfrac{G\,m_1\,m_2}{r^2}$ (7.15) $G = 6.67\times10^{-11}$ N·m²/kg²	Calculates the mutual attractive forces between two masses.
Acceleration due to Gravity at an Altitude h	$a_g = \dfrac{G\,M_E}{(R_E + h)^2}$ (7.18)	Calculates the acceleration due to gravity at an altitude of h above the Earth surface.
Gravitational Potential Energy	$U = -\dfrac{G\,m_1\,m_2}{r}$ (7.19)	Defines gravitational potential energy for a two mass system ($U_\infty = 0$).
Kepler's Law of Periods	$T^2 = K\,r^3$ (7.23) $K = 2.97\times10^{-19}$ s²/m³ for Sun	Relates period with radius for objects orbiting the Sun.
Escape Speed	$v_{esc} = \sqrt{\dfrac{2\,G\,M_E}{R_E}}$ $= \sqrt{2\,g\,R_E}$ (7.24)	Gives the escape speed of the Earth.
Energy of Orbiting Satellites Orbiting Earth	$E = -\dfrac{G\,m\,M_E}{2\,r}$ (7.27) $K = \lvert E\rvert$ (7.29)	Calculates the total energy and kinetic energy of objects orbiting the Earth.

V. Solutions of Selected Exercises and Paired Exercises

5. (a) $(10°)\times\dfrac{\pi\text{ rad}}{180°} = 0.175$ rad.

(b) $(270°)\times\dfrac{\pi\text{ rad}}{180°} = 4.71$ rad.

(c) $(45°)\times\dfrac{\pi\text{ rad}}{180°} = 0.785$ rad.

(d) $(450°)\times\dfrac{\pi\text{ rad}}{180°} = 7.85$ rad.

6. (a) $(\pi/25\text{ rad})\times\dfrac{180°}{\pi\text{ rad}} = 7.20°$.

(b) $(\pi/10\text{ rad})\times\dfrac{180°}{\pi\text{ rad}} = 18.0°$.

(c) $(1.25\text{ rad})\times\dfrac{180°}{\pi\text{ rad}} = 71.6°$.

(d) $(4\pi\text{ rad})\times\dfrac{180°}{\pi\text{ rad}} = 720°$.

12. (a) $s = r\,\theta = (0.50\text{ km})(300°)\times\dfrac{\pi\text{ rad}}{180°} = = 2.6$ km.

(b) $s = (0.50\text{ km})(4)(2\pi\text{ rad}) = 13$ km.

16. Since $C = 2\pi\,r = (6.28\text{ rad})r$, the answer is no. So we can cut 6 such pieces and one 0.28 rad piece.

28. (a) $\omega = \dfrac{\Delta\theta}{\Delta t} = \dfrac{(24 \text{ rev})(2\pi \text{ rad/rev})}{(3.0 \text{ min})(60 \text{ s/min})} = 0.84 \text{ rad/s}.$

(b) $v = r\,\omega$, $v_4 = (4.0 \text{ m})(0.838 \text{ rad/s}) = 3.4 \text{ m/s}$ and $v_5 = (5.0 \text{ m})(0.838 \text{ rad/s}) = 4.2 \text{ m/s}.$

35. (a) $f = 1 \text{ rev/d} = \dfrac{1 \text{ rev}}{24(3600 \text{ s})} = 1.16\times10^{-5} \text{ Hz}.$

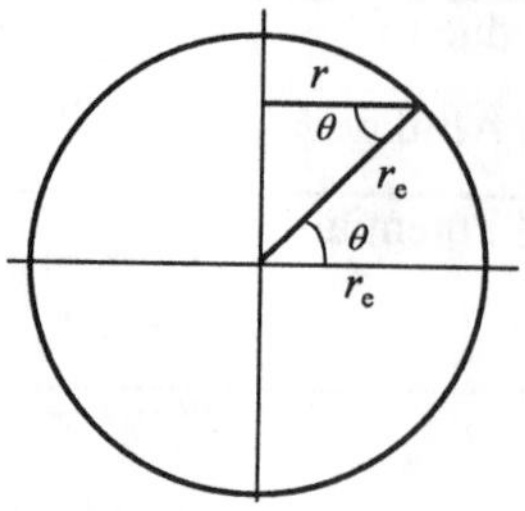

(b) $\omega = \dfrac{\Delta\theta}{\Delta t} = \dfrac{2\pi \text{ rad}}{24(3600 \text{ s})} = 7.27\times10^{-5}$ rad/s in the direction of geographic north..

(c) The angular velocity of any point on the Earth is the same (it is one rev per day).

$r = r_e \cos\theta$, so $\dfrac{v_{45}}{v_{eq}} = \dfrac{r\omega_{45}}{r_e\,\omega_{eq}} = \dfrac{r_e \cos 45°}{r_e} = \sin 45° = 70.7\%.$

Therefore $\dfrac{\omega_{45}}{\omega_{eq}} = 100\%.$

48. (a) $v = \dfrac{\Delta d}{\Delta t}\dfrac{2\pi r}{\Delta t} = \dfrac{2\pi(1.50 \text{ m})}{1.20 \text{ s}} = 7.85 \text{ m/s}.$

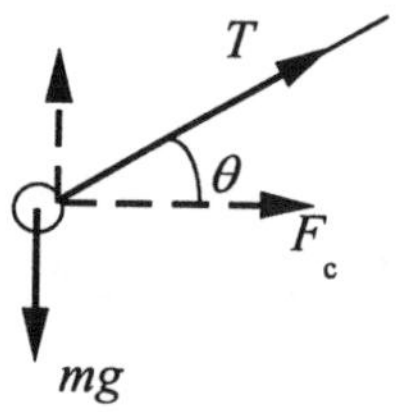

(b) $a_c = \dfrac{v^2}{r} = \dfrac{(7.85 \text{ m/s})^2}{1.50 \text{ m}} = 41.1 \text{ m/s}^2.$

(c) No, the string can not be exactly horizontal. There must be something upward (a component of the tension) to balance the downward weight.

52. The Earth rotates once a day. $\omega = \dfrac{\Delta\theta}{\Delta t} = \dfrac{2\pi \text{ rad}}{(24)(3600 \text{ s})} = 7.27\times10^{-5} \text{ rad/s}.$

$a_c = r\omega^2 = (6.4\times10^6 \text{ m} + 3.56\times10^7 \text{ m})(7.27\times10^{-5} \text{ rad/s})^2 = 0.22 \text{ m/s}^2$ and gravity causes this.

64. (a) Given: $\omega_o = 0$, $\omega = \dfrac{v}{r} = \dfrac{2.20 \text{ m/s}}{17.5 \text{ m}} = 0.126 \text{ rad/s}$, $t = 15.0$ s. Find: α.

$\alpha = \dfrac{\omega - \omega_o}{t} = \dfrac{0.126 \text{ rad/s} - 0}{15.0 \text{ s}} = 8.40\times10^{-3} \text{ rad/s}^2.$

(b) After reaching the constant operating speed, $\alpha = 0$ and so $a_t = r\,\alpha = 0$.

78. $a_g = \dfrac{G M_e}{(R_e + h)^2} = \dfrac{(6.67\times10^{-11} \text{ N·m}^2/\text{kg}^2)(6.0\times10^{24} \text{ kg})}{(6.4\times10^6 \text{ m} + 8.8\times10^3 \text{ m})^2} = 9.74 \text{ m/s}^2.$

81. $F_1 = F_3 = \dfrac{G\,m^2}{d^2} = \dfrac{(6.67\times10^{-11}\ \text{N·m}^2/\text{kg}^2)(2.5\ \text{kg})^2}{(1.0\ \text{m})^2} = 4.17\times10^{-10}\ \text{N},$

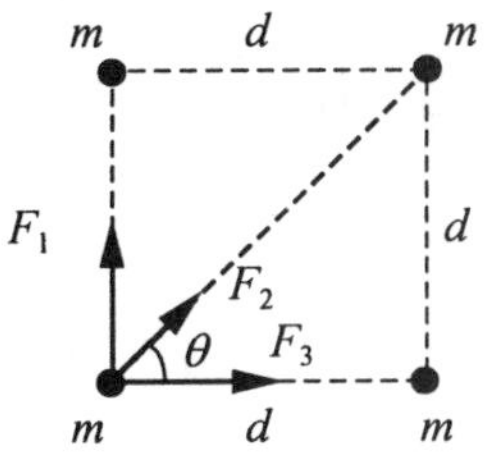

The diagonal distance is $\sqrt{(1.0\ \text{m})^2 + (1.0\ \text{m})^2} = \sqrt{2}\ \text{m}$,

so $F_3 = \dfrac{(6.67\times10^{-11}\ \text{N·m}^2/\text{kg}^2)(2.5\ \text{kg})^2}{(\sqrt{2}\ \text{m})^2} = 2.08\times10^{-10}\ \text{N}.$

From symmetry the net force is

$F = \sqrt{(4.17\times10^{-10}\ \text{N})^2 + (4.17\times10^{-10}\ \text{m})^2} + 2.08\times10^{-10}\ \text{N} = 8.0\times10^{-10}\ \text{N},$

toward opposite corner.

82. When the Earth's gravitational force equals the lunar gravitational force, we have

$F_\text{e} = \dfrac{G\,M_\text{e}\,m}{x^2} = F_\text{m} = \dfrac{G\,M_\text{m}\,m}{(3.8\times10^8\ \text{m} - x)^2}.$

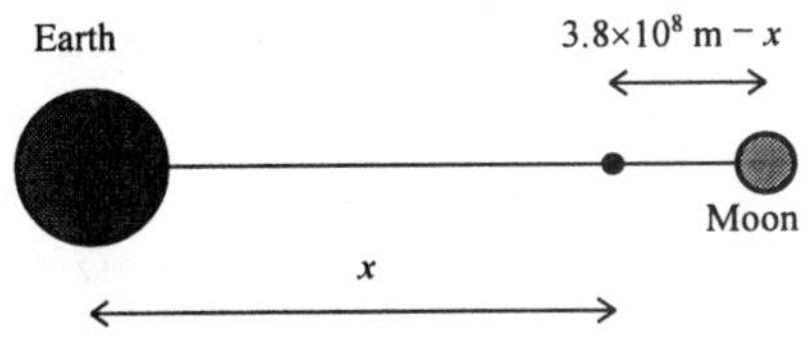

Taking the square root on both sides gives

$\dfrac{\sqrt{M_\text{e}}}{x} = \dfrac{\sqrt{M_\text{m}}}{3.8\times10^8\ \text{m} - x},$

or $\sqrt{7.4\times10^{22}\ \text{kg}}\ x = \sqrt{6.0\times10^{24}\ \text{kg}}\ (3.8\times10^8\ \text{m} - x).$ Solve for $x = 3.4\times10^8$ m from Earth.

No, there are still other gravitational forces from the other planets and the Sun.

90. (a) $U_\text{tot} = -\dfrac{G\,m_1\,m_2}{r_{12}} - \dfrac{G\,m_1\,m_3}{r_{13}} - \dfrac{G\,m_2\,m_3}{r_{23}}$

$= -(6.67\times10^{-11}\ \text{N·m}^2/\text{kg}^2)\left[\dfrac{(1.0\ \text{kg})^2}{0.80\ \text{m}} + \dfrac{(1.0\ \text{kg})^2}{0.80\ \text{m}} + \dfrac{(1.0\ \text{kg})^2}{0.80\ \text{m}}\right] = -2.5\times10^{-10}\ \text{J}.$

(b) From symmetry, the force at the center is zero. So the force per unit mass is also 0.

100. Replace the mass of the Sun with the mass of the Earth.

$K == \dfrac{4\pi^2}{G\,M_\text{e}} = \dfrac{4\pi^2}{(6.67\times10^{-11}\ \text{N·m}^2/\text{kg}^2)(6.0\times10^{24}\ \text{kg})} = 9.86\times10^{-14}\ \text{s}^2/\text{m}.$

$T =$ 1day = 24(3600 s) = 86400 s (synchronous satellite). From $T^2 = K\,r^3 = K\,(R_\text{e} + h)^3$,

we have $h = \sqrt[3]{\dfrac{T^2}{K}} - R_\text{e} = \sqrt[3]{\dfrac{(86400\ \text{s})^2}{9.86\times10^{-14}\ \text{s}^2/\text{m}}} - 6.4\times10^6\ \text{m} = 3.6\times10^7\ \text{m}.$

107. The Moon's gravitational attraction is greater on the water on the near side than on the Earth and produces one bulge for one tide;. the attraction is greater on the Earth than on the water on the far side and so the Earth moves toward the Moon and leaves the water behind for another bulge.

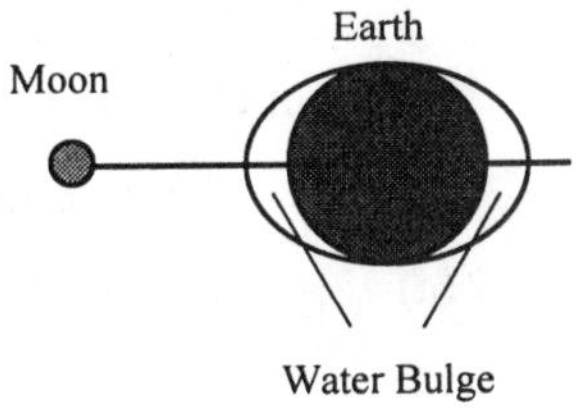

108. The centripetal force is provided by the combination of

$T - mg\cos\theta = m\frac{v^2}{r}$, where $\cos\theta = \frac{h}{l}$ and v could be found from energy conservation. When the girl is h below the original position,

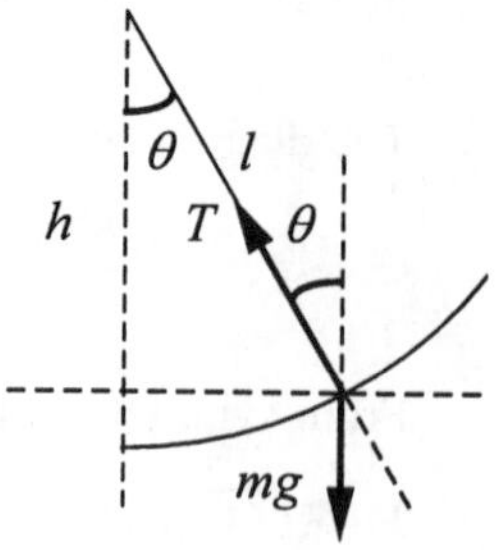

we have $\frac{1}{2}m(0)^2 + mgh = \frac{1}{2}mv^2 + mg(0)$, so $v^2 = 2gh$.

Therefore $T = mg\cos\theta + m\frac{v^2}{r} = \frac{mgh}{l} + \frac{m(2gh)}{l} = \frac{3mgh}{l}$.

(a) $h = 0$, so $T = 0$.

(b) $h = 12\ m - 5.0\ \text{m} = 7.0\ \text{m}$, so $T = \frac{3(60\ \text{kg})(9.80\ \text{m/s}^2)(7.0\ \text{m})}{10\ \text{m}} = 1.2\times10^3\ \text{N}$.

(c) $h = 12\ \text{m} - 2.0\ \text{m} = 10\ \text{m}$, so $T = \frac{3(60\ \text{kg})(9.80\ \text{m/s}^2)(10\ \text{m})}{10\ \text{m}} = 1.8\times10^3\ \text{N}$.

114. $K + U = 0$, so $\frac{1}{2}mv_e^2 - \frac{GM_e m}{R_e + h} = 0$.

Therefore $v_e = \sqrt{\frac{2GM_e}{R_e + h}} = \sqrt{\frac{2(6.67\times10^{-11}\ \text{N}\cdot\text{m}^2/\text{kg}^2)(6.0\times10^{24}\ \text{kg})}{6.4\times10^6\ \text{m} + 750\times10^3\ \text{m}}} = 1.1\times10^4\ \text{m/s}$.

120. Centripetal force is supplied by the tension in the sling. When released, the rock goes tangential with a much higher speed than with just the arm. (Sling increases the radius and $v = r\,\omega$.)

VI. Practice Quiz

1. Convert 50 revolutions per minute to rad/s.

(a) 314 rad/s (b) 5.2 rad/s (c) 0.83 rad/s (d) 157 rad/s (e) 2.6 rad/s

2. The cutting cord on a gas-powered weed cutter is 0.16 m in length. If the motor rotates at the rate of 20 rev/s, what is the approximate tangential velocity of the end of the cord?

(a) 20 m/s (b) 25 m/s (c) 35 m/s (d) 65 m/s (e) 630 m/s

3. A 0.300-kg mass, attached to the end of a 0.75-m string, is whirled around in a smooth level table. If the maximum tension that the string can withstand is 250 N then what maximum tangential speed can the mass have if the string is not to break?

(a) 22.4 m/s (b) 25 m/s (c) 19.4 m/s (d) 32.7 m/s (e) 275 m/s

4. How many radians does a 0.300-m radius automobile tire rotate after starting from rest and accelerating at a constant angular acceleration of 2.00 rad/s^2 over a 5.00 s interval?

(a) 0.500 rad (b) 1.00 rad (c) 2.00 rad (d) 12.5 rad (e) 25.0 rad

5. The gravitation attractive force between two masses is F. If the masses are moved to half of their initial distance, what is the gravitational attractive force?

(a) $F/4$ (b) $F/2$ (c) F (d) $2F$ (e) $4F$.

6. A hypothetical planet has a mass of half that of the Earth and a radius of twice that of the Earth. What is the acceleration due to gravity on the planet in terms of g?

(a) g (b) $g/2$ (c) $g/4$ (d) $g/8$ (e) $g/16$

7. A wheel starts from rest and reaches an angular speed of 6.0 rad/s while turning through 2.0 revolutions. What is the angular acceleration of the wheel?

(a) 0.24 rad/s^2 (b) 1.4 rad/s^2 (c) 3.0 rad/s^2 (d) 9.0 rad/s^2 (e) 18 rad/s^2

8. An object moves with a constant speed of 20 m/s on a circular track of radius 100 m. The acceleration of the object is

(a) zero. (b) 0.20 rad/s^2. (c) 0.40 m/s^2. (d) 2.0 m/s^2. (e) 4.0 rad/s^2.

9. The minute hand in a clock has a length of 0.30 m. What distance does the tip of the minute hand sweep through in 15 minutes?

(a) 0.12 m (b) 0.47 m (c) 0.36 m (d) 0.24 m (e) 27 m

10. How many revolutions per minute must a circular, rotating spacestation (r = 1000 m) rotate to produce an artificial gravity of 9.80 m/s^2?

(a) 0.95 rpm (b) 0.85 rpm (c) 0.75 rpm (d) 0.65 rpm (e) 0.55 rpm

Answers to Practice Quiz:

1. b 2. a 3. b 4. e 5. e 6. d 7. b 8. e 9. b 10. a

CHAPTER 8

Rotational Motion and Equilibrium

I. Chapter Objectives

Upon completion of this chapter, you should be able to:

1. distinguish between pure translational and pure rotational motions of a rigid body and state the condition(s) for rolling without slipping.
2. define torque, apply the conditions for mechanical equilibrium, and describe the relationship between the location of the center of gravity and stability.
3. describe what is meant by the moment of inertia of a rigid body and apply the rotational form of Newton's second law to physical situations.
4. discuss, explain, and use the rotational forms of work, kinetic energy, and power.
5. define angular momentum and apply the conservation of angular momentum to physical situations.

II. Key Terms

Upon completion of this chapter, you should be able to define and/or the explain following key terms:

rigid body	**center of gravity**
translational motion	**equilibrium**
rotational motion	**stable**
instantaneous axis of rotation	**unstable**
moment (lever arm)	**moment of inertia**
torque	**parallel axis theorem**
translational equilibrium	**rotational**
concurrent forces	**work**
rotational equilibrium	**power**
mechanical equilibrium	**kinetic energy**
static equilibrium	**angular momentum conservation**

The definitions and/or explanations of the most important key terms can be found in the following section: **III. Chapter Summary and Discussion.**

III. Chapter Summary and Discussion

1. Rigid Bodies, Translation, and Rotation

A **rigid body** is an object or system of particles in which the interparticle distances (distances between particles) are fixed and remain constant. In pure **translational motion**, every particle of an object or system of particles has the same instantaneous velocity, which means there is no rotation. In pure **rotational motion**, all particles of an object or system of particles have the same instantaneous angular velocity.

Generally, rigid body motion is a combination of translational and rotational motions. If an object is rolling without slipping, the following conditions must be satisfied:

$$s = r\,\theta, \quad \text{or} \quad v = r\,\omega, \quad \text{or} \quad a = r\,\alpha.$$

2. Torque, Equilibrium, and Stability

The magnitude of the **torque** (τ) produced by a force about an axis of rotation is given by $\tau = r_\perp F$. The distance $r_\perp = r \sin\theta$ is called the **moment arm** or **lever arm**, which is the *perpendicular* distance from the axis of rotation to the line of the action of the force. The SI unit of torque is m·N (the same as for work N·m or J, but which we write in reverse for distinction). By convention, counterclockwise torque is positive and clockwise torque is negative.

Note: The moment arm is the *perpendicular* distance from the axis of rotation to the *line of the action* of the force, which is usually not the same as simply the distance from the axis to the force.

An object is in **translational equilibrium** if the net force on it is zero, and an object is in **rotational equilibrium** if the net torque is zero. If an object is in both translational equilibrium and rotational equilibrium, it is said to be in **mechanical equilibrium**. So the conditions for an object in mechanical equilibrium are $\Sigma \mathbf{F}_i = 0$ and $\Sigma\tau_i = 0$. A special case of mechanical equilibrium is **static equilibrium** in which objects are at rest and remain that way.

Note: When applying $\Sigma\tau_i = 0$, you can choose anywhere as the axis of rotation because the object is not rotating about any point. However, some choices of axes of rotation can greatly simplify the solutions of such problems. Generally, the axis of rotation should be chosen through a point through which some unknown force(s) act to make the moment of arm(s) of the force(s) zero, therefore the torque(s) zero.

Example 8.1 The bolts on a car wheel require tightening to a torque of 90 m·N. If a 30 cm long wrench is used, what is the magnitude of the force required

(a) when the force is perpendicular to the wrench,

(b) when the force is 53° to the wrench as shown.

Solution: Given: $r = 0.30$ m, $\tau = 90$ m·N, (a) $\theta = 90°$, (b) $\theta = 53°$.

Find: F in (a) and (b).

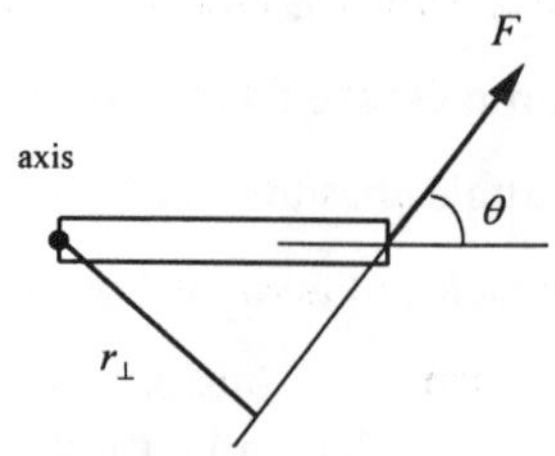

(a) $r_\perp = r \sin \theta = (0.30 \text{ m}) \sin 90° = 0.30$ m.

From $\tau = r_\perp F$, we have $F = \dfrac{\tau}{r_\perp} = \dfrac{90 \text{ m·N}}{0.30 \text{ m}} = 3.0 \times 10^2$ N.

(b) $r_\perp = (0.30 \text{ m}) \sin 53° = 0.24$ m. $F = \dfrac{90 \text{ m·N}}{0.24 \text{ m}} = 3.8 \times 10^2$ N.

Why does it need more force in (b) than in (a)?

Example 8.2 A uniform board of weight 40 N supports two children weighing 500 N and 350 N, respectively. If the support is at the center of the board and the 500-N child is 1.5 m from the center. What is the position of the 350-N child?

Solution: Since the board is uniform, its center of mass and center of gravity are at the center. Choose the support as the axis of rotation since there is an unknown force N (although we can find it easily with $\Sigma \mathbf{F}_i = 0$). The weight of the 500-N child produces a ccw torque, the weight of the 350-N child gives a cw torque, the weight of the board and the normal force by the support have no contribution to torque because their moment arms are zero.

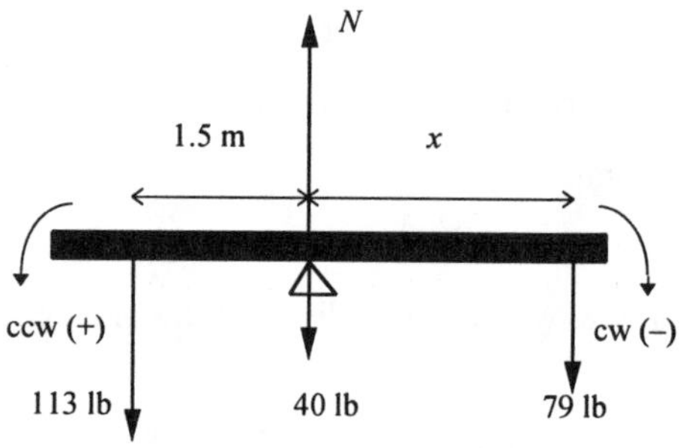

From the condition for rotational equilibrium

$\Sigma \tau_i = (500 \text{ N})(1.5 \text{ m}) - (350 \text{ N})\, x + N\,(0) = 0.$ Solve for $x = 2.1$ m.

Example 8.3 A 10-m uniform beam of weight 100 N is supported by two ropes at the ends as shown. If a 400-N person sits at 2.0 m from one end of the beam, what are the tensions in the ropes?

Solution: Since the beam is uniform, its center of mass and center of gravity are at its geometrical center, which is 5.0 m from one end. We first draw the free-body diagram of the beam. Where should we choose the axis of rotation? What if we chose the center of the beam? If we do so, we will have two unknowns (T_1 and T_2) in our $\Sigma\tau_i = 0$ equation. We cannot solve for two unknowns within just one equation (Although we can get another equation from $\Sigma \mathbf{F}_i = 0$. Then we have to solve simultaneous equations.) We can simplify the solution of this problem by choosing our axis either at T_1 or T_2. Let's try T_2 first, i.e., the axis of rotation is at the right end of the beam.

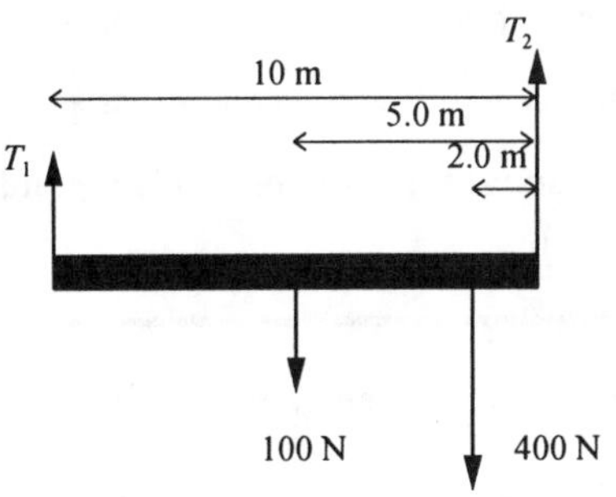

About the right end, T_1 has a cw torque, the 100 N and 400 N forces have ccw torques, and T_2 has zero torque since its moment arm is zero.

From the condition for rotational equilibrium, we have

$\Sigma\tau_i = -T_1\,(10\text{ m}) + (100\text{ N})(5.0\text{ m}) + (400\text{ N})(2.0\text{ m}) + T_2\,(0) = 0.$ Solve for $T_1 = 130\text{ N}.$

To find T_2, we can choose the left end as the axis of rotation and repeat our calculation. Or we can use the condition for translational equilibrium.

$\Sigma \mathbf{F}_i = T_1 + T_2 - 100\text{ N} - 400\text{ N} = 0.$ So $T_2 = 500\text{ N} - T_1 = 500\text{ N} - 130\text{ N} = 370\text{ N}.$

If an object is in **stable equilibrium**, any small displacement results in a restoring force or torque, which tends to return the object to its original equilibrium. For an object in **unstable equilibrium**, any small displacement from equilibrium results in a force or torque that tends to take the object farther away from its equilibrium position. An object will be in stable equilibrium as long as its center of gravity lies above and inside its original base of support., i.e., line of action of the weight force of the center of gravity intersects the original base of support.

3. Rotational Dynamics

The torque on a particle due to a constant force is $\tau = m\,r^2\,\alpha$. Similarly for a rotating body about a fixed axis, $\tau = \Sigma\,(m_i r_i^2)\alpha = I\,\alpha$, where $I = \Sigma\, m_i r_i^2$ is called the **moment of inertia** (rotational analog of mass). The moments of inertia of many continuous objects (objects treated as an infinite number of particles, each with a very small mass) have to be determined with calculus. The results are listed in Figure 8.15 on page 261 in the text.

The moment of inertia about an axis parallel to an axis through its center of mass is given by the **parallel axis theorem**: $I = I_{CM} + M\,d^2$, where I_{CM} is the moment of inertia about an axis through the center of mass, M is the mass of the object, and d is the distance between the parallel axes.

In Chapter 7, we discussed the fact that the kinematic equations for angular motion are identical to those for linear motion if the linear quantities are replaced by their angular counterparts (x by θ, v by ω, and a by α). In dynamics, the rotational counterparts of force and mass are torque and moment of inertia. Therefore we can write the **rotational form of Newton's second law** as $\tau = I\,\alpha$ (for linear motion, it is $\mathbf{F} = m\,\mathbf{a}$).

Example 8.4 The moment of inertia of a sphere rotating about an axis through its center is $\frac{2}{5}MR^2$, where M is the mass of the sphere and R is the radius of the sphere. What is the moment of inertia of the sphere about an axis through the contact point when it is rolling on a horizontal surface?

Solution: When the sphere is rolling, the axis of rotation is though the point where the sphere makes contact with the surface. The distance from this axis to an axis through the center of the sphere is R, the radius of the sphere. From the parallel axis theorem, we have $I = I_{CM} + M\,d^2 = \frac{2}{5}MR^2 + MR^2 = \frac{7}{5}MR^2$.

Example 8.5 A solid cylinder of mass 10 kg is pivoted about a frictionless axis through the center O. A rope wrapped around the outer radius R_1 = 1.0 m, exerts a force F_1 = 5.0 N to the right. A second rope wrapped around another section of radius R_2 = 0.50 m exerts a force F_2 = 6.0 N downward.

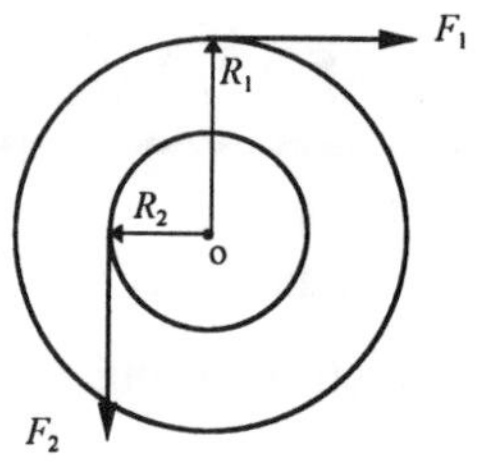

(a) What is the angular acceleration of the disk?

(b) If the disk starts from rest, how many radians does it rotate through in the first 5.0 s?

Solution: Given: M = 10 kg, F_1 = 5.0 N, R_1 = 1.0 m, F_2 = 6.0 N, R_2 = 0.50 m, t = 5.0 s.

Find: (a) α (b) θ.

(a) About an axis through O, the torque by F_1 is cw and the torque by F_2 is ccw. From Figure 8.15, the moment of inertia of a disk is given by $I = \frac{1}{2}MR_2^2 = \frac{1}{2}(10\text{ kg})(1.0\text{ m})^2 = 5.0\text{ kg·m}^2$.

The net torque on the disk is $\tau_{net} = -F_1R_1 + F_2R_2 = -(5.0\text{ N})(1.0\text{ m}) + (6.0\text{ N})(0.50\text{ m}) = -2.0\text{ m·N}$.

From Newton's second law, we have $\alpha = \frac{\tau_{net}}{I} = \frac{-2.0\text{ m·N}}{5.0\text{ kg·m}^2} = -0.40\text{ rad/s}^2$. The negative sign indicates that the disk will accelerates clockwise.

(b) From rotational kinematics, $\theta = \omega_o\,t + \frac{1}{2}\alpha\,t^2 = 0 + \frac{1}{2}(-0.40\text{ rad/s}^2)(5.0\text{ s})^2 = -5.0\text{ rad}$.

Again, the negative sign indicates that the angular displacement is counterclockwise.

---

4. Rotational Work and Kinetic Energy

As was discussed earlier, the rotational counterparts of x, v, a, F, and m are θ, ω, α, τ, and I. The equations in linear motion are still valid if we replace the linear quantities by their rotational counterparts. The original linear equations are given in parentheses (a detailed comparison can be found in Table 8.1 on page 267 in the text).

Work:	$W = \tau\,\theta$	$(W = F\,d = F\,x)$
Kinetic Energy:	$K = \frac{1}{2} I\,\omega^2$	$(K = \frac{1}{2} m\,v^2)$
Power:	$P = \tau\,\omega$	$(P = F\,v)$
Work-Energy Theorem:	$W = \Delta K = \frac{1}{2} I\,\omega^2 - \frac{1}{2} I\,\omega_o^2$	$(W = \Delta K = \frac{1}{2} m\,v^2 - \frac{1}{2} m\,v_o^2)$

The kinetic energy of a **rolling body** (without slipping) relative to an axis through the contact point is the sum of the rotational kinetic energy about an axis through the center of mass and the translational kinetic energy of the center of mass and, i.e., $K = \frac{1}{2} I_{CM}\,\omega^2 + \frac{1}{2} M\,v_{CM}^2$.

Example 8.6 A solid sphere of mass 1.0 kg and radius 0.010 m rolls with a speed of 10 m/s. How high up an inclined plane can it climb before coming to rest?

Solution: Given: $M = 1.0$ kg, $R = 0.010$ m, $v_{CM}^2 = 10$ m/s.

Find: h.

We use the conservation of energy to solve this problem. Choose the bottom of the inclined plane as the reference level for gravitational potential energy ($U_o = 0$ at $h_o = 0$). The mechanical energy at the bottom should be equal to the mechanical energy at the top. When the sphere is at the bottom of the sphere, its potential energy is zero and it has both translational and rotational kinetic energies. When the sphere reaches the maximum height, its speed is zero and the only form of energy it has is potential. Using $v_{CM} = R\omega$ and $I_{CM} = \frac{2}{5} M\,R^2$ we have

$$K = \frac{1}{2} I_{CM}\,\omega^2 + \frac{1}{2} M\,v_{CM}^2 = \frac{1}{2}\left(\frac{2}{5} M\,R^2\right)\frac{v_{CM}^2}{R^2} + \frac{1}{2} M\,v_{CM}^2 = \frac{7}{10} M\,v_{CM}^2 = Mgh.$$

So $$h = \frac{7\,v_{CM}^2}{10\,g} = \frac{7\,(10\text{ m/s})^2}{10(9.80\text{ m/s}^2)} = 7.1\text{ m}.$$

Example 8.7 A disk of mass 10 kg and radius 0.20 m is accelerated by a motor from rest to an angular speed of 20 rad/s during a 0.40 s interval.

(a) How much work is required?

(b) What is the power output of the motor?

Solution: Given: $M = 10$ kg, $R = 0.20$ m, $\omega_o = 0$, $\omega = 20$ rad/s, $t = 0.40$ s.

Find: (a) W (b) P.

(a) The moment of inertia of a disk is given by $I = \frac{1}{2}M R^2 = \frac{1}{2}(10 \text{ kg})(0.20 \text{ m})^2 = 0.20 \text{ kg·m}^2$.

From work-energy theorem, we have

$W = \Delta K = \frac{1}{2}I\,\omega^2 - \frac{1}{2}I\,\omega_o^2 = \frac{1}{2}I\,\omega^2 = \frac{1}{2}(0.20 \text{ kg·m}^2)(20 \text{ rad/s})^2 = 40 \text{ J}.$

(b) $P = \dfrac{W}{t} = \dfrac{40 \text{ J}}{0.40 \text{ s}} = 100 \text{ W}.$

---

5. Angular Momentum

The definition of angular momentum is analogous to linear momentum. It is defined as the product of moment of inertia and angular velocity. $\mathbf{L} = I\,\omega$ (it is $p = m\,v$ for linear momentum). The SI unit of angular momentum is kg·m²/s. The vector direction of angular momentum is determined by the right hand rule as applied to angular velocity in Chapter 7. The rotational form of Newton's second law can also be written in terms of angular momentum: $\tau = \dfrac{\Delta \mathbf{L}}{\Delta t}$.

In the absence of an external, unbalanced torque, the total angular momentum of a system is conserved (remains constant). Mathematically, this can be written as

$$I\,\omega = I_o\,\omega_o.$$

Example 8.8 A figure skater rotating at 5.00 rad/s with arms extended has a moment of inertia of 2.25 kg·m². If the arms are pulled in so the moment of inertia decreases to 1.80 kg·m², what is the final angular speed?

Solution: Given: $I_o = 2.25$ kg·m² $I = 1.80$ kg·m²

$\omega_o = 5.00$ rad/s

Find: ω.

From the conservation of angular momentum, $I\,\omega = I_o\,\omega_o$, we have

$$\omega = \frac{I_o\,\omega_o}{I} = \frac{(2.25 \text{ kg·m}^2)(5.00 \text{ rad/s})}{1.80 \text{ kg·m}^2} = 6.25 \text{ rad/s}.$$

The figure skater rotates faster after pulling the arms in.

Why does the moment of inertia decrease when the arms are pulled in? Why are there no net external torque on the skater?

IV. Mathematical Summary

Condition for Rolling without Slipping	$v = r\,\omega$ (8.1) (or $s = r\,\theta$ or $a = r\,\alpha$)	Defines the condition for rolling without slipping.
Torque (magnitude)	$\tau = r_\perp F = r\,F \sin\theta$ (8.2)	Defines the magnitude of torque.
Conditions Equilibrium	$\Sigma \mathbf{F}_i = 0$ and $\Sigma \tau_i = 0$ (8.3)	Define the conditions for translational and rotational mechanical equilibrium.
Torque on a particle (magnitude)	$\tau = m\,r^2\,\alpha$ (8.4)	Calculates the magnitude of torque on a particle.
Moment of Inertia	$I = \Sigma m_i\,r_i^2$ (8.6)	Defines the moment of inertia.
Newton's Second Law (magnitude)	$\tau = I\,\alpha$ (8.7)	Gives the rotational form of Newton's second law
Parallel Axis Theorem	$I = I_{CM} + M\,d^2$ (8.8)	Calculates the moment of inertia about an axis parallel to one though the center of mass.
Rotational Work	$W = \tau\,\theta$ (8.9)	Defines rotational work.
Rotational Power	$P = \tau\,\omega$ (8.10)	Defines Rotational power.
Work-Energy Theorem	$W = \Delta K = \frac{1}{2}I\omega^2 - \frac{1}{2}I\omega_o^2$ (8.11)	Relates work with change in kinetic energy.
Rotational Kinetic Energy	$K = \frac{1}{2}I\,\omega^2$ (8.12)	Defines rotational kinetic energy.
Kinetic Energy of a Rolling Object	$K = \frac{1}{2}I_{CM}\,\omega^2 + \frac{1}{2}M\,v_{CM}^2$ (8.13)	Gives the total kinetic energy of a rolling object.
Angular Momentum of a Particle (magnitude)	$L = m\,r_\perp^2\,\omega$ (8.14)	Defines the angular momentum of a particle in circular motion.
Angular Momentum of a Rigid Body	$\mathbf{L} = I\,\omega$ (8.16)	Defines the angular momentum of a rigid body.
Torque in Terms of Angular Momentum	$\tau = \dfrac{\Delta \mathbf{L}}{\Delta t}$ (8.17)	Expresses Newton's second law in terms of angular momentum.
Conservation of Angular Momentum	$I\,\omega = I_o\,\omega_o$ (8.18)	Gives the law of the conservation of angular momentum when the net external torque is zero.

V. Solutions of Selected Exercises and Paired Exercises

8. Dividing $s = r\,\theta$ (the condition of without slipping) by a time interval t gives $\frac{s}{t} = r\frac{\theta}{t}$, or $v = r\,\omega$.

Dividing $v = r\,\omega$ by a time interval t again yields $\frac{v}{t} = r\frac{\omega}{t}$, or $a = r\,\alpha$.

10. From $v = r\,\omega$, we have $\omega = \frac{v}{r} = \frac{0.25 \text{ m/s}}{0.15 \text{ m}} = 1.7$ rad/s.

28. Choose the joint (where F_j is) as the axis of rotation.

$\Sigma\tau = F_m$ (0.18 m) sin 15° − (3.0 kg)(9.80 m/s²)(0.26 m) = 0, so $F_m = 1.6\times10^2$ N.

32. (a) The center of gravity (CG) of the first book is at the center. So for the last book not to fall, its CG can not displace more than 12.5 cm relative to the CG of the first book (within the base). The CG of each successive book on the top is moved 1.5 cm relative to that of the one below. Therefore the number of books which can be on top of the first one is $\frac{12.5 \text{ cm}}{1.5 \text{ cm}} = 8.33$.

Thus we can stack a total of 1 + 8 = 9 books.

(b) The total height is 9(5.0) cm = 45 cm. So the CM is at $\frac{45 \text{ cm}}{2} = 22.5$ cm.

49. (a) $I = \Sigma\, mr^2 = (2.0 \text{ kg})(0.30 \text{ m})^2 + (4.0 \text{ kg})(0.75 \text{ m})^2 = 2.4 \text{ kg}\cdot\text{m}^2$.

(b) $X_{CM} = \frac{\Sigma_i\, m_i\, x_i}{M} = \frac{(2.0 \text{ kg})(0.30 \text{ m}) + (4.0 \text{ kg})(0.75 \text{ m})}{2.0 \text{ kg} + 4.0 \text{ kg}} = 0.60 \text{ m}$.

So $I = (2.0 \text{ kg})(0.60 \text{ m} - 0.30 \text{ m})^2 + (4.0 \text{ kg})(0.75 \text{ m} - 0.60 \text{ m})^2 = 0.27 \text{ kg}\cdot\text{m}^2$.

(c) $I = I_{CM} + M\,d^2 = 0.27 \text{ kg}\cdot\text{m}^2 + (6.0 \text{ kg})(0.60 \text{ m})^2 = 2.4 \text{ kg}\cdot\text{m}^2$.

52. $I_{CM} = \frac{2}{5} M_e R_e^2$. $I = I_{CM} + M_e\, d^2 = \frac{2}{5} M_e R_e^2 + M_e R_e^2 = \frac{7}{5} M_e R_e^2$.

So $\frac{I}{I_{CM}} = 3.5$, or 3.5 times.

54. First find the angular acceleration from dynamics. Use the result from Exercise 8.52.

$$\alpha = \frac{\tau}{I} = \frac{\tau}{\frac{7}{5} M R^2} = \frac{10 \text{ m}\cdot\text{N}}{\frac{7}{5}(15 \text{ kg})(0.15 \text{ m})^2} = 21.2 \text{ rad/s}^2.$$

From $\omega^2 = \omega_o^2 + 2\alpha\theta$, we have $\theta = \frac{\omega^2 - \omega_o^2}{2\alpha} = \frac{(7.5 \text{ rad/s})^2 - (3.0 \text{ rad/s})^2}{2(21.2 \text{ rad/s}^2)} = 1.1$ rad.

62. The hoop rotates about an instantaneous axis of rotation through the point of contact (point O).

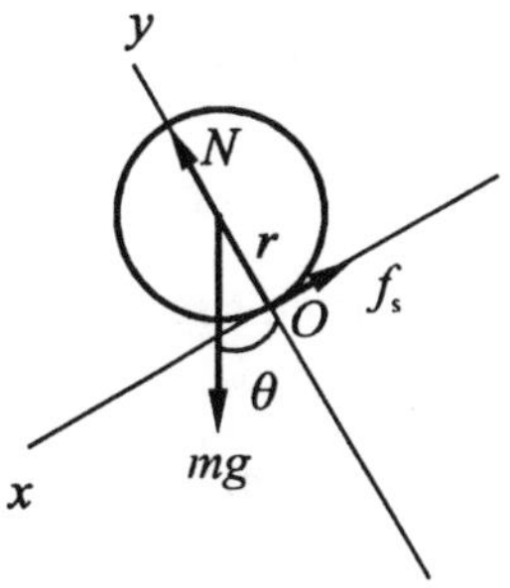

The moment of inertia about this axis is

$I = I_{CM} + M d^2 = M R^2 + M R^2 = 2M R^2$.

The torque by gravity is $\tau = MgR \sin \theta$.

So $a = R\,\alpha = R\dfrac{\tau}{I} = R\dfrac{MgR \sin \theta}{2M R^2} = \dfrac{g \sin \theta}{2}$

$= \dfrac{(9.80 \text{ m/s}^2) \sin 15°}{2} = 1.3 \text{ m/s}^2$.

72. $K = \frac{1}{2} m v^2 + \frac{1}{2} I \omega^2 = \frac{1}{2} m v^2 + \frac{1}{2} I \dfrac{v^2}{R^2}$. So $K_h = \frac{1}{2} m v_h^2 + \frac{1}{2} m R^2 \dfrac{v_h^2}{R^2} = m v_h^2$,

$K_c = \frac{1}{2} m v_c^2 + \frac{1}{2} \times \frac{1}{2} m R^2 \dfrac{v_c^2}{R^2} = \dfrac{3}{4} m v_c^2$, and $K_s = \frac{1}{2} m v_s^2 + \frac{1}{2} \times \frac{2}{5} m R^2 \dfrac{v_s^2}{R^2} = \dfrac{7}{10} m v_s^2$.

Since they are all released from the same height, the K's are the same. That means the v is the greatest for the sphere and smallest for the hoop. Therefore the sphere gets to the bottom first and the hoop last.

$v_s = \sqrt{\dfrac{10\,K}{7\,m}}$, $v_c = \sqrt{\dfrac{4\,K}{3\,m}}$, and $v_h = \sqrt{\dfrac{K}{m}}$.

89. (a) From angular momentum conservation: $I\,\omega = I_o\,\omega_o$,

so $M R_a^2\,\omega_a = M R_p^2\,\omega_p$, or $R_a\,v_a = R_p\,v_p$. Since $R_p < R_a$, $v_p > v_a$.

(b) $\dfrac{v_p}{v_a} = \dfrac{R_a}{R_p} = \dfrac{1.52}{1.47} = 1.03$.

90. Orbital: The Earth can be considered as a particle to the sun.

$\omega_o = \dfrac{2\pi \text{ rad}}{(365)(24)(3600 \text{ s})} = 1.99\times10^{-7}$ rad/s, and $I_o = M R^2$,

so $L_o = I_o\,\omega_o = (6.0\times10^{24} \text{ kg})(1.5\times10^{11} \text{ m})^2\,(1.99\times10^{-7} \text{ rad/s}) = 2.69\times10^{40} \text{ kg·m}^2\text{/s}$.

Spin: The Earth has to be considered as a sphere.

$\omega_s = \dfrac{2\pi \text{ rad}}{(24)(3600 \text{ s})} = 7.27\times10^{-5}$ rad/s, and $I_s = \dfrac{2}{5} M r^2$,

so $L_s = \dfrac{2}{5}(6.0\times10^{24} \text{ kg})(6.4\times10^{6} \text{ m})^2\,(7.27\times10^{-5} \text{ rad/s}) = 7.15\times10^{33} \text{ kg·m}^2\text{/s}$.

Therefore $\dfrac{L_o}{L_s} = \dfrac{2.69\times10^{40}}{7.15\times10^{33}} = 3.8\times10^{6}$.

They are not in the same direction because the Earth's axis is tilted.

103. Both rotate about an instantaneous axis though the contact point.

From the parallel-axis theorem: $I_h = I_{CM} + M d^2 = M R^2 + M R^2 = 2M R^2$,

$$I_c = \frac{1}{2}M(R_1^2 + R_2^2) + M R_2^2 = \frac{1}{2}M(R_1^2 + 2R_2^2).$$

From Newton's second law: $Mg\,(R \sin\theta) = I\,\alpha = I\dfrac{a}{R}$,

so for the hoop: $Mg\,r \sin\theta = 2MR^2\dfrac{a_h}{R}$, ☞ $a_h = \dfrac{g \sin\theta}{2}$,

for the annular cylinder: $Mg\,R_2 \sin\theta = \frac{1}{2}M(R_1^2 + 2R_2^2)\dfrac{a_c}{R_2}$,

or $a_c = \dfrac{2g\sin\theta}{R_1^2/R_2^2 + 2} = \dfrac{g\sin\theta}{R_1^2/2R_2^2 + 1}$.

Since $R_2 > R_1$, $R_1^2/2R_2^2 < 1$, or $R_1^2/2R_2^2 + 1 < 2$, so $a_c > a_h$.

Therefore the hoop rolls (accelerates) more slowly.

112. On the verge of the unstable equilibrium, the weight goes though the lower corner of the trailer.

So $\tan\theta = \dfrac{L/2}{h} = \dfrac{3.66\text{ m}}{2(3.58\text{ m})} = 0.511$,

therefore $\theta = 27°$.

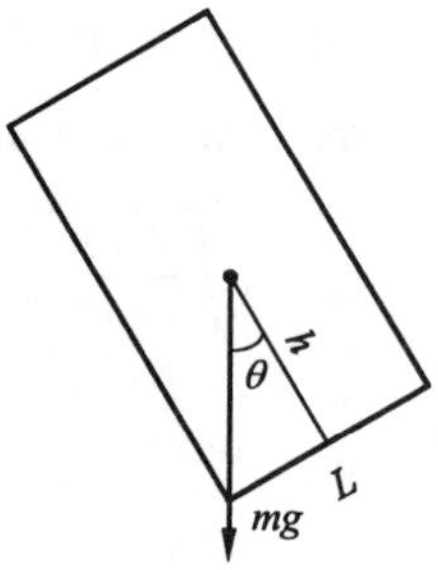

VI. Practice Quiz

1. A meter stick is supported by a knife edge at the 50 cm mark and has masses of 0.40 kg and 0.60 kg hanging at the 20 cm and 80 cm marks, respectively. Where (at what mark) should a third mass of 0.30 kg be hung to keep the stick balanced?

(a) 20 cm (b) 25 cm (c) 30 cm (d) 50 cm (e) 70 cm

2. A rotating flywheel can be used to store energy. If it is required to store 1.00×10^6 J of energy when rotating at 400 rad/s, what is the moment of inertia of the wheel?

(a) 6.25 kg·m^2 (b) 12.5 kg·m^2 (c) 25 kg·m^2 (d) 50 kg·m^2 (e) 5000 kg·m^2

3. The Earth moves about the Sun in an elliptical orbit. As the Earth moves closer to the Sun, then which of the following best describes the orbiting speed of the Earth about the Sun?

(a) increases (b) decreases (c) conserves (d) remains constant (e) none of the above

4. A wheel with moment of inertia of 2.00 kg·m^2 has a net torque of 2.50 m·N applied to it. What angular acceleration does it experience?

(a) 0.200 rad/s^2 (b) 1.25 rad/s^2 (c) 2.00 rad/s^2 (d) 2.50 rad/s^2 (e) 5.00 rad/s^2

5. A sphere of radius 0.20 m rolls without slipping. If the sphere rotates through 3.0 revolutions, how far does it roll?

(a) 0.60 m (b) 1.2 m (c) 1.9 m (d) 3.8 m (e) 7.6 m

6. The rotational analog of force is

(a) moment of inertia. (b) kinetic energy. (c) moment arm. (d) work. (e) torque.

7. A sphere of mass 10 kg and radius 0.010 m is released from the top of a 1.0-m high inclined plane. What is the speed of the sphere when it reaches the bottom of the inclined plane?

(a) 3.7 m/s (b) 4.4 m/s (c) 5.6 m/s (d) 6.3 m/s (e) 7.2 m/s

8. If a constant net torque is applied to an object, that object will

(a) rotate with a constant angular velocity. (b) rotate with a constant angular acceleration.
(c) have an increasing moment of inertia. (d) have a decreasing moment of inertia.
(e) none of the above.

9. A 4.00-m long rod is hinged at one end. The rod is initially held in the horizontal position, and then released as the free end is allowed to fall. What is the angular acceleration as it is released? (The moment of inertia of a rod about one end is $ML^2/3$.)

(a) 2.45 rad/s^2 (b) 3.68 rad/s^2 (c) 4.90 rad/s^2 (d) 6.75 rad/s^2 (e) 7.35 rad/s^2

10. A 500-N person stands on a uniform board of weight 100 N and length 8.0 m. The board is supported at each end. If the support force at the right end is three times that at the left end, how far from the right end is the person?

(a) 4.0 m (b) 2.0 m (c) 1.6 m (d) 2.0 m (e) 6.4 m

Answers to Practice Quiz:

1. c 2. b 3. a 4. b 5. d 6. e 7. a 8. b 9. b 10. c

CHAPTER 9

Solids and Fluids

I. Chapter Objectives

Upon completion of this chapter, you should be able to:

1. distinguish between stress and strain and use elastic moduli to compute dimensional changes.
2. explain the pressure-depth relationship, state Pascal's principle and describe how it is used in practical applications.
3. relate the buoyant force and Archimedes' principle and tell whether an object will float in a fluid based on relative densities.
4. describe the source of surface tension and its effects and explain why liquids rise in small tubes by capillary action.
5. identify the simplifications used in describing fluid flow and use the continuity equation and Bernoulli's equation to explain common effects.

*6. define the coefficient of viscosity, calculate flow rate using Poiseuille's law, and interpret a fluid's Reynolds number with regard to turbulent flow.

II. Key Terms

Upon completion of this chapter, you should be able to define and/or explain the following key terms:

fluid	**atmosphere (atm)**	**cohesive forces**
stress	**Pascal's principle**	**capillary action**
strain	**absolute pressure**	**ideal fluid**
elastic moduli	**gauge pressure**	**streamlines**
elastic limit	**barometer**	**equations of continuity**
Young's modulus	**buoyant force**	**flow rate equation**
shear modulus	**Archimedes' principle**	**Bernoulli's equation**
bulk modulus	**specific gravity**	***viscosity**
pressure	**surface tension**	***Poiseuille's law**
pascal (Pa)	**adhesive forces**	***Reynolds number**

The definitions and/or explanations of the most important key terms can be found in the following section: **III. Chapter Summary and Discussion**.

III. Chapter Summary and Discussion

1. Solids and Elastic Moduli

All materials are elastic to some degree and can be deformed. **Stress** is the quantity that describes the force causing the deformation and **strain** is a relative measure of how much deformation a given stress produces. Quantitatively, stress is the applied force per unit cross-sectional area, and strain is the ratio of the change in dimensions to the original dimensions.

$$\text{Stress} = \frac{F}{A},\quad \text{tensile strain (in length)} = \frac{\Delta L}{L_o},\quad \text{shear strain (in area)} = \frac{x}{h},\quad \text{volume strain} = \frac{\Delta V}{V_o}.$$

In general, stress is proportional to strain up to the elastic limit. The constant of proportionality, which depends on the nature of the material, is called the **elastic modulus** and is defined as the ratio of stress to strain,

$$\text{elastic modulus} = \frac{\text{stress}}{\text{strain}}.$$

There are three types of elastic moduli, Young's modulus Y, the shear modulus S, and the bulk modulus B.

$$Y = \frac{F/A}{\Delta L/L_o},\quad S = \frac{F/A}{x/h} \approx \frac{F/A}{\phi},\quad B = \frac{F/A}{-\Delta V/V_o} = -\frac{\Delta p}{\Delta V/V_o}.$$ The SI units of moduli are N/m^2 or pascal (Pa).

The compressibility is the inverse of the bulk modulus, $k = \dfrac{1}{B}$.

Example 9.1 An aluminum wire 2.0 m in length and 2.0 mm in diameter supports a 10.0 kg mass.

(a) What is the stress in the wire?

(b) What is the elongation of the wire?

Solution: Given: $F = w = mg = (10.0\text{ kg})(9.80\text{ m/s}^2) = 98\text{ N}$, $L_o = 2.0\text{ m}$,

$r = d/2 = 1.0\text{ mm} = 1.0\times10^{-3}\text{ m}$, $Y = 7.0\times10^{10}\text{ N/m}^2$ (from Table 9.1)

Find: (a) stress = F/A (b) ΔL.

(a) $$\text{stress} = \frac{F}{A} = \frac{F}{\pi r^2} = \frac{98\text{ N}}{\pi(1.0\times10^{-3}\text{ m})^2} = 3.1\times10^7\text{ N/m}^2.$$

(b) From $Y = \dfrac{F/A}{\Delta L/L_o}$, we have $$\Delta L = \frac{F/A}{Y L_o} = \frac{3.12\times10^7\text{ N/m}^2}{(7.0\times10^{10}\text{ N/m}^2)(2.0\text{ m})} = 2.2\times10^{-4}\text{ m} = 0.22\text{ mm}.$$

Example 9.2 A shear force of 4.0×10^2 N is applied to one face of an aluminum cube with sides of 30 cm. What is the resulting relative displacement?

Solution: Given: $F = 4.0\times10^2$ N, $h = 0.30$ m, $S = 2.5\times10^{10}$ N/m^2 (from Table 9.1)

Find: x.

The area of a side is $A = h^2$.

From $S = \dfrac{F/A}{x/h}$, $x = \dfrac{F/A}{S/h} = \dfrac{F\,h}{A\,S} = \dfrac{F\,h}{h^2\,S} = \dfrac{F}{h\,S} = \dfrac{4.0\times10^2\text{ N}}{(0.20\text{ m})(2.5\times10^{10}\text{ N/m}^2)} = 8.0\times10^{-8}$ m.

2. Fluids: Pressure and Pascal's Principle

A **fluid** is a substance that flows and cannot support a shear. Both liquids and gases are fluids.

Pressure is defined as the force per unit area, $P = \dfrac{F}{A}$, and has units of N/m^2 or pascal (Pa).

Absolute pressure on an object submerged at a depth h below the surface of a fluid is given by the **pressure-depth equation**, $p = p_o + \rho g h$, where p_o is the pressure on the fluid surface and ρ is the density of the fluid. **Gauge pressure** (static pressure) is the difference between the absolute pressure and the atmospheric pressure, i.e., $p_{\text{gauge}} = p - p_a = \rho g h$, where $p_a = 1.01\times10^5$ Pa. When a gauge is used to measure pressure, it measures the gauge pressure.

Pascal's principle states that any external pressure allied to an enclosed fluid is transmitted undiminished to every point in the fluid and to the walls of its container. Hydraulic press and lift are good practical examples of this principle.

Example 9.3 (a) What is the absolute pressure at a location 15.0 m below the surface of sea?

(b) What is the gauge pressure there?

Solution: Given: $p_o = p_a = 1.01\times10^5$ Pa, $h = 15.0$ m, $\rho = 1.03\times10^3$ kg/m^3 (sea water from Table 9.2)

Find: (a) p (b) $p - p_a$.

(a) $p = p_o + \rho g h = 1.01\times10^5\text{ Pa} + (1.03\times10^3\text{ kg/m}^3)(9.80\text{ m/s}^2)(15.0\text{ m}) = 2.52\times10^5$ Pa.

(b) $p_{\text{gauge}} = p - p_a = 2.52\times10^5\text{ Pa} - 1.01\times10^5\text{ Pa} = 1.51\times10^5$ Pa.

Example 9.4 In a hydraulic garage lift, the small piston has a radius of 5.0 cm and the large piston has a radius of 15 cm. What force must be applied on the small piston in order to lift a car weighing 20,000 N on the large piston?

Solution: According to Pascal's principle, the external pressure exerted at the small piston is transmitted undiminished to the large piston. Therefore $p_1 = \frac{F_1}{A_1} = p_2 = \frac{F_2}{A_2}$.

Given: $r_1 = 0.050$ m, $r_2 = 0.15$ m, $F_2 = 20{,}000$ N

Find: F_1.

From $\frac{F_1}{A_1} = \frac{F_2}{A_2}$, we have $F_1 = \left(\frac{A_1}{A_2}\right) F_2 = \left(\frac{\pi r_1^2}{\pi r_2^2}\right) F_2 = \left(\frac{r_1}{r_2}\right)^2 F_2 = \left(\frac{0.050}{0.15}\right)^2 (2.0\times10^4 \text{ N}) \approx 2{,}200$ N.

Note: What a wonderful device! The radius of the large piston is three times the radius of the small piston, (0.15 m)/(0.050 m) = 3. However, the force required on the small piston is only 1/9 of the force on the large piston, (20,000 N)/(2,200 N) = 1/9. Why?

3. Buoyancy and Archimedes' Principle

An object submerged (either partially or completely) in a fluid will experience an upward buoyant force exerted by the fluid. **Archimedes' principle** states that the buoyant force is equal to the weight of the fluid displaced., i.e., if an object displaces 3.0 N of fluid, it will experience a 3.0 N buoyant force. Mathematically, the principle is expressed as $F_B = w_f = m_f g = (\rho_f V_f)g = \rho_f V_f g$, where ρ_f is the density of the fluid, V_f is the volume of fluid displaced (which is the same as the volume fraction of object submerged *in* the fluid).

Archimedes' principle is very useful in determining the volume of objects (therefore density). That was what Archimedes allegedly used to determine that the crown of his king was not made of pure gold.

Density and floating:

(1) An object will float in a fluid if the average density of the object is less than the density of the fluid.

(2) An object will sink in a fluid if the average density of the object is greater than the density of the fluid.

(3) An object will be in equilibrium at any submerged depth in a fluid if the average densities of the object and fluid are equal.

The **specific gravity** of a substance is the ratio of the density of the substance to the density of water, and so is a pure number (dimensionless).

Example 9.5 A boat approximating a rectangular box measures 5.0 m long, 1.0 wide and 0.50 m high. How many kilograms of mass can be loaded onto the boat before it sinks in a lake? (Neglect the mass of the boat itself.)

Solution: Given: $L = 5.0$ m, $W = 1.0$ m, $H = 0.50$ m, $\rho_{\text{water}} = 1.0\times10^3$ kg/m^3.

Find: m.

Density approach: On the verge of sinking, the density of the load is equal to the density of water.

From $\rho_{\text{water}} = \rho = \dfrac{m}{V} = \dfrac{m}{LWH}$, we have

$m = \rho_{\text{water}} LWH = (1.0\times10^3 \text{ kg/m}^3)(5.0 \text{ m})(1.0 \text{ m})(0.50\text{m}) = 2.5\times10^3$ kg (that is equivalent to 5500 lb!)

Force approach: On the verge of sinking, the buoyant force is equal to the weight of the load.

$F_B = \rho_f V_f g = w = mg$, so $m = \rho_f V_f = (1.0\times10^3 \text{ kg/m}^3)(5.0 \text{ m})(1.0\text{m})(0.50 \text{ m}) = 2.5\times10^3$ kg.

Example 9.6 A bargain hunter bought a pure (24-karat) "gold" crown at a flea market. After getting it home, she suspends the crown on a scale and finds its weight to be 7.84 N. She then weighs the crown while it is immersed in water and finds that the scale reads 6.86 N. Is the crown made of pure gold? (Archimedes' problem!)

Solution: We can use many different physical quantities to identify matter, since many physical quantities are unique to different elements. For example, gold and aluminum each have their own unique density. So if we can determine the density of the crown, we can determine whether it is made of pure gold. When the crown is submerged in water, the volume of water displaced is equal to the volume of the crown. The apparent weight decrease is due to the buoyant force.

Given: $w = 7.84$ N, w (in water) = 6.86 N, $\rho_{\text{water}} = 1.0\times10^3$ kg/m^3.

Find: ρ.

First find the volume of the crown with Archimedes' principle.

$F_B = 7.84 \text{ N} - 6.86 \text{ N} = 0.98 \text{ N} = \rho_f V_f g$, so $V_f = \dfrac{F_B}{\rho_f g} = \dfrac{0.98 \text{ N}}{(1.0\times10^3 \text{ kg/m}^3)(9.80 \text{ m/s}^2)} = 1.00\times10^{-4} \text{ m}^3$.

Therefore $\rho = \dfrac{m}{V} = \dfrac{w/g}{V} = \dfrac{w}{gV} = \dfrac{7.84 \text{ N}}{(9.80 \text{ m/s}^2)(1.00\times10^{-4} \text{ m}^3)} = 8.0\times10^3 \text{ kg/m}^3$.

Looking up Table 9.2, this crown is not made of gold (the density of gold is 19.3×10^3 kg/m^3).

---

4. Surface Tension and Capillary Action

The **surface tension** of a liquid is caused by the inward pull on the surface molecules which causes the surface of the liquid to contract and be stretched. Surface tension is defined as $\gamma = \dfrac{F}{L}$, where F is the force and L is the length of the object in fluid. The SI unit of surface tension is N/m.

Adhesive forces (or adhesion) are attractive forces between unlike molecules and **cohesive forces** (or cohesion) are attractive forces between like molecules. A relative measure of these is given by the **contact angle,** ϕ, which is the angle between a surface and a line tangent to a drop of liquid on the surface.

Capillary action, the rise of a liquid of density ρ in a small diameter tube, is the result of both surface tension and adhesion. The rise height is given by $h = \dfrac{2\gamma \cos\phi}{\rho g r}$, where r is the radius of the tube.

Example 9.7 To what height would water at 0°C rise in a glass capillary tube with a diameter of

(a) 0.50 mm?

(b) 1.00 mm?

Solution: Given: $\gamma = 0.076$ N/m (from Table 9.3), $\phi = 0°$ (from Table 9.4), $\rho = 1.0 \times 10^3$ kg/m^3,

(a) $r = d/2 = 0.25$ mm (b) $r = 0.50$ mm.

Find: h.

$$\text{(a) } h = \frac{2\gamma \cos\phi}{\rho g r} = \frac{2(0.076 \text{ N/m}) \cos 0°}{(1.0\times 10^3 \text{ kg/m}^3)(9.80 \text{ m/s}^2)(0.25\times 10^{-3} \text{ m})} = 0.062 \text{ m} = 6.2 \text{ cm}.$$

$$\text{(b) } h = \frac{2(0.076 \text{ N/m}) \cos 0°}{(1.0\times 10^3 \text{ kg/m}^3)(9.80 \text{ m/s}^2)(0.50\times 10^{-3} \text{ m})} = 0.031 \text{ m} = 3.1 \text{ cm}.$$

---

5. Fluid Dynamics and Bernoulli's Equation

Ideal fluid flow is a steady, irrotational, nonviscous, and incompressible flow. For the conservation of mass, the **equation of continuity** for fluid flow is $\rho A v$ = constant, where ρ is the density, v is the speed, and A is the cross-sectional area. If the fluid is incompressible (ρ = constant), then $A v$ = constant, or $A_1 v_1 = A_2 v_2$, which is called the **flow rate equation**, since the quantity measures flow rate (volume per unit time).

Bernoulli's equation is the direct result of the application of conservation of energy to fluid. It is written as $p + \frac{1}{2}\rho v^2 + \rho g y$ = constant. If the flow is at a constant height y = constant, then $p + \frac{1}{2}\rho v^2$ = constant. In verbal statement, this is "the higher the speed, the lower the pressure, or vice versa". This equation is very successful in explaining physical phenomena such as the curved baseball.

Example 9.8 An ideal fluid flows at 12 m/s in a horizontal circular pipe. If the pipe widens to twice its original radius, what is the flow speed in the section?

Solution: Given: $v_1 = 12$ m/s, $r_2 = 2r_1$.

Find: v_2.

From the flow rate equation, $A_1 v_1 = A_2 v_2$, we have

$$v_2 = \frac{A_1 v_1}{A_2} = \frac{\pi r_1^2 v_1}{\pi r_2^2} = \frac{r_1^2 v_1}{(2\,r_1^2)^2} = \frac{v_1}{4} = \frac{12 \text{ m/s}}{4} = 3.0 \text{ m/s}.$$

Note: The pipe widens to twice its original radius and the speed decreases to 1/4 of its original value. Why?

Example 9.9 In Example 9.8, if the fluid is water and the pressure at the smaller section is 3.0×10^4 Pa, what is the pressure at the larger section?

Solution: Since the fluid is flowing horizontally, y is constant.

Given: $\rho = 1.0\times10^3$ kg/m^3, $p_1 = 3.0\times10^4$ Pa, $v_1 = 12$ m/s, $v_2 = 3.0$ m/s (Example 9.8).

Find: p_2.

From Bernoulli's equation, $p_1 + \frac{1}{2}\rho v_1^2 = p_2 + \frac{1}{2}\rho v_2^2$, we have

$$p_2 = p_1 + \tfrac{1}{2}\rho\,(v_1^2 - v_2^2) = 3.0\times10^4 \text{ Pa} + \tfrac{1}{2}(1.0\times10^3 \text{ kg/m}^3)[(12 \text{ m/s})^2 - (3.0 \text{ m/s})^2] = 9.8\times10^4 \text{ Pa}.$$

So the speed is lower and pressure is higher in the larger section.

---

*6. Viscosity, Poiseuille's Law, and Reynolds Number

Coefficient of viscosity (or **viscosity**), η, is a measure of the fluid's internal resistance to flow. Its definition is $\eta = \dfrac{F\,h}{A\,v}$, where F/A is the shear stress needed to maintain the flow between two parallel planes separated by a distance h at a relative velocity of magnitude v. Its SI unit is Pa·s.

The average **flow rate**, $Q = A\,v$ (volume/time, m^3/s), depends on the characteristics of the fluid and the pipe as well as on the pressure difference between the ends of the pipe. **Poiseuille's law** gives a relationship for the flow rate.

The **Reynolds number** is a dimensionless number which gives an indication of if turbulence occurs in fluid flow. If the number is about 2000 or more (≥ 2000), the flow is turbulent.

IV. Mathematical Summary

Stress	$\text{stress} = \frac{F}{A}$ (9.1)	Defines stress
Strain (tensile)	$\text{strain} = \frac{\Delta L}{L_o} = \frac{L - L_o}{L_o}$ (9.2)	Defines tensile strain
Young's Modulus	$Y = \frac{F/A}{\Delta L/L_o}$ (9.4)	Relates Young's modulus with stress and strain
Shear Modulus	$S = \frac{F/A}{x/h} \approx \frac{F/A}{\phi}$ (9.5)	Relates shear modulus with stress and shear strain
Bulk Modulus	$B = \frac{F/A}{-\Delta V/V_o} = -\frac{\Delta p}{\Delta V/V_o}$ (9.6)	Relates bulk modulus with stress and volume strain
Compressibility	$k = \frac{1}{B}$ (9.7)	Defines compressibility
Pressure	$p = \frac{F}{A}$ (9.8)	Defines pressure in terms of force and area
Pressure-Depth Equation	$p = p_o + \rho g h$ (9.10)	Expresses pressure as a function of depth in fluid
Barometric (Atmospheric) pressure	$p_a = \rho g h$ (9.13)	Computes atmospheric pressure from barometric measurements
Archimedes' Principle	$F_b = m_f g = \rho_f g V_f$ (9.14)	Calculates buoyant force
Surface Tension	$\gamma = \frac{F}{L}$ (9.16)	Defines surface tension
Capillary Height	$h = \frac{2\gamma \cos\phi}{\rho g r}$ (9.17)	Computes capillary height in a capillary tube
Equation of Continuity	$\rho_1 A_1 v_1 = \rho_2 A_2 v_2$ or $\rho A v = \text{constant}$ (9.18)	Relates flow characteristics for ideal fluid
Flow Rate Equation	$A_1 v_1 = A_2 v_2$ or $Av = \text{constant}$ (9.19)	Rewrites equation of continuity for an incompressible fluid
Bernoulli's Equation	$p_1 + \frac{1}{2}\rho v_1^2 + \rho g y_1 = p_2 + \frac{1}{2}\rho v_2^2 + \rho g y_2$ (9.20)	Applies the conservation of energy to a fluid
*Coefficient of Viscosity	$\eta = \frac{F h}{A v}$ (9.21)	Defines the coefficient of viscosity
*Flow Rate	$Q = A\bar{v} = \frac{\Delta V}{\Delta t}$ (9.22)	Defines flow (volume) rate

V. Solutions of Selected Exercises and Paired Exercises

7. (a) The compression stress $= \dfrac{F}{A} = \dfrac{(500\text{ N})\sin 37°}{(0.040\text{ m})(0.040\text{ m})} = 1.9\times10^5\text{ N/m}^2$.

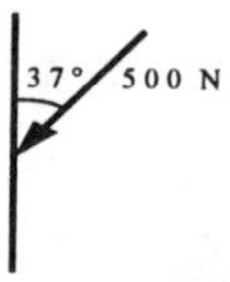

(b) The shear stress $= \dfrac{(500\text{ N})\cos 37°}{(0.040\text{ m})(0.040\text{ m})} = 2.5\times10^5\text{ N/m}^2$.

9. Stress $= \dfrac{F}{A} = \dfrac{mg}{\pi r^2} = \dfrac{(6.0\text{ kg})(9.80\text{ m/s}^2)}{\pi(0.50\times10^{-3}\text{ m})^2} = 7.49\times10^7\text{ N/m}^2$,

strain $= \dfrac{\Delta L}{L_o} = \dfrac{1.4\times10^{-3}\text{ m}}{2.0\text{ m}} = 7.0\times10^{-4}$. So $Y = \dfrac{\text{stress}}{\text{strain}} = \dfrac{7.49\times10^7\text{ N/m}^2}{7.0\times10^{-4}} = 1.1\times10^{11}\text{ N/m}^2$.

12. It is a compression strain.

Strain $= -\dfrac{\text{stress}}{Y} = -\dfrac{F/A}{Y} = -\dfrac{F}{YA} = -\dfrac{mg}{YA} = -\dfrac{(12.0\times10^3\text{ kg})(9.80\text{ m/s}^2)}{(20\times10^{10}\text{ N/m}^2)(0.200\text{ m})(0.150\text{ m})} = -1.96\times10^{-5}$.

Also Strain $= \dfrac{\Delta L}{L_o}$. So $L = L_o + \Delta L = (1 - 1.96\times10^{-5})(4.00\text{ m}) = 3.99992\text{ m}$.

25. (a) The pressure inside the can is equal to the atmospheric pressure outside. When the liquid is poured from an unvented can, a partial vacuum develops inside and the pressure difference causes the pouring to be difficult. By opening the vent, you are allowing air to go into the can and the pressures are equalized so the liquid can be easily poured.

(b) When you squeeze a medicine dropper before inserting it into a liquid, you are forcing the air out and reducing the pressure inside the dropper. When you release the top with the dropper in a liquid, the liquid rises in the dropper due to atmospheric pressure.

(c) When we inhale, the lungs expand so the pressure decreases. This rushes the air into the lungs. When we exhale, the lungs contract and the pressure increases. This pushes the air out.

36. $\Delta p = \rho gh = (1.29\text{ kg/m}^3)(9.80\text{ m/s}^2)(30\text{ m}) = 379\text{ Pa}$.

So the fractional decrease is $\dfrac{\Delta p}{p_a} = \dfrac{379\text{ Pa}}{1.013\times10^5\text{ Pa}} = 3.74\times10^{-3} = 0.37\%$.

42. (a) From Pascal's principle: $p_i = p_o = \dfrac{F_o}{A_o} = \dfrac{mg}{A_o} = \dfrac{(1.4\times10^3\text{ kg})(9.80\text{ m/s}^2)}{0.20\text{ m}^2} = 6.9\times10^4\text{ Pa}$.

(b) $F_i = p_i A_i = (6.86\times10^4\text{ Pa})(\pi)(0.025\text{ m})^2 = 1.3\times10^2\text{ N}$.

(c) Since the fluid is enclosed, $h_i A_i = h_o A_o$.

So $\Delta U = F_o h_o = (p_o A_o)h_o = p_i(A_o h_o) = p_i(A_i h_i) = F_i h_i = W$.

54. (a) He did it by water displacement.

(b) $\rho = \dfrac{m}{V} = \dfrac{0.750\text{ kg}}{3.980\times10^{-5}\text{ m}^3} = 18.8\times10^3\text{ kg/m}^3 < \rho_g = 19.3\times10^3\text{ kg/m}^3$. So the answer is no.

60. For the open cube to float, weight must be equal to the buoyant force.

$w = mg = F_b = \rho_f g V_f = \rho_f g L^3$, where V_f is the volume of the open cube.

So $L = \sqrt[3]{\dfrac{m}{\rho_f}} = \sqrt[3]{\dfrac{\rho V}{\rho_f}} = \sqrt[3]{\dfrac{(7.8\times10^3\text{ kg/m}^3)(1.0\text{ m})^3}{1000\text{ kg/m}^3}} = 2.0\text{ m}$,

where V is the volume of the solid cube.

72. From $h = \dfrac{2\gamma\cos\phi}{\rho g r}$, we have

$$d = 2r = 2\frac{2\gamma\cos\phi}{\rho g h} = 2\frac{2(0.073\text{ N/m})\cos 0°}{(1000\text{ kg/m}^3)(9.80\text{ m/s}^2)(0.050\text{ m})} = 6.0\times10^{-4}\text{ m}.$$

86. From the equation of continuity, $\rho_A A_A v_A = \rho_B A_B v_B$ and $A_A = A_B$,

we have $\dfrac{\rho_B}{\rho_A} = \dfrac{v_A}{v_B} = 3$.

So B by 3 times.

98. From $Q = \dfrac{\pi r^4 \Delta p}{8\eta L}$, we have

$$\Delta p = \frac{8\eta L Q}{\pi r^4} = \frac{8(1.00\times10^{-3}\text{ Pl})(6.0\text{ m})(40\text{ L/min})(10^{-3}\text{ m}^3\text{/L})(1\text{ min/60 s})}{\pi(0.015\text{ m})^4} = 2.0\times10^2\text{ Pa}.$$

104. From $F_b = \rho_f g V$, the scale reading is

$T = mg - F_b = \rho V g - F_b = (\rho - \rho_f)Vg = (7.8\times10^3\text{ kg/m}^3 - 0.68\times10^3\text{ kg/m}^3)(9.80\text{ m/s}^2)(0.36\text{ m}^3)$

$= 2.5\times10^4\text{ N}$.

119. (a) When it is in the water, the force applied is

$F = mg - F_b = \rho V g - \rho_f g V = (\rho - \rho_f)gV$

$= (7.8\times10^3\text{ kg/m}^3 - 1.0\times10^3\text{ kg/m}^3)(9.80\text{ m/s}^2)(0.25\text{ m})(0.20\text{ m})(10\text{ m}) = 3.3\times10^4\text{ N}$.

(b) When it is out of water, the force applied is

$F = mg = \rho V g = (7.8\times10^3\text{ kg/m}^3)(9.80\text{ m/s}^2)(0.25\text{ m})(0.20\text{ m})(10\text{ m}) = 3.8\times10^4\text{ N}$.

VI. Practice Quiz

1. What is the absolute pressure at the bottom of a 5.0 m deep swimming pool? (Note: the pressure contribution from the atmosphere is 1.01×10^5 Pa.)

   (a) 0.49×10^5 Pa (b) 0.72×10^5 Pa (c) 1.01×10^5 Pa (d) 1.5×10^5 Pa (e) 2.51×10^5 Pa

2. A 4.00-kg cylinder made of solid iron is supported by a string while submerged in water. What is the tension in the string? (The density of iron is 7.86×10^3 kg/m^3.)

   (a) 0 N (b) 2.50 N (c) 19.6 N (d) 23.7 N (e) 34.2 N

3. When you blow some air above a strip of paper, the paper rises. This is because

   (a) the air above the paper moves faster and the pressure is higher.

   (b) the air above the paper moves faster and the pressure is lower.

   (c) the air above the paper moves slower and the pressure is lower.

   (d) the air above the paper moves slower and the pressure is higher.

   (e) none of the above applies.

4. An ideal fluid flows through a pipe made of two sections with diameters of 2.0 and 6.0 cm, respectively. What is the speed ratio of the flow through the 6.0 cm section to that through the 2.0 cm section?

   (a) 1/9 (b) 1/3 (c) 1 (d) 3 (e) 9

5. A 15,000-N car on a hydraulic lift rests on a cylinder with a piston of diameter 0.40 m. What is the pressure applied to the liquid in order to lift the car?

   (a) 0.75×10^5 N/m^2 (b) 1.0×10^5 N/m^2 (c) 1.2×10^5 N/m^2 (d) 1.5×10^5 N/m^2 (e) 2.0×10^5 N/m^2

6. Which of the following is associated with the law of conservation of energy for fluids?

   (a) Archimedes' principle (b) Bernoulli's principle (c) Pascal's principle

   (d) Poiseuille's law (e) equation of continuity

7. A mass *m* is hung from identical wires made of aluminum, brass, copper, and steel. Which wire will stretch the least?

   (a) aluminum (b) brass (c) copper (d) steel (e) all the same.

8. Water flows at a speed of 8.0 m/s in a horizontal pipe 2.0 cm in diameter. The water then enters a horizontal pipe 4.0 cm in diameter. What is the difference in pressure between the two segments?

   (a) 1.0×10^4 N/m^2 (b) 2.0×10^4 N/m^2 (c) 3.0×10^4 N/m^2 (d) 4.0×10^4 N/m^2 (e) 5.0×10^4 N/m^2

9. In some places, the Greenland ice sheet is 1.0 kilometer thick. Estimate the gauge pressure underneath the ice.

(a) 2.5×10^{6} N/m^2 (b) 4.5×10^{6} N/m^2 (c) 9.0×10^{6} N/m^2 (d) 9.0×10^{5} N/m^2 (e) 2.5×10^{5} N/m^2

10. If the density of gold is 19.3×10^{3} kg/m^3, what buoyant force does a 0.60-kg gold crown experience when it is immersed in water?

(a) 3.0×10^{-5} N (b) 3.0×10^{-4} N (c) 3.0×10^{-2} N (d) 0.30 N (e) 3.0 N

Answers to Practice Quiz:

1.d 2.e 3.b 4.a 5.c 6.b 7.d 8.c 9.c 10.d

CHAPTER 10

Temperature

I. Chapter Objectives

Upon completion of this chapter, you should be able to:

1. distinguish between temperature and heat.
2. explain how a temperature scale is constructed and convert temperatures from one scale to another.
3. describe the ideal gas law and explain how it is used to determine absolute zero.
4. calculate the thermal expansions of solids and liquids.
5. relate kinetic theory and temperature and explain the process of diffusion and Graham's law.

II. Key Terms

Upon completion of this chapter, you should be able to define and/or explain the following key terms:

temperature
heat
internal energy
thermometer
thermal expansion
Fahrenheit temperature scale
Celsius temperature scale
Boyle's law
Charles' law
ideal (perfect) gas law
absolute zero
Kelvin temperature scale
kelvin
triple point of water
Avogadro's number
mole
thermal coefficient of linear expansion
thermal coefficient of area expansion
thermal coefficient of volume expansion
kinetic theory of gases
diffusion
Graham's law
osmosis

The definitions and/or explanations of the most important key terms can be found in the following section: **III. Chapter Summary and Discussion.**

III. Chapter Summary and Discussion

1. Temperature and Heat

Temperature is a relative measure, or indication, of hotness and coldness. In kinetic theory of gases, temperature is a measure of the average kinetic energy of the molecules.

Heat is energy that is transferred from one object to another because of a temperature difference. The total energy (kinetic plus potential) of all molecules of a body or system is the **internal energy**.

When heat is transferred between two objects, whether or not they are touching, they are said to be in **thermal contact**. When there is no longer a net heat transfer between objects in thermal contact, they are at the same temperature and are said to be in **thermal equilibrium**.

2. The Celsius and Fahrenheit Temperature Scales

The two most common temperature scales are **Celsius temperature scale** and **Fahrenheit temperature scale**. On the Celsius scale, water freezes at 0°C and boils at 100°C; while on the Fahrenheit scale, water freezes at 32°F and boils at 212°F. The conversions between the two scales are $T_F = \frac{9}{5} T_C + 32$, or $T_C = \frac{5}{9}(T_F - 32)$.

Example 10.1 What is the temperature 50.0°C on the Fahrenheit scale?

Solution: Given: $T_C = 50.0°C$. Find: T_F.

$T_F = \frac{9}{5} T_C + 32 = \frac{9}{5}(50) + 32 = 122°F$.

Example 10.2 The temperature changes from 35°F during the night to 75°F during the day. What is the temperature change on the Celsius scale?

Solution: Although we can convert both 35°F and 75°F to their Celsius temperatures and then calculate the change, we can take advantage of the conversion equations to calculate the temperature change ΔT. From $T_C = \frac{5}{9}(T_F - 32)$, we can see that for every Fahrenheit degree increase, the Celsius temperature increases by 5/9 of a degree, or $\Delta T_C = \frac{5}{9} \Delta T_F$.

Given: $\Delta T_F = 75°F - 35°F = 40$ F°.

[Note: the unit for Fahrenheit temperature change is F° (Fahrenheit degree), not °F (degrees Fahrenheit).]

Find: ΔT_C.

$\Delta T_C = \frac{5}{9} \Delta T_F = \frac{5}{9}(40) = 22$ C°. (Note: the unit for Celsius temperature change is C°, not °C.)

3. Ideal Gas Law and Absolute Temperature

An ideal gas is monatomic (molecule has only one atom), low density, and low pressure gas. The **ideal** (or perfect) **gas law** relates the pressure, volume and the temperature of the gas, $pV = Nk_B T$, where N is the number of molecules and k_B is the Boltzmann's constant which has a value of 1.38×10^{-23} J/K. It describes real, low-density gases fairly well. Two other forms of the ideal gas law are $pV = nRT$ and $\frac{p_1 V_1}{T_1} = \frac{p_2 V_2}{T_2}$, where n is the number of moles of molecules and R is the universal gas constant which has values of 8.31 J/mol·K or 0.0821 L·atm/mol·K. One mole of substance contains $N_A = 6.02 \times 10^{23}$ molecules (Avogadro's number).

A gas can be used to measure temperature as a function of time at a constant volume. Extrapolation to zero pressure defines absolute zero temperature. **Absolute zero** is the foundation for the **Kelvin temperature scale,** which uses absolute zero and the **triple point of water** as fixed points. The conversion between Kelvin and Celsius temperatures is $T_K = T_C + 273.15$, or more commonly to three significant figures, $T_K = T_C + 273$.

Note: The temperature in the ideal gas law *must* be the Kelvin temperature.

Example 10.3 What is −40°F on the Kelvin scale?

Solution: Since there is no direct conversion given between Fahrenheit and Kelvin temperatures, we first convert Fahrenheit to Celsius.

Given: $T_F = -40°C$. Find: T_K.

$T_C = \frac{5}{9}(T_F - 32) = \frac{5}{9}(-40 - 32) = \frac{5}{9}(-72) = -40°C$.

(Wow! The Celsius and Fahrenheit have the same nominal reading at this temperature.)

Next, $T_K = T_C + 273 = -40 + 273 = 233$ K.

Example 10.4 A gas has a volume of 0.20 m³, a temperature of 30°C, and a pressure of 1.0 atm (one atmosphere of pressure). It is heated to 60°C and compressed to a volume of 0.15 m³. Find the new pressure in atmospheres.

Solution: Given: $T_1 = 30°C$, $T_2 = 60°C$,

$V_1 = 0.20\ m^3$, $V_2 = 0.15\ m^2$,

$p_1 = 1.0$ atm.

Find: p_2 (in atm).

The temperatures are given on the Celsius scale, we first convert them to Kelvin temperatures.

$T_1 = 30°C = 30 + 273 = 303$ K and $T_2 = 60°C = 60 + 273 = 333$ K.

From $\frac{p_1 V_1}{T_1} = \frac{p_2 V_2}{T_2}$, we have $p_2 = \frac{V_1}{V_2}\frac{T_2}{T_1} p_1 = \frac{0.20\ m^3}{0.15\ m^2}\frac{333\ K}{303\ K}(1.0\ atm) = 1.5$ atm.

Note: Leaving the pressure in units of atm gives the new pressure in atm because of the ratio form of the ideal gas law. Also, the temperatures used in the ideal gas law *must* be the Kelvin temperatures. What answer would you get if you use Celsius temperature?

Example 10.5 An ideal gas in a container of volume of 100 cm^3 at 20°C has a pressure of 100 N/m^2. Determine the number of gas molecules in the container.

Solution: Given: $V = 100\ cm^3$, $T = 20°C$, $p = 100\ N/m^2$.

Find: N.

The volume is not given in m^3 (the standard SI unit). We first convert volume to m^3 and temperature to Kelvin. Although 1 m = 100 cm, 1 $m^3 \neq 100\ cm^3$. Actually $1\ m^3 = (1\ m)^3 = (100\ cm)^3 = 10^6\ cm^3$.

$$V = (100\ cm^3)\frac{1\ m^3}{10^6\ cm^3} = 1.0 \times 10^{-4}\ m^3.$$

$$T = T_C + 273 = 20 + 273 = 293\ K.$$

From $pV = Nk_B T$, we have $N = \frac{pV}{k_B T} = \frac{(100\ N/m^2)(1.0\times10^{-4}\ m^3)}{(1.38\times10^{-23}\ J/K)(293\ K)} = 2.5\times10^{18}$ molecules.

---

4. Thermal Expansion

The **thermal expansion** of a material is characterized by its coefficient of expansion. For solids, the **thermal coefficient of linear expansion,** α, applies to one-dimensional length changes; the **thermal coefficient of area expansion** (approximately 2α) applies to two-dimensional area changes; and the **thermal coefficient of volume expansion** (approximately 3α) applies to three-dimensional volume changes. For fluids with no definite shape, only volume expansion is applicable and a special thermal coefficient of volume expansion β is used. The equations are:

Linear: $\frac{\Delta L}{L_o} = \alpha\,\Delta T.$ Area: $\frac{\Delta A}{A_o} = 2\alpha\,\Delta T.$ Volume: $\frac{\Delta V}{V_o} = 3\alpha\,\Delta T$ (or $= \beta\,\Delta T$ for fluids).

Example 10.6 A steel bridge is 1000 m long at −20°C in winter. What is its length when the temperature rises to 40°C in summer? (For steel, $\alpha = 11\times10^{-6}$ /C° and ignore significant figures.)

Solution: Given: $L_o = 1000$ m, $T_i = -20$°C, $T_F = 40$°C, $\alpha = 11\times10^{-6}$ /C°.

Find: L.

$\Delta T = T_f - T_i = 40°\text{C} - (-20°\text{C}) = 60$ C°.

From $\dfrac{\Delta L}{L_o} = \alpha\Delta T$, we have $\Delta L = \alpha\Delta T L_o$.

So $L = L_o + \Delta L = L_o + \alpha\Delta T L_o = L_o(1 + \alpha\Delta T) = (1000\text{ m})[1 + (11\times10^{-6}\text{ /C°})(60\text{ C°})] = 1000.66$ m.

Note: Because of expansion, engineers must leave gaps between steel sections.

Example 10.7 A 500-mL glass beaker of water is filled to the rim at a temperature of 0°C. How much water will overflow if the water is heated to a temperature of 95°C? (Ignore the expansion of the beaker, why?)

Solution: Given: $V_o = 500$ mL, $T_i = 0$°C, $T_f = 95$°C, $\beta = 2.1\times10^{-4}$ /C° (from Table 10.1).

Find: ΔV.

$\Delta T = T_f - T_i = 95°\text{C} - 0°\text{C} = 95$ C°.

From $\dfrac{\Delta V}{V_o} = \beta\Delta T$, we have $\Delta V = \beta\Delta T\, V_o = (2.1\times10^{-4}\text{ /C°})(95\text{ C°})(500\text{ mL}) = 10$ mL.

Note: In this Example, we can ignore the expansion of the beaker because glass has a much smaller coefficient of volume expansion (approximately 10×10^{-6} /C°).

5. The Kinetic Theory of Gases

The **kinetic theory of gases** uses statistical methods to derive the ideal gas law from mechanical principles. Some of the important conclusions of this theory are:

(1) Temperature is a measure of the average kinetic energy of the molecules, $\frac{1}{2}mv_{\text{rms}}^2 = \frac{3}{2}k_B T$, where v_{rms} is the root-mean-square speed of the molecules and m is the mass of the molecule.

(2) The ideal gas law can be expressed in terms of the root-mean-square speed of the molecules, $pV = \frac{1}{3}Nmv_{\text{rms}}^2$.

(3) The total internal energy of an ideal gas is given by $U = \frac{3}{2}Nk_B T = \frac{3}{2}nRT$.

Diffusion is a process of random molecular mixing in which particular molecules move from a region of higher concentration to one of lower concentration. **Osmosis** is the diffusion of a liquid across a permeable membrane because of a concentration gradient (difference).

Example 10.8 Calculate the rms (root–mean–square) speed of a hydrogen molecule and an oxygen molecule when the temperature is 300 K. (The masses of hydrogen and oxygen molecules are 3.3×10^{-27} kg and 5.3×10^{-26} kg, respectively.)

Solution: Given: $T = 300$ K, $m_H = 3.3\times10^{-27}$ kg, $m_O = 5.3\times10^{-26}$ kg.

Find: v_{rms} for H_2 and O_2.

From $\frac{1}{2} m v_{rms}^2 = \frac{3}{2} k_B T$, we have $v_{rms} = \sqrt{\frac{3\,k_B\,T}{m}}$.

For H_2: $v_{rms} = \sqrt{\frac{3(1.38\times10^{-23}\text{ J/K})(300\text{ K})}{3.3\times10^{-27}\text{ kg}}} = 1.9\times10^3$ m/s (about 4,000 mi/h).

For O_2: $v_{rms} = \sqrt{\frac{3(1.38\times10^{-23}\text{ J/K})(300\text{ K})}{5.3\times10^{-26}\text{ kg}}} = 4.8\times10^2$ m/s.

Why does more massive molecule gas gave smaller v_{rms}?

Example 10.9 If the temperature of a gas increases from 20°C to 40°C, by what factor does the rms (root–mean–square) speed increases?

Solution: The temperatures *must* be in Kelvins!

Given: $T_1 = 20°\text{C} = 20 + 273 = 293$ K, $T_2 = 40°\text{C} = 40 + 273 = 313$ K.

Find: $\frac{(v_{rms})_2}{(v_{rms})_1}$.

From $\frac{1}{2} m v_{rms}^2 = \frac{3}{2} k_B T$, we have $v_{rms} = \sqrt{\frac{3\,k_B\,T}{m}}$.

So $\frac{(v_{rms})_2}{(v_{rms})_1} = \sqrt{\frac{T_2}{T_1}} = \sqrt{\frac{313\text{ K}}{293\text{ K}}} = 1.03$, or it increases by only 3%.

Note: Since k_B and m are constant, they are divided out in the ratio $\frac{(v_{rms})_2}{(v_{rms})_1}$. The Celsius temperature doubles in this Example, the rms speed does not double (it increases by only 3%), nor does the Kelvin temperature.

IV. Mathematical Summary

Celsius-Fahrenheit Conversion	$T_F = \frac{9}{5} T_C + 32$ (10.1) $T_C = \frac{5}{9}(T_F - 32)$ (10.2)	Relates Celsius temperature and Fahrenheit temperature
Boyle's Law	$p\,V =$ constant (10.3) or $p_1 V_1 = p_2 V_2$ (10.3)	Relates pressure and volume for constant temperature
Charles' Law	$\frac{V}{T} =$ constant (10.4) or $\frac{V_1}{T_1} = \frac{V_2}{T_2}$	Relates volume and absolute temperature for constant volume
Ideal (or perfect) Gas Law	$p\,V = N k_B T$ (10.5) or $\frac{p_1 V_1}{T_1} = \frac{p_2 V_2}{T_2}$ or $p\,v = n\,R\,T$	Relates pressure, volume, and absolute temperature $k_B = 1.38 \times 10^{-23}$ J/K $R = 0.0821$ L·atm/mol·K = 8.31 J/K·mole Avogadro's number: $N_A = 6.02 \times 10^{23}$ molecules/mole
Kelvin-Celsius Conversion	$T_K = T_C + 273.15$ (10.6) or $T_K = T_C + 273$	Relates Kelvin temperature and Celsius temperature
Thermal Expansion of Solids: Linear	$\frac{\Delta L}{L_o} = \alpha \Delta T$ (10.8) or $L = L_o(1 + \alpha \Delta T)$ (10.9)	Calculates the fractional change in length in terms of the coefficient of linear expansion and change in temperature
Thermal Expansion of Solids: Area	$\frac{\Delta A}{A_o} = 2\alpha \Delta T$ (10.10) or $A = A_o(1 + 2\alpha \Delta T)$	Calculates the fractional change in area in terms of the coefficient of linear expansion and change in temperature
Thermal Expansion of Solids: Volume	$\frac{\Delta V}{V_o} = 3\alpha \Delta T$ (10.11) or $V = V_o(1 + 3\alpha \Delta T)$	Calculates the fractional change in volume in terms of the coefficient of linear expansion and change in temperature
Thermal Volume Expansion of Fluids	$\frac{\Delta V}{V_o} = \beta \Delta T$ (10.12)	Calculates the fractional change in volume in terms of the coefficient of volume expansion and change in temperature
Results of Kinetic Theory of Gases	$p\,V = \frac{1}{3} N m\, v_{rms}^2$ (10.13) $\frac{1}{2} m\, v_{rms}^2 = \frac{3}{2} k_B T$ (10.14) $U = \frac{3}{2} N k_B T = \frac{3}{2} nRT$ (10.15)	Gives the results of the kinetic theory of gases
Diffusion Rate Ratio (Graham's Law)	$\frac{R_1}{R_2} = \sqrt{\frac{m_2}{m_1}}$ (10.16)	Relates the ratio of diffusion rates to the ratio of masses for constant temperature and pressure

V. Solutions of Selected Exercises and Paired Exercises

8. $T_F = \frac{9}{5}\,T_C + 32 = \frac{9}{5}\,(39.4°\text{C}) + 32 = 103°\text{F}.$

14. (a) The x intercept is found at $T_F = \frac{9}{5}\,T_C + 32 = 0,$ so $T_C = -18°\text{C}.$

(b) $T_C = \frac{5}{9}\,(T_F - 32).$ So the slope is $\frac{5}{9}$ and the intercept is $-32°\text{F}.$

24. $500°\text{F} = \frac{5}{9}\,(500 - 32)\ °\text{C} = 260°\text{C} = 533\ \text{K}.$ So the answer is 500°F.

30. $T_1 = 92°\text{F} = \frac{5}{9}\,(92 - 32)\ °\text{C} = 33.3°\text{C} = 306.3\ \text{K}, \quad T_2 = 32°\text{F} = 0°\text{C} = 273\ \text{K}.$

From $\frac{p_1V_1}{T_1} = \frac{p_2V_2}{T_2},$ we have $V_2 = \frac{p_1V_1T_2}{T_1p_2} = \frac{(20.0\ \text{lb/in}^2)(0.20\ \text{m}^3)(273\ \text{K})}{(306.3\ \text{K})(14.7\ \text{lb/in}^2)} = 0.24\ \text{m}^3.$

Here ib/in² can be used since it is in a ratio.

44. $\frac{\Delta L}{L_o} = \alpha\,\Delta T = (12\times10^{-6}\ \text{C}^{\circ -1})(-5.0°\text{C} - 20°\text{C}) = -3.0\times10^{-4},$ or 0.03%.

52. $A_o = \pi r^2 = \pi(4.00\ \text{cm})^2 = 50.27\ \text{cm}^2.$

$\Delta A = 2\alpha A_o\,\Delta T = 2(24\times10^{-6}\ \text{C}^{\circ -1})(50.27\ \text{cm}^2)(150°\text{C} - 20°\text{C}) = 0.314\ \text{cm}^2.$

So the final area is $50.27\ \text{cm}^2 + 0.314\ \text{cm}^2 = 50.6\ \text{cm}^2.$

58. (a) $V = V_o + \Delta V = V_o\,(1 + 3\alpha\,\Delta T) = (0.100\ \text{m})^3\,[1 + 3(24\times10^{-6}\ \text{C}^{\circ -1})(300°\text{C} - 20°\text{C})]$

$= 1.0202\times10^{-3}\ \text{m}^3.$

(b) $L = \sqrt[3]{V} = \sqrt[3]{1.02016\times10^{-3}\ \text{m}^3} = 0.10068\ \text{m}.$

(c) Yes. Each edge expands according to $L = L_o + \Delta L = L_o\,(1 + \alpha\,\Delta T)$, within rounding error.

66. From $v_{rms} = \sqrt{\frac{3\,k_B\,T}{m_o}},$ we have $\frac{(v_{rms})_2}{(v_{rms})_1} = \sqrt{\frac{T_2}{T_1}} = \sqrt{\frac{373\ \text{K}}{298\ \text{K}}} = 1.119.$

So the percentage change is $\frac{(v_{rms})_2 - (v_{rms})_1}{(v_{rms})_1} = \frac{(v_{rms})_2}{(v_{rms})_1} - 1 = 0.119 = 12\%.$

72. $U = \frac{3}{2} N k_B T.$ If U is doubled, the Kelvin temperature is doubled.

So $K' = 2K = 2(298 \text{ K}) = 596 \text{ K} = 323°\text{C}.$

84. From $\frac{p_o V_o}{T_o} = \frac{p V}{T}$, we have $\frac{V}{V_o} = \frac{T p_o}{T_o p}$.

Also $V = V_o + \Delta V = V_o (1 + \beta \Delta T)$, or $\frac{V}{V_o} = 1 + \beta \Delta T = 1 + \beta (T - T_o)$.

So $1 + \beta (T - T_o) = \frac{T p_o}{T_o p}$, or $\beta = \frac{T p_o - T_o p}{T_o p (T - T_o)}$.

92. (a) The piston advances according to ideal gas law.

(b) From $\frac{p_o V_o}{T_o} = \frac{p V}{T}$, we have $\frac{V}{V_o} = \frac{T p_o}{T_o p} = \frac{T}{T_o} = \frac{313 \text{ K}}{283 \text{ K}} = 1.106.$

So the fractional change is $\frac{V - V_o}{V_o} = \frac{V}{V_o} - 1 = 0.106.$

95. $\Delta V = \beta V_o \Delta T = (1.8 \times 10^{-4} \text{ C°}^{-1})(0.200 \text{ cm}^3)(30 \text{ C°}) = 1.08 \times 10^{-3} \text{ cm}^3.$

From $V = A h = \pi r^2 h$, we have $\Delta h = \frac{\Delta V}{\pi r^2} = \frac{1.08 \times 10^{-3} \text{ cm}^3}{\pi (0.0325 \text{ cm})^2} = 3.3 \text{ cm}.$

VI. Practice Quiz

1. Oxygen condenses into a liquid at approximately 90 K. What temperature, in degrees Fahrenheit, does this correspond to?

(a) –118°F (b) –193°F (c) –217°F (d) –265°F (e) –297°F

2. A balloon has a volume of 1.0 m³. As it rises in the Earth's atmosphere, its volume expands. What will be its new volume if its original temperature and pressure are 20°C and 1.0 atm, and its final temperature and pressure are –40°C and 0.10 atm?

(a) 1.0 m³ (b) 2.0 m³ (c) 4.0 m³ (d) 6.0 m³ (e) 8.0 m³

3. The absolute (Kelvin) temperature of an ideal gas is directly proportional to which of the following properties, when taken as an average, of the molecules of the gas?

(a) speed (b) momentum (c) mass (d) kinetic energy (e) potential energy

4. A fixed container holds oxygen and nitrogen gases at the same temperature. Which of the following statements is correct?

(a) The oxygen molecules have the greater kinetic energy.

(b) The nitrogen molecules have the greater kinetic energy.

(c) The oxygen molecules have the greater speed.

(d) The nitrogen molecules have the greater speed.

(e) They have the same speed and kinetic energy.

5. A steel beam is 15.0 m long at a temperature of $-15°C$. What is its expansion when the temperature is 35°C?

(a) 5.50×10^{-4} m (b) 1.65×10^{-4} m (c) 3.85×10^{-4} m (d) 2.20×10^{-4} m (e) 1.10×10^{-3} m.

6. An ideal gas has a pressure of 2.5 atm, a volume of 1.0 L at a temperature of 30°C. How many molecules are in the gas?

(a) 6.1×10^{23} (b) 6.0×10^{22} (c) 2.4×10^{22} (d) 2.4×10^{23} (e) 6.0×10^{25}

7. A brass cube, 10 cm on a side, is heated with a temperature change of 200 C°. By what percentage does its volume change?

(a) 5.7×10^{-3} % (b) 0.57 % (c) 1.1 % (d) 1.1×10^{-3} % (e) 0.10 %

8. Between 0°C and 4°C, the thermal coefficient of volume expansion of water is

(a) positive. (b) negative. (c) zero. (d) becoming less dense. (e) becoming more dense.

9. A molecule has a rms speed of 500 m/s at 20°C. What is its rms speed at 80°C?

(a) 500 m/s (b) 1,000 m/s (c) 2,000 m/s (d) 550 m/s (e) 600 m/s

10. Which one of the choices below best describes a system made of ice, water, and steam existing together?

(a) absolute zero (b) triple point (c) freezing point (d) boiling point (e) sublimation point

Answers to Practice Quiz:

1. e 2. e 3. d 4. d 5. a 6. b 7. c 8. b 9. d 10. b

CHAPTER 11

Heat

I. Chapter Objectives

Upon completion of this chapter, you should be able to:

1. distinguish the various units of heat and define the mechanical equivalent of heat.
2. describe specific heat and its cause, and explain how the specific heat of materials are obtained using calorimetry.
3. compare and contrast the three common phases of matter, and relate latent heat to phase changes.
4. describe the three methods of heat transfer and give practical and/or environmental examples of each.

II. Key Terms

Upon completion of this chapter, you should be able to define and/or explain the following key terms:

heat
calorie (cal)
kilocalorie (kcal)
British thermal unit (Btu)
freezing point
liquid phase
gaseous (vapor) phase
boiling point
phase diagrams
evaporation
conduction
thermal conductors
thermal insulators
emissivity
black body
mechanical equivalent of heat
specific heat
solid heat
solid phase
melting point
condensation point
sublimation
latent heat
latent heat of fusion
latent heat of vaporization
thermal conductivity
convection
radiation
infrared radiation
Stefan's law

The definitions and/or explanations of the most important key terms can be found in the following section: **III. Chapter Summary and Discussion**.

III. Chapter Summary and Discussion

1. Units of Heat

Heat is energy that is transferred from one object to another because of a temperature difference. The SI unit of heat is joule (J).

Other units of heat commonly used are the **calorie** (cal) and the **kilocalorie** (kcal, food Calorie, note the capital C for "large" calorie). A kilocalorie is the amount of heat required to raise the temperature of 1 kg of water by 1 C° (from 14.5°C to 15.5°C).

A **British thermal unit (Btu)** is the amount heat needed to raise the temperature of 1 lb of water by 1 F° (from 63°F to 64°F).

The relationship between unit of heat and the standard SI unit (J) is called the **mechanical equivalent of heat**: 1 kcal = 4186 J (or 1 cal = 4.186 J).

2. Specific Heat

Specific heat is the amount of heat required to raise the temperature of 1 kg of substance by 1 C°. Among the substances known to us, water has a relatively large specific heat (more heat required per kg/C°) value of 1 cal/g·C°, or 1 kcal/kg·C°, or 4186 J/kg·C°. The high specific heat of water explains why the climate doesn't vary much with season in places near a large body of water such as coastal California.

The heat needed to change the temperature of a mass m of ΔT is then $Q = cm\Delta T$, where Q is heat and c is the specific heat.

Within an isolated system, the heat lost by an object will be gained by other objects of the system. According to the conservation of energy, heat lost = heat gained. This is the basis of **calorimetry** (the method of mixing).

Note: Since heat lost is a negative quantity, when applying calorimetry, you should write $-Q_{\text{lost}} = Q_{\text{gained}}$, (that is, equate magnitude).

Example 11.1 A 2.0-kg aluminum block is originally at 10°C. If 140 kJ of heat energy are added to the block, what is its final temperature?

Solution: Given: $m = 2.0$ kg, $c = 920$ J/kg·C° (Table 11.1), $Q = 140$ kJ $= 140\times10^3$ J, $T_i = 10$°C.

Find: T_f.

From $Q = cm\Delta T$, we have $\Delta T = \dfrac{Q}{c\,m} = \dfrac{140\times10^3 \text{ J}}{(920 \text{ J/kg·C°})(2.0 \text{ kg})} = 76$ C°.

So $T_f = T_i + \Delta T = 10$°C + 76 C° = 86°C.

Example 11.2 A 0.600-kg piece of metal is heated to 100°C and placed in an aluminum can of mass 0.200 kg which contains 0.500 kg of water initially at 17.3°C. The final equilibrium temperature of the mixture is 20.2°C. What kind of metal is it?

Solution: Here we want to identify the metal. If we can find its specific heat (remember it is specific to the substance), we can identify the metal by looking up Table 11.1. Use the following subscripts:

u = unknown metal, w = water, a = aluminum, i = initial, and f = final.

Given: $m_u = 0.600$ kg, m_a 0.200 kg, $m_w = 0.500$ kg, $t_{iu} = 100$°C, $t_{fu} = 20.2$°C,

$t_{ia} = t_{iw} = 17.3$°C, $T_{fa} = T_{fw} = 20.2$°C, $c_w = 4186$ J/kg·C°, $c_a = 920$ J/kg·C°.

Find: c_u.

In the mixing process, metal loses heat, and aluminum and water gain heat.

The heat lost by metal is $Q_{lost} = c_u m_u \Delta T_u = c_u$ (0.600 kg)(20.2°C − 100°C) = −(47.9 kg·C°)c_u.

The heat gained by aluminum and water is $Q_{gained} = c_a m_a \Delta T_a + c_w m_w \Delta T_w$

= (920 J/kg/C°)(0.200 kg)(20.2°C − 17.3°C) + (4186 J/kg·C°)(0.500 kg)(20.2°C − 17.3°C) = 6.60×10^3 J.

From calorimetry, $-Q_{lost} = Q_{gained}$, we have −[−(47.9 kg·C°)c_u] = 6.60×10^3 J.

Solving for c_u, $c_u = 138$ J/kg·C°.

Looking in Table 11.1, we conclude that the metal is lead.

---

3. Phase Change and Latent Heat

Matter normally exist in three phases: solid, liquid, and gas. **Latent heat** is the heat involved in a phase change and does not go into changing the temperature, but in breaking or forming molecular bonds.

The latent of heat for a solid-liquid phase change is called the **latent heat of fusion** (L_f), which is the heat required to change 1 kg of solid to liquid while the temperature remains at the **freezing point**. The latent of heat for a liquid-gas phase change is called the **latent heat of vaporization** (L_v), which is the heat required to change 1 kg of liquid to gas while the temperature remains at the **boiling point**. Therefore, the heat involved for a mass m is given by $Q = m\,L$, where L is the latent heat.

Example 11.3 How much heat energy is needed to change 10 kg of ice at –20°C to steam at 120°C?

Solution: Ice at –20°C first needs to be raised to 0°C, and then melted to 0°C liquid water. The water at 0°C needs to be raised to 100°C, where it is vaporized to steam. Finally, the temperature of steam needs to be raised to 120°C.

Use the following subscripts: i = ice, w = water, and s = steam.

Given: $m = 10$ kg, $c_i = 2100$ J/kg·C°, $c_w = 4186$ J/kg·C°, $c_s = 2010$ J/kg·C°, $L_f = 3.33\times10^5$ J/kg, $L_v = 22.6\times10^5$ J/kg.

Find: Q_{tot} (total heat).

The heat required to raise the temperature of ice from –20°C to 0°C is

$$Q_1 = c_i\, m_i\, \Delta T_i = (2100 \text{ J/kg·C°})(10 \text{ kg})[0°\text{C} - (-20°\text{C})] = 4.2\times10^5 \text{ J}.$$

The heat required to melt ice at 0°C to water at 0°C is

$$Q_2 = m\, L_f = (10 \text{ kg})(3.33\times10^5 \text{ J/kg}) = 3.33\times10^6 \text{ J}.$$

The heat required to raise the temperature of water from 0°C to 100°C is

$$Q_3 = c_w\, m_w\, \Delta T_w = (4186 \text{ J/kg·C°})(10 \text{ kg})(100°\text{C} - 0°\text{C}) = 4.186\times10^6 \text{ J}.$$

The heat required to vaporize water at 100°C to steam at 100°C is

$$Q_4 = m\, L_v = (10 \text{ kg})(22.6\times10^5 \text{ J/kg}) = 22.6\times10^6 \text{ J}.$$

The heat required to raise the temperature of steam from 100°C to 120°C is

$$Q_5 = c_s\, m_s\, \Delta T_s = (2010 \text{ J/kg·C°})(10 \text{ kg})(120°\text{C} - 100°\text{C}) = 4.02\times10^5 \text{ J}.$$

So the total heat required is $Q_{tot} = \Sigma Q_i = Q_1 + Q_2 + Q_3 + Q_4 + Q_5 = 3.1\times10^7$ J.

4. Heat Transfer

There are three mechanisms of heat transfer: **conduction, convection**, and **radiation**. \

The **thermal conductivity** (k) characterizes the heat-conducting ability of a material. Its SI unit is J/m·s·C°. Materials with large k values are called **thermal conductors** while materials with small k values are called **thermal insulators**. The heat conduction rate (heat flow rate) is given by $\frac{\Delta Q}{\Delta t} = \frac{kA\Delta T}{d}$, where A is the cross-sectional area, d the thickness, and ΔT is the temperature difference.

Convection involves mass transfer (conduction does not). For example, when cold water is in contact with a hot object, such as the bottom of a pot on a stove, the object transfers heat to the water by conduction, and the warmer water carries the heat away with it by convection as it rises (less dense than cold water at the top).

Radiation is the transfer of energy by electromagnetic waves and requires no transfer medium (both conduction and convection require a transfer medium). The rate at which an object radiates energy is given by $P = \sigma A e T^4$, where P is the power radiated in watts (W) or joules/s (J/s), e is the **emissivity** which characterizes the material and has dimensionless value between 0 and 1, A is the surface area of the object, T is the absolute temperature of the object, and $\sigma = 5.67\times10^{-8}$ W/m²·K⁴ is called the *Stefan-Boltzmann constant*. A good emitter of radiation is also a good absorber. A perfect emitter and absorber ($e = 1$) is called a black body.

If an object is at temperature T and its surroundings are at a temperature of T_s, then the net rate of energy loss or gain due to radiation alone is given by $P_{net} = \sigma A e (T_s^4 - T^4)$.

Note: The temperatures used in the radiation equations *must* be in Kelvin.

Example 11.4 A window glass 0.50-cm thick has dimensions of 3.0 m by 1.5 m. If the outside temperature is −10°C and the inside temperature 20°C,

(a) what is the rate of heat conduction through the window?

(b) How much heat flows through the window in 1.0 h due to conduction only?

Solution: Given: $d = 0.50$ cm, $A = L\,W = (3.0\text{ m})(1.5\text{ m}) = 4.5\text{ m}^2$, $\Delta T = 20°\text{C} - (-10°\text{C}) = 30\text{ C}°$,

$k = 0.084$ J/m·s·C° (Table 11.4).

Find: (a) $\Delta Q/\Delta t$ (b) ΔQ.

$$\text{(a)}\ \frac{\Delta Q}{\Delta t} = \frac{kA\Delta T}{d} = \frac{(0.084\text{ J/m·s·C°})(4.5\text{ m}^2)(30\text{ C°})}{0.50\times10^{-2}\text{ m}} = 2.26\times10^3\text{ J/s} = 2.3\times10^3\text{ W} = 2.3\text{ kW}.$$

$$\text{(b)}\ \Delta Q = \frac{\Delta Q}{\Delta t}\Delta t = (2.26\times10^3\text{ J/s})(1.0\text{ h})(3600\text{ s/h}) = 8.1\times10^6\text{ J}.$$

Example 11.5 A radiator has an emissivity of 0.70 and its exposed area is 1.50 m². The temperature of the radiator is 100°C and the surrounding temperature is 20°C. What is the net heat flow rate from the body?

Solution: Given: $e = 0.70$, $A = 1.50\text{ m}^2$, $T = 100°\text{C} = 273 + 100 = 373$ K,

$T_s = 20°\text{C} = 273 + 20 = 293$ K.

Find: P_{net}.

$$P_{net} = \sigma A e (T_s^4 - T^4) = (5.67\times10^{-8}\text{ W/m}^2\text{·K}^4)(1.50\text{ m}^2)(0.70)[(293\text{ K})^4 - (373\text{ K})^4]$$

$= -7.1\times10^2$ J/s $= -7.1\times10^2$ W. The negative sign indicates that the radiator is losing energy.

IV. Mathematical Summary

Mechanical Equivalent of Heat	1 kcal = 4186 J 1 cal = 4.186 J	Gives the conversion between calorie and joules
Specific Heat (capacity)	$Q = mc\Delta T$ (11.1)	Relates heat with mass, specific heat, and change in temperature
Latent Heat	$L = mL$ (11.2)	Relates heat with latent heat and mass (for water) $L_f = 3.3\times10^5$ J/kg (80 kcal/kg) $L_v = 22.6\times10^5$ J/kg (540 kcal/kg)
Thermal Conduction	$\dfrac{\Delta Q}{\Delta t} = \dfrac{kA\Delta T}{d}$ (11.3)	Calculates the heat conduction rate
Stefan's Law	$P = \sigma A e T^2$ (11.5)	Computes the heat power
Radiant Power Loss or Gain	$P_{net} = \sigma A e(T_s^4 - T^4)$ (11.6)	Computes the net heat power

V. Solutions of Selected Exercises and Paired Exercises

6. (a) $1 \text{ Cal} = (1000 \text{ cal}) \times \dfrac{1 \text{ Btu}}{252 \text{ cal}} = 3.97 \text{ Btu} < 4 \text{ Btu}.$

(b) $210 \text{ Cal} = (210\times10^3 \text{ cal}) \times \dfrac{1 \text{ Btu}}{252 \text{ cal}} = 8.33\times10^2 \text{ Btu}.$

14. From $Q = mc\Delta T$, we have $\Delta T = \dfrac{Q}{cm} = \dfrac{(42 \text{ cal})(4.19 \text{ J/cal})}{(920 \text{ J/kg}\cdot\text{C}°)(5.0\times10^{-3} \text{ kg})} = 38.2°\text{C}.$

So the final temperature is 38.2°C + 20°C = 58°C.

24. 1 cm^3 of water has a mass of 1 g. Ice gains heat and water and the cup loses heat. Let T be the final temperature. There is a phase change (fusion for ice). The heat gained by ice is

$Q_i = m_i L_f + c_w m_i \Delta T_i = (0.050 \text{ kg})(3.3\times10^5 \text{ J/kg}) + (4186 \text{ J/kg}\cdot\text{C}°)(0.050 \text{ kg})(T - 0°\text{C})$

$= 1.65\times10^4 \text{ J} + (209 \text{ J/C}°)T.$

The heat lost by the water and the aluminum cup is

$Q_L = (4186 \text{ J/kg.C}°)(0.300 \text{ kg})(T - 25°\text{C}) + (920 \text{ J/kg}\cdot\text{C}°)(0.100 \text{ kg})(T - 25°\text{C}) = (1348 \text{ J/C}°)(T - 25°\text{C}).$

From Calorimetry: $-Q_{lost} = Q_{gained}$, we have $-(1348 \text{ J/C}°)(T - 25°\text{C}) = 1.65\times10^4 \text{ J} + (209 \text{ J/C}°)T.$

Solve for T, $T = 11°\text{C}.$

35. The difference in heat needed for the two processes is the heat needed for melting the ice at 0°C and evaporating the water at 100°C. The heat needed to raise the temperature of water from 0°C to 100°C is the same for both processes.

So: $\Delta Q = m\,L_f + m\,L_v = (1.0\text{ kg})(3.3\times10^5\text{ J/kg} + 22.6\times10^5\text{ J/kg}) = 2.6\times10^6\text{ J}$ more.

40. 1 liter of water has a mass of 1 kg. The heat needed to be removed from water is

$Q_{\text{lost}} = mc\Delta T = (4186\text{ J/kg·C°})(1.0\text{ kg})(20°\text{C} - 100°\text{C}) = -3.35\times10^5\text{ J}.$

The ice and the melt water have to absorb that much heat. Let the mass of ice be m.

$Q_{\text{gained}} = m\,L_f + mc\Delta T = m\,(3.3\times10^5\text{ J/kg}) + m(4186\text{ J/kg.C°})(20°\text{C} - 0°\text{C}) = (4.14\times10^5\text{ J/kg})m.$

From calorimetry: $-Q_{\text{lost}} = Q_{\text{gained}}$, we have $m = \dfrac{3.35\times10^5\text{ J}}{4.14\times10^5\text{ J/kg}} = 0.81\text{ kg}.$

53. (a) This convects the heat from the hot soup to the cooler air.

(b) No. The ice blocks air flow and cooling of the air. (Also, the air conditioner is less efficient and runs more, increasing electric cost.)

62. The normal body temperature is 37°C.

$$\frac{\Delta Q}{\Delta t} = \frac{kA\Delta T}{d} = \frac{(0.20\text{ J/m·s.C°})(1.5\text{ m}^2)(37°\text{C} - 33°\text{C})}{0.040\text{ m}} = 30\text{ J/s} = 7.2\times10^{-3}\text{ kcal/s}.$$

72. $P_{\text{net}} = \sigma A\,e(T_s^4 - T^4) = (5.67\times10^{-8}\text{ W/m}^2\text{·K}^4)(4\pi)(0.075\text{ m})^2\,(0.58)[(273 + 300)^4\text{ K}^4 - (273 + 20)^4\text{ K}^4]$

$= 233\text{ J/s}.$ In 1.0 min, $\Delta Q = P_{\text{net}}\,t = (233\text{ J/s})(60\text{ s}) = 1.4\times10^4\text{ J}.$

80. 1 mL of water has a mass of 1 g. Let the initial temperature of the shot be T.

The heat lost by the shot is $Q_{\text{copper}} = cm\Delta T = (390\text{ J/kg·C°})(0.150\text{ kg})(28°\text{C} - T) = (58.5\text{ J/C°})(28°\text{C} - T)$,

the heat gained by water is $Q_{\text{water}} = (4186\text{ J/kg·C°})(0.200\text{ kg})(28°\text{C} - 25°\text{C}) = 2512\text{ J}$,

the heat gained by the cup is $Q_{\text{cup}} = (920\text{ J/kg·C°})(0.375\text{ kg})(28°\text{C} - 25°\text{C}) = 1035\text{ J}.$

From Calorimetry: $-Q_{\text{lost}} = Q_{\text{gained}}$, we have $-(58.5\text{ J/C°})(28°\text{C} - T) = 2512\text{ J} + 1035\text{ J}.$

Solving for T, $T = 88.7°\text{C}.$

90. The heat required to raise the temperature of ice from −10°C to 0°C is

$Q_1 = cm\Delta T = (2100\text{ J/kg·C°})(0.75\text{ kg})[0°\text{C} - (-10°\text{C})] = 1.575\times10^4\text{ J},$

the heat required to change 0°C ice to 0OC water is

$Q_2 = m\,L_f = (0.75\text{ kg})(3.3\times10^5\text{ J/kg}) = 2.475\times10^5\text{ J},$

the heat required to raise the temperature if water from 0°C to 20°C is

$Q_3 = (4186\text{ J/kg·C°})(0.75\text{ kg})(20°\text{C} - 0°\text{C}) = 6.279\times10^4\text{ J}.$

So the total heat required is $\Sigma Q = 3.26\times10^5\text{ J} = 78\text{ kcal}.$

VI. Practice Quiz

1. In a cloud formation, water vapor condenses into water droplets which get bigger and bigger until it rains. This will cause the temperature of the air in the clouds to

   (a) get warmer. (b) get cooler. (c) stay constant. (d) there is no air in the clouds. (e) it freezes.

2. A machine gear consists of 0.10 kg of iron and 0.16 kg of copper. How much heat is added to the gear if its temperature increases by 35 C°. (The specific heats for iron and copper are 460 and 390 J/kg·C°.)

   (a) 9.1×10^2 J (b) 3.8×10^3 J (c) 4.0×10^3 J (d) 4.4×10^3 J (e) none of the above

3. How much heat is required to boil away 0.500 kg of water at 50°C?

   (a) 1.05×10^5 J (b) 1.67×10^5 J (c) 1.13×10^6 J (d) 1.23×10^6 J (e) 2.71×10^5 J

4. A 0.10 kg piece of cooper, initially at 95°C, is dropped into 0.20 kg of water contained in a 0.28-kg aluminum calorimeter; the water and calorimeter are initially at 15°C. What is the final temperature of the system?

   (a) 19.2 °C (b) 18.3°C (c) 17.8°C (d) 23.7°C (e) 25.0°C

5. If the absolute temperature of a radiator is doubled, by what factor does the radiating power change?

   (a) 2 (b) 4 (c) 8 (d) 16 (e) 32

6. The reason ocean temperatures do not vary drastically is that

   (a) water has a relatively high rate of conduction. (b) water is a good radiator.

   (c) water is a poor heat conductor. (d) water has a lower specific heat.

   (e) water has a relatively high specific heat.

7. A 5000-W heater is used to heat water. How long will it take the heater to heat 20 kg of water from 20°C to 100°C? (Assume no heat loss.)

   (a) 2 minutes (b) 12 minutes (c) 22 minutes (d) 32 minutes (e) 42 minutes

8. On a cold day, a piece of metal feels much colder to the touch than a piece of wood. This is due to the difference in which of the following physical properties?

   (a) density (b) specific heat (c) latent heat (d) thermal conductivity (e) temperature

9. Equal masses of water at 20°C and 80°C are mixed. What is the final temperature of the mixture?

(a) 30°C (b) 40°C (c) 50°C (d) 60°C (e) 70°C

10. A person makes ice tea by adding ice to 1.8 kg of hot tea (water), initially at 80°C. How many kilograms of ice, initially at 0°C, are required to bring the mixture to 10°C?

(a) 1.0 kg (b) 1.2 kg (c) 1.5 kg (d) 1.7 kg (e) 1.9 kg

Answers to Practice Quiz:

1.a 2.b 3.d 4.c 5.d 6.e 7.b 8.d 9.c 10.c

CHAPTER 12

Thermodynamics

I. Chapter Objectives

Upon completion of this chapter, you should be able to:

1. define thermodynamic systems and states of systems, and explain how processes affect such systems.
2. explain the relationship of internal energy, heat, and work expressed by the first law, and analyze various "iso-" processes.
3. state and explain the second law of thermodynamics in several forms, and explain the concept of entropy.
4. explain the concept of a heat engine and compute thermal efficiency, and explain the concept of a heat pump and compute coefficient of performance.
5. explain how the Carnot cycle applies to heat engines, compute the ideal efficiency, and state the third law of thermodynamics.

II. Key Terms

Upon completion of this chapter, you should be able to define and/or explain the following key terms:

thermodynamics
system
thermally isolated system
completely isolated system
heat reservoir
equations of state
process
irreversible process
reversible process
first law of thermodynamics
isobaric process
isometric process
isothermal process
adiabatic process
second law of thermodynamics
entropy
isentropic process
heat engine
thermal efficiency
heat pump
coefficient of performance
Carnot cycle
Carnot (ideal) efficiency
third law of thermodynamics

The definitions and/or explanations of the most important key terms can be found in the following section: **III. Chapter Summary and Discussion**.

III. Chapter Summary and Discussion

1. Thermodynamic Systems, States, and Processes

Thermodynamics deals with the transfer or action (dynamics) of heat. A **system** is a definite quantity of matter enclosed by boundaries or surfaces, either real or imaginary. An *open system* or *closed system* refers to whether or not mass can be transferred into or out of a system.

There is no heat transfer in or out of a **thermally isolated system**. A **completely isolated system** has no energy exchange (that is, no interaction at all) with its surroundings.

A **heat reservoir** is a system with an unlimited heat capacity for which the addition or removal of heat causes no change in temperature.

A **thermodynamic equation of state** is a mathematical relationship of the thermodynamic or state variables, such as pressure, volume and temperature for an ideal gas, $pV = Nk_BT$. A **process** is a change in the state or the thermodynamic variables of a system. A **reversible process** is one with a known path and an **irreversible process** is one for which intermediate steps are nonequilibrium states. Irreversible does not mean that the system can't be taken back to the initial state; it only means the process path can't be retraced, because of the nonequilibrium conditions.

2. The First Law of Thermodynamics

The **first law of thermodynamics** is simply a statement of the conservation of energy for thermodynamic systems. It relates the change in internal energy, heat, and work: $Q = \Delta U + W$, where Q is heat, ΔU is the change in internal energy, and W is work.

Note: Be aware of the strict sign convention for work and heat: $+Q$, heat *added*; $-Q$, heat *removed*; $+W$, work done *by* system; $-W$, work done *on* system.

The work done by an expanding gas at constant pressure is given by $W = p\Delta V = p(V_2 - V_1)$. For processes in which the pressure is not a constant, work can be calculated graphically by computing the area under the curve in a pressure versus volume graph.

"Iso-" processes are ones in which one of the thermodynamic variables is held constant. For example: **isobaric** (constant pressure), **isometric** (constant volume), **isothermal** (constant temperature or constant internal energy for an ideal gas), **adiabatic** ($Q = 0$ or no heat transfer).

Example 12.1 A system consists of 3.0 kg of water at 80°C. 25 J of work is done on the system by stirring with a paddle wheel, while 63 J of heat is removed. What is the change in internal energy of the system?

Solution: Be careful about the sign convention for heat and work.

Given: $W = -25$ J (done on system), $Q = -63$ J (removed).

Find: ΔU.

From $Q = \Delta U + W$, we have $\Delta U = Q - W = -63 \text{ J} - (-25 \text{ J}) = -63 \text{ J} + 25 \text{ J} = -38 \text{ J}$.

The internal energy of the system decreases.

Example 12.2 During an isothermal process, 5.0 J of heat is removed from an ideal gas.

(a) What is the change in internal energy?

(b) What is the work done in the process?

Solution: Given: isothermal process, $Q = -5.0$ J (removed).

Find: (a) ΔU (b) W.

(a) In an isothermal process, the temperature and internal energy of an ideal gas is constant because for ideal gas, the internal energy is proportional to temperature. So $\Delta U = 0$

(b) From $Q = \Delta U + W$, we have $W = Q - \Delta U = -5.0 \text{ J} - 0 = -5.0 \text{ J}$. That is, 5.0 J of work is done on the system (it is compressed).

Example 12.3 In an isometric process, the internal energy of a system decreases by 50 J.

(a) What is the work done?

(b) What is the heat exchange?

Solution: Given: isometric process, $\Delta U = -50$ J (decrease).

Find: (a) W (b) Q.

(a) In an isometric process, the volume is a constant, so $\Delta V = 0$.

Therefore, $W = p\Delta V = 0$.

(b) $Q = \Delta U + W = -50 \text{ J} + 0 = -50 \text{ J}$. That is, 50 J of heat is removed from the system.

---

3. The Second Law of Thermodynamics and Entropy

The **second law of thermodynamics** specifies the direction in which a process can naturally or spontaneously take place. Many equivalent statements are used to describe the second law according to the situation or application.

Heat will not flow spontaneously from a colder body to a warmer body. This does not mean heat cannot flow from a warmer body to a colder body. All it says is that it cannot take place spontaneously.

Heat energy cannot be transformed completely into mechanical work. This does not mean energy is not conserved. All it says is that heat energy cannot be 100% converted to mechanical work.

It is impossible to construct an operational perpetual motion engine.

The total entropy of the universe increases in every natural process.

Entropy is a measure of the disorder of a system and is related to heat and temperature. The change in entropy (ΔS) when an amount of heat (Q) is added to (or removed from) a system by a reversible process at a constant temperature (T) is given by $\Delta S = \frac{Q}{T}$, where T is the Kelvin temperature and the units for entropy are joules per kelvin (J/K).

Example 12.4 1.0 kg of steam at 100°C condenses to water at 100°C. What is the change in entropy in the process?

Solution: Given: $m = 1.0$ kg, $T = 100°\text{C} = 100 + 273 = 373$ K, $L_v = 22.6\times10^5$ J/kg.

Find: ΔS.

The heat removed in the process is $Q = -m\,L_v = -(1.0\text{ kg})(22.6\times10^5\text{ J/kg}) = -22.6\times10^5$ J.

So $\Delta S = \frac{Q}{T} = \frac{-22.6\times10^5\text{ J}}{373\text{ K}} = -6.1\times10^3$ J/K. The entropy decreases (more order in final state).

Example 12.5 Show that it is a violation of the second law of thermodynamics to have some heat , say 10 J, to transfer spontaneously from a cold reservoir at 300 K to a hot reservoir at 350 K.

Solution: According to the second law of thermodynamics, the total entropy of an isolated system increases for every natural process. That is, the change in entropy for all natural process must be greater than zero.

Given: $Q_{\text{cold}} = -10$ J, $Q_{\text{hot}} = 10$ J, $T_{\text{cold}} = 300$ K, $T_{\text{hot}} = 350$ K.

Find: ΔS.

For the cold reservoir: $\Delta S_{\text{cold}} = \dfrac{Q_{\text{cold}}}{T_{\text{cold}}} = \dfrac{-10\text{ J}}{300\text{ K}} = -3.33\times10^{-2}\text{ J/K}.$

For the hot reservoir: $\Delta S_{\text{hot}} = \dfrac{Q_{\text{hot}}}{T_{\text{hot}}} = \dfrac{10\text{ J}}{350\text{ K}} = 2.86\times10^{-2}\text{ J/K}.$

The total change in entropy for the system (both reservoirs) is

$\Delta S = \Delta S_{\text{cold}} + \Delta S_{\text{hot}} = -3.33\times10^{-2}\text{ J/K} + 2.86\times10^{-2}\text{ J/K} = -4.7\times10^{-3}\text{ J/K}.$

The total change in entropy for the process is negative, that is, the total entropy decreases. This is a violation of the second law of thermodynamics so the process described in the problem cannot spontaneously take place in isolation.

---

4. Heat Engines and Heat Pumps

A **heat engine** is any cyclic device that converts heat to work. It absorbs heat (Q_{hot}) from a high-temperature reservoir, does work (W), and exhausts heat (Q_{cold}) to a low-temperature reservoir. After one cycle, every thermodynamic variable returns to its original value so $\Delta U = 0$. Therefore, from the first law of thermodynamics, $W = Q_{\text{hot}} - Q_{\text{cold}}$. The **thermal efficiency** of a heat engine is defined as the ratio of its work output and its heat input: $\varepsilon_{\text{th}} = \dfrac{W}{Q_{\text{in}}} = \dfrac{Q_{\text{hot}} - Q_{\text{cold}}}{Q_{\text{hot}}} = 1 - \dfrac{Q_{\text{cold}}}{Q_{\text{hot}}}$.

A **heat pump** is a device that transfers heat energy from a low-temperature reservoir to a high-temperature reservoir. To do this, there must be work input, according to the second law of thermodynamics. The **coefficient of performance** (cop) for a refrigerator (or air conditioner) is expressed as $\text{cop}_{\text{ref}} = \dfrac{Q_{\text{in}}}{W} = \dfrac{Q_{\text{cold}}}{Q_{\text{hot}} - Q_{\text{cold}}}$. The coefficient of performance for a heat pump in heating mode is defined as $\text{cop}_{\text{hp}} = \dfrac{Q_{\text{out}}}{W} = \dfrac{Q_{\text{hot}}}{Q_{\text{hot}} - Q_{\text{cold}}}$. For either case, $Q_{\text{hot}} = W + Q_{\text{cold}}$, according to the first law of thermodynamics.

Example 12.6 A heat engine has an efficiency of 35.0% and extracts 150 J of heat from a hot reservoir per cycle.

(a) How much work does it perform in each cycle?

(b) How much heat does it exhaust in each cycle?

Solution: Given: $\varepsilon_{\text{th}} = 35.0\% = 0.350$, $Q_{\text{hot}} = 150\text{ J}$.

Find: (a) W (b) Q_{cold}.

(a) From $\varepsilon_{\text{th}} = \dfrac{W}{Q_{\text{hot}}}$, we have $W = \varepsilon_{\text{th}}\, Q_{\text{hot}} = (0.350)(150\text{ J}) = 52.5\text{ J}$.

(b) $Q_{\text{hot}} = W + Q_{\text{cold}}$, so $Q_{\text{cold}} = Q_{\text{hot}} - W = 150\text{ J} - 52.5\text{ J} = 97.5\text{ J}$.

Example 12.7 A refrigerator removes heat from the freezing compartment of a freezer at the rate of 20 kJ per cycle and ejects 24 kJ per cycle into a room.

(a) How much work input is required in each cycle?

(b) What is the coefficient of performance of the refrigerator?

Solution: Given: $Q_{\text{cold}} = 20 \text{ kJ} = 20\times 10^3 \text{ J}$, $Q_{\text{hot}} = 24 \text{ kJ} = 24\times 10^3 \text{ J}$.

Find: (a) W (b) cop_{ref}.

(a) $Q_{\text{hot}} = W + Q_{\text{cold}}$, so $W = Q_{\text{hot}} - Q_{\text{cold}} = 24\times 10^3 \text{ J} - 20\times 10^3 \text{ J} = 4.0\times 10^3 \text{ J}$.

(b) $\text{cop}_{\text{ref}} = \dfrac{Q_{\text{cold}}}{W} = \dfrac{20\times 10^3 \text{ J}}{4.0\times 10^3 \text{ J}} = 5.0$,

or $\text{cop}_{\text{ref}} = \dfrac{Q_{\text{cold}}}{Q_{\text{hot}} - Q_{\text{cold}}} = \dfrac{20\times 10^3 \text{ J}}{24\times 10^3 \text{ J} - 20\times 10^3 \text{ J}} = 5.0$.

5. The Carnot Cycle and Ideal Heat Engines

The ideal **Carnot cycle** for a heat engine consists of two isothermal and two adiabatic processes. This is the most efficient engine, and the **Carnot efficiency** gives an unattainable, upper limit of efficiency: The efficiency is expressed as $\varepsilon_{\text{C}} = 1 - \dfrac{T_{\text{cold}}}{T_{\text{hot}}} = \dfrac{T_{\text{hot}} - T_{\text{cold}}}{T_{\text{hot}}}$, where T_{cold} and T_{hot} are the Kelvin temperatures of the low-temperature and high-temperature reservoirs, respectively.

Note: The temperatures in the Carnot efficiency calculation *must* be in Kelvin.

The third law of thermodynamics states that it is impossible to reach a temperature of absolute zero.

Example 12.8 What is the Carnot efficiency of a gasoline engine which operates between 430 K and room temperature, 300 K?

Solution: Given: $T_{\text{hot}} = 430 \text{ K}$, $T_{\text{cold}} = 300\text{K}$.

Find: ε_{C}.

$\varepsilon_{\text{C}} = 1 - \dfrac{T_{\text{cold}}}{T_{\text{hot}}} = 1 - \dfrac{300 \text{ K}}{430 \text{ K}} = 0.30 = 30\%$. Remember this is the unattainable, upper limit! No gasoline engine can have a higher efficiency if it is operating between these two temperatures!

IV. Mathematical Summary

First Law of Thermodynamics	$Q = \Delta U + W$ (12.1)	Relates heat, change in internal energy, and work $+Q$, heat *added* to system $-Q$, heat *removed* from system $+W$, work done *by* system $-W$, work done *on* system
Work Done by Gas (constant pressure)	$W = p\Delta V = p(V_2 - V_1)$ (12.2)	Calculates the work done by an expanding gas at constant pressure
Change in Entropy (constant temperature)	$\Delta S = \frac{Q}{T}$ (12.4)	Defines the change in entropy at constant temperature
Thermal Efficiency	$\varepsilon_{th} = \frac{W}{Q_{in}} = \frac{Q_{hot} - Q_{cold}}{Q_{hot}}$ $= 1 - \frac{Q_{cold}}{Q_{hot}}$ (12.6)	Defines and calculates the thermal efficiency of a heat engine
Coefficient of Performance for Refrigerator or Air Conditioner	$\text{cop}_{ref} = \frac{Q_{in}}{W}$ $= \frac{Q_{cold}}{Q_{hot} - Q_{cold}}$ (12.7)	Defines and calculates the coefficient of performance of a refrigerator or air conditioner
Coefficients of Performance of a Heat Pump in Heating Mode	$\text{cop}_{hp} = \frac{Q_{out}}{W}$ $= \frac{Q_{hot}}{Q_{hot} - Q_{cold}}$ (12.8)	Defines and calculates the coefficient of performance for a heat pump in heating mode. A heat pump in cooling mode works like a refrigerator or air conditioner
Carnot Efficiency	$\varepsilon_C = 1 - \frac{T_{cold}}{T_{hot}}$ $= \frac{T_{hot} - T_{cold}}{T_{hot}}$ (12.10)	Defines and calculates the Carnot efficiency for an ideal heat engine

V. Solutions of Selected Exercises and Paired Exercises

3. For an ideal gas, $pV = Nk_B T = nRT$. So a plot of p vs. V will not be a straight line.

4. (a) isothermal

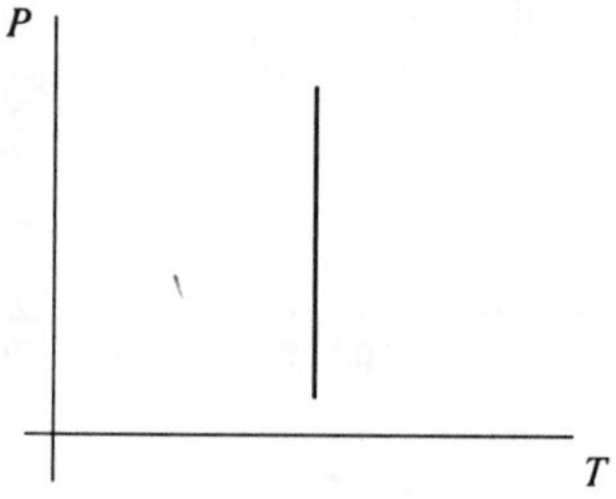

(b) isobaric

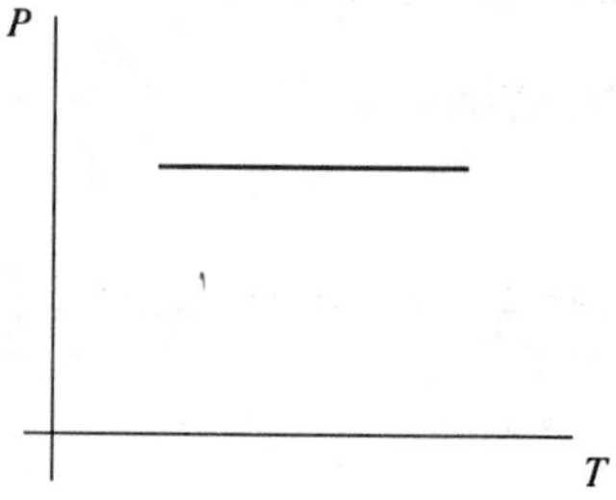

(c) isometric

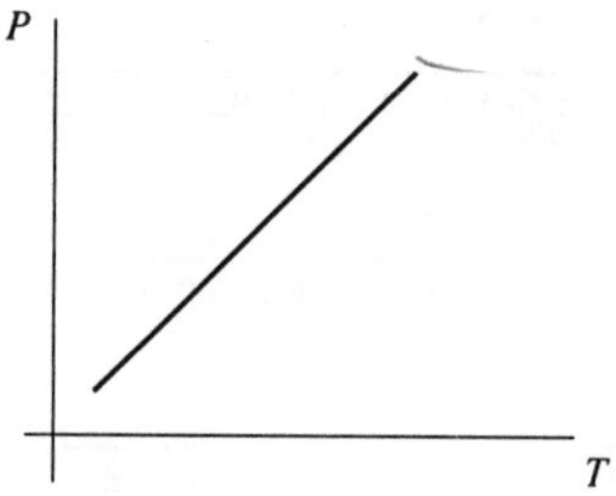

(d) adiabatic

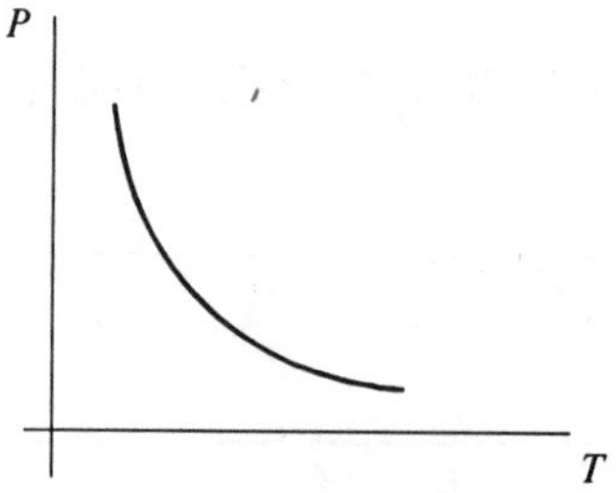

5. (a) isothermal

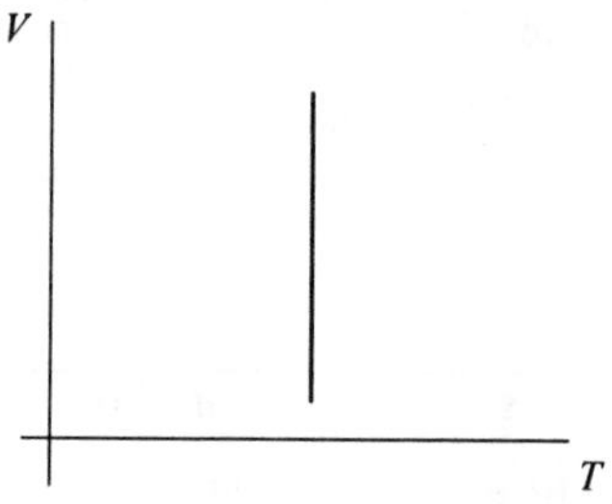

(b) isobaric

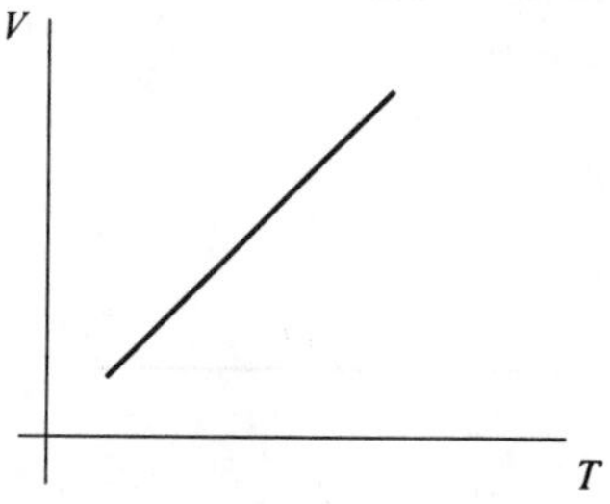

(c) isometric

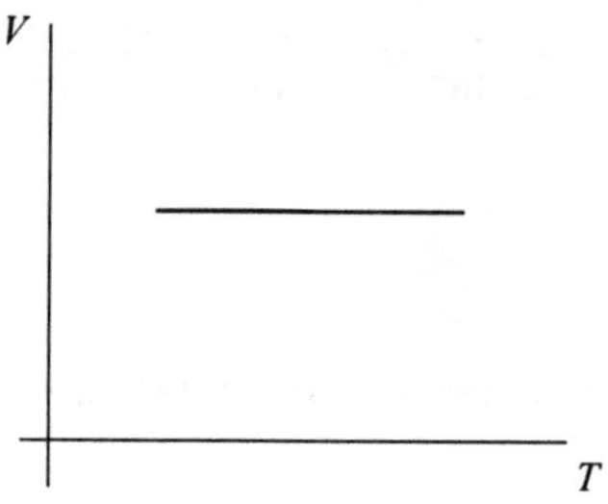

(d) adiabatic

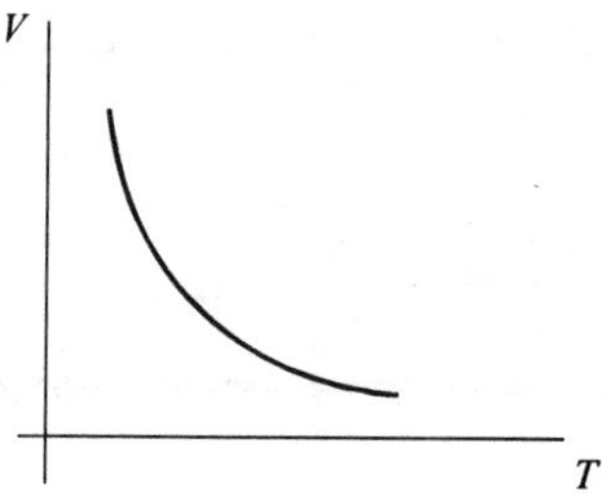

12. Since the container is rigid, $\Delta V = 0$, so $W = 0$.

$Q = \Delta U + W = \Delta U = 3.0\text{ kcal} = 1.3\times10^4\text{ J}$.

24. The work done equals the area under the curve (the area of the trapezoid). The work is negative for compression (work is done on the system).

$W = \frac{1}{2}(2.0\times10^5 \text{ Pa} + 5.0\times10^5 \text{ Pa})(0.50 \text{ m}^3) = -1.8\times10^5 \text{ J}.$

34. For an isothermal process, U is a constant or $\Delta U = 0$. So $Q = \Delta U + W = W = 3.0\times10^3$ J.

Therefore $\Delta S = \dfrac{Q}{T} = \dfrac{3.0\times10^3 \text{ J}}{273 \text{ K}} = 11$ J/K.

40. $\Delta S = \Delta S_1 + \Delta S_2 = \dfrac{Q_1}{T_1} + \dfrac{Q_2}{T_2} = \dfrac{-1.5\times10^3 \text{ J}}{473 \text{ K}} + \dfrac{1.5\times10^3 \text{ J}}{333 \text{ K}} = 1.33$ J/K.

A negative ΔS would mean the universe is becoming more ordered (impossible).

54. (a) $\varepsilon_{\text{th}} = 1 - \dfrac{Q_{\text{cold}}}{Q_{\text{hot}}} = 1 - \dfrac{100 \text{ kcal}}{150 \text{ kcal}} = 33\%.$

(b) $W = Q_{\text{hot}} - Q_{\text{cold}} = 50 \text{ kcal} = 2.1\times10^5$ J.

64. (a) $Q_{\text{hot}} = m\,L_{\text{v}} = (8.00 \text{ kg})(22.6\times10^5 \text{ J/kg}) = 1.81\times10^7$ J.

From $\varepsilon_{\text{th}} = \dfrac{W}{Q_{\text{hot}}}$, we have $W = \varepsilon_{\text{th}}\,Q_{\text{hot}} = 0.300(1.81\times10^7 \text{ J}) = 5.42\times10^6$ J.

(b) $Q_{\text{cold}} = Q_{\text{hot}} - W = 1.81\times10^7 \text{ J} - 5.42\times10^6 \text{ J} = 1.27\times10^7$ J.

76. (a) From $\varepsilon_{\text{th}} = 1 - \dfrac{Q_{\text{cold}}}{Q_{\text{hot}}}$, we have $Q_{\text{hot}} = \dfrac{Q_{\text{cold}}}{1-\varepsilon} = \dfrac{1200 \text{ J}}{1-0.40} = 2000$ J.

(b) From $\varepsilon_{\text{C}} = 1 - \dfrac{T_{\text{cold}}}{T_{\text{hot}}}$, we have $T_{\text{hot}} = \dfrac{T_{\text{cold}}}{1-\varepsilon_{\text{c}}} = \dfrac{(273+50) \text{ K}}{1-0.40} = 538 \text{ K} = 265°\text{C}.$

80. From $\varepsilon_{\text{c}} = 1 - \dfrac{T_{\text{cold}}}{T_{\text{hot}}}$, we have $T_{\text{cold}} = (1-\varepsilon_{\text{c}})T_{\text{hot}} = (1-0.40)(273+350) \text{ K} = 374 \text{ K} = 101°\text{C}.$

90. (a) $\text{cop}_{\text{c}} = \dfrac{Q_{\text{hot}}}{Q_{\text{hot}} - Q_{\text{cold}}} = \dfrac{T_{\text{hot}}}{T_{\text{hot}} - T_{\text{cold}}}.$

(b) The efficiency improves as the temperature difference between the two reservoirs decreases.

101. (a) With the piston locked in place, $\Delta V = 0$.

So $\Delta U = Q - W = Q - p\Delta V = Q = (3.5 \text{ kcal})(4190 \text{ J/kcal}) = 1.5\times10^4$ J.

(b) 1 liter of water has a mass of 1 kg. $\Delta T = \dfrac{Q}{cm} = \dfrac{1.47\times10^4 \text{ J}}{(4190 \text{ J/kg}\cdot\text{C}°)(2.0 \text{ kg})} = 1.75 \text{ C}°.$

So the final temperature is $20°\text{C} + 1.75 \text{ C}° = 22°\text{C}$.

104. 1 liter of water has a mass of 1 kg and 1 liter of ethyl alcohol has a mass of 0.80 kg (the density of ethyl alcohol is 0.80 times that of water). Assume the final temperature is T.

$Q_{\text{lost}} = cm\Delta T = (4190\ \text{J/kg}\cdot\text{C}°)(0.50\ \text{kg})(T - 25.0°\text{C}) = (2095\ \text{J/C}°)(T - 25.0°\text{C})$,

$Q_{\text{gained}} = (2430\ \text{J/kg}\cdot\text{C}°)(0.40\ \text{kg})(T - 18.0°\text{C}) = (972\ \text{J/C}°)(T - 18.0°\text{C})$.

From calorimetry: $-Q_{\text{lost}} = Q_{\text{gained}}$, we have $-(2095\ \text{J/C}°)(T - 25.0°\text{C}) = (972\ \text{J/C}°)(T - 18.0°\text{C})$.

Solving for T, $T = 22.78°\text{C}$. So $Q_{\text{gained}} = 4.648\times10^3$ J.

Use average temperature in entropy calculations.

$$\Delta S = \Delta S_w + \Delta S_e = \frac{-4.648\times10^3\ \text{J}}{(273 + 23.89)\ \text{K}} + \frac{4.648\times10^3\ \text{J}}{(273 + 20.39)\ \text{K}} = 0.187\ \text{J/K}.$$

VI. Practice Quiz

1. An ideal gas undergoes an *adiabatic* process in doing 25 J of work,. What is the change in internal energy?
(a) 0 (b) 50 J (c) 25 J (d) −50 J (e) −25 J

2. A heat engine receives 6000 J of heat from its combustion process and loses 4000 J through the exhaust and friction. What is its efficiency?
(a) 33% (b) 40% (c) 60% (d) 67% (e) 73%

3. According to the second law of thermodynamics, for any process that may occur within an isolated system, which of the following choices applies?
(a) entropy remains constant (b) entropy increases (c) entropy decreases
(d) both (a) and (b) (e) both (a) and (c)

4. The efficiency of a Carnot engine is 35.0%. What is the temperature of the cold reservoir if the temperature of the hot reservoir is 500 K?
(a) 175 K (b) 325 K (c) 269 K (d) 231 K (e) 345 K

5. In a isometric process, there is no change in
(a) pressure. (b) volume. (c) temperature. (d) heat. (e) internal energy.

6. When 0.50 kg of ice melts, the change in entropy is
(a) 4.6×10^2 J/K. (b) $1.7\times10^3\times10^2$ J/K. (c) 8.4×10^2 J/K. (d) 6.1×10^2 J/K. (e) 3.3×10^3 J/K.

7. The work done on an ideal gas system in an isothermal process is −400 J. What is the change in internal energy?

(a) zero (b) −400 J (c) −800 J (d) 400 J (e) 800 J

8. If a heat engine has an efficiency of 30% and its power output is 600 W. What is the rate of heat input from the high-temperature reservoir?

(a) zero (b) 1.8 kW (c) 2.0 kW (d) 2.4 kW (e) 3.0 kW

9. During each cycle of operation, a refrigerator absorbs 230 J of heat from the freezer compartment and expels 356 J of heat to the room. How much work input is required in each cycle?

(a) zero (b) 460 J (c) 712 J (d) 586 J (e) 126 J

10. What is the work done by the gas as it expands from pressure p_1 and volume V_1 to pressure p_2 and volume V_2 along a straight line path as shown?

(a) zero (b) $\frac{1}{2}(p_1 + p_2)(V_1 + V_2)$ (c) $\frac{1}{2}(p_1 - p_2)(V_1 + V_2)$

(d) $\frac{1}{2}(p_1 + p_2)(V_1 - V_2)$ (e) $\frac{1}{2}(p_1 - p_2)(V_1 - V_2)$

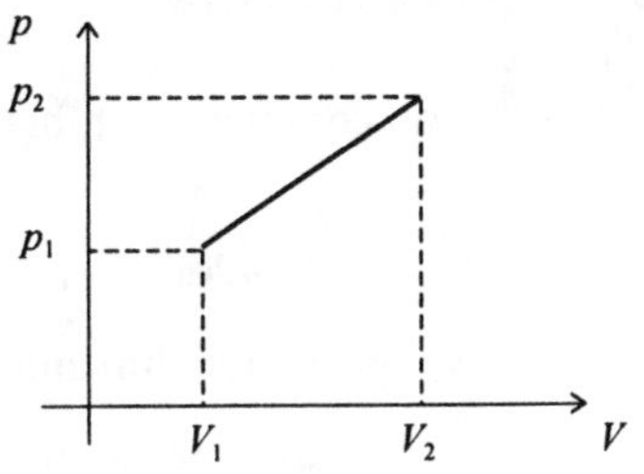

Answers to Practice Quiz:

1. e 2. a 3. d 4. b 5. b 6. d 7. a 8. c 9. e 10. d

CHAPTER 13

Vibrations and Waves

I. Chapter Objectives

Upon completion of this chapter, you should be able to:

1. describe simple harmonic motion and relate energy and speed in such motion.
2. write the equation of motion for SHM and explain what is meant by phase and phase differences.
3. describe wave motion in terms of various parameters and identify different types of waves.
4. explain various wave phenomena.
5. describe the formation and characteristics of standing waves, and explain the phenomenon of resonance.

II. Key Terms

Upon completion of this chapter, you should be able to define and/or explain the following key terms:

Hooke's law
simple harmonic motion (SHM)
displacement
amplitude
period
frequency
hertz (Hz)
equation of motion
phase constant
damped harmonic motion
wave
wave motion
wavelength (λ)
wave speed
transverse wave
longitudinal wave
principle of superposition
constructive interference
destructive interference
total constructive interference
total destructive interference
reflection
refraction
diffraction
standing wave
node
antinode
natural (resonant) frequencies
fundamental frequency
harmonic series
resonance

The definitions and/or explanations of the most important key terms can be found in the following section: **III. Chapter Summary and Discussion**.

III. Chapter Summary and Discussion

1. Simple Harmonic Motion

The motion of an oscillating or vibrating object depends on the restoring forces that make it go back and forth. The simplest type of force is spring force which is described by Hooke's law, $\mathbf{F} = -k\,\mathbf{x}$, where k is the spring constant, and the negative sign indicates that the force, **F**, is opposite to the displacement, **x**.

Motion under the influence of the type of force described by Hooke's law is called **simple harmonic motion (SHM)**, because this force is the simplest restoring force and because the motion can be described by harmonic functions (sines and cosines). SHM is described by the following parameters:

amplitude (A):	the magnitude of the maximum displacement of a mass from its equilibrium position,
period (T):	the time for one complete oscillation,
frequency (f):	the number of cycles per second.

The frequency and period are related by $f = \frac{1}{T}$. The SI unit of frequency is 1/s, or hertz (Hz). We sometimes descriptively say cycles/s (cycles per second, cps).

The total mechanical energy of an object in SHM is directly proportional to the square of its amplitude. For example, for a mass m oscillating on a spring of spring constant k, $E = \frac{1}{2}kA^2$. The maximum speed of the mass is directly proportional to the amplitude, $v_{max} = \sqrt{\frac{k}{m}}\,A$, because $E = \frac{1}{2}kA^2 = \frac{1}{2}mv_{max}^2$.

Example 13.1 A 0.50-kg mass is attached to a spring of spring constant 20 N/m along a horizontal, frictionless surface. The object oscillates in simple harmonic motion and has a speed of 1.5 m/s at the equilibrium position.

(a) What is the total energy of the system?

(b) What is the amplitude?

(c) At what location are the values for the potential and kinetic energies the same?

Solution: Given: $m = 0.50$ kg, $v_{max} = 1.5$ m/s, $k = 20$ N/m.

Find: (a) E (b) A (c) x when $K = U$.

(a) At the equilibrium position, $x = 0$, so the potential energy is zero. Therefore, the total energy is equal to kinetic energy. $E = K = \frac{1}{2}mv_{max}^2 = \frac{1}{2}(0.50\text{ kg})(1.5\text{ m/s})^2 = 0.56$ J.

(b) At the amplitude ($x = \pm A$), $v = 0$, so the total energy is equal to the potential energy.

$$E = U = \tfrac{1}{2}kx^2 = \tfrac{1}{2}kA^2, \quad \text{therefore} \quad A = \sqrt{\frac{2E}{k}} = \sqrt{\frac{2(0.56\text{ J})}{20\text{ N/m}}} = 0.24\text{ m}.$$

(c) When $K = U$, $E = K + U = U + U = 2U = 2\tfrac{1}{2}kx^2 = kx^2$,

$$\text{so } x = \sqrt{\frac{E}{k}} = \sqrt{\frac{0.56\text{ J}}{20\text{ N/m}}} = 0.17\text{ m}.$$

Example 13.2 A mass is attached to a spring of spring constant 60 N/m along a horizontal, frictionless surface. The spring is initially stretched by a force of 5.0 N on the mass and let go. It takes the mass 0.50 s to go back to its equilibrium position when it is oscillating.

(a) What is the amplitude?

(b) What is the period?

(c) What is the frequency?

Solution: Given: $k = 60$ N/m, $F = 5.0$ N, $t = 0.50$ s.

Find: (a) A (b) T (c) f.

(a) The initial stretch is the maximum distance from the equilibrium position. From Hooke's law, $F = kx$, we have $A = x_{max} = \dfrac{F}{k} = \dfrac{5.0\text{ N}}{60\text{ N/m}} = 0.083$ m.

(b) The motion from the amplitude to the equilibrium position is only 1/4 of a complete oscillation. So the period is $T = 4t = 4(0.50\text{ s}) = 2.0$ s.

(c) $f = \dfrac{1}{T} = \dfrac{1}{2.0\text{ s}} = 0.50$ Hz.

---

2. Equation of Motion

The **equation of motion** of an object in SHM has the general form $y = A\sin(\omega t + \delta)$, where A is the amplitude, ω is the angular frequency of the motion ($\omega = 2\pi f = \frac{2\pi}{T}$), and δ is the **phase constant**. The amplitude and the phase constant depend on the initial conditions of the situation. The frequency f depends on the intrinsic properties of the system (the stiffness and the inertia). From the equation of motion, we can express the velocity and acceleration of an object in SHM as

$$v = A\,\omega\cos(\omega t + \delta) \quad \text{and} \quad a = -\omega^2 A\sin(\omega t + \delta) = -\omega^2 y.$$

For a mass-spring system, $T = 2\pi\sqrt{\frac{m}{k}}$, and $f = \frac{1}{2\pi}\sqrt{\frac{k}{m}}$, or $\omega = \sqrt{\frac{k}{m}}$, where m is the mass of the object and k is the spring constant.

For a simple pendulum, $T = 2\pi\sqrt{\frac{L}{g}}$, and $f = \frac{1}{2\pi}\sqrt{\frac{g}{L}}$, where L is the length of the pendulum and g is the acceleration due to gravity. (Note: the mass of the pendulum is not in the equation so the period is independent of mass.)

Without a driving force and with frictional losses, the amplitude and the energy of an oscillator will decrease with time, giving rise to **damped harmonic motion**.

Example 13.3 An object with a mass of 4.0 kg is attached to a spring with a spring constant of 10 N/m. The object is displaced by 5.0 cm from the equilibrium position and let go.

(a) What is the amplitude A?

(b) What is the period T?

(c) What is the frequency f?

Solution: Given: $m = 4.0$ kg, $k = 10$ N/m, initial $x = 5.0$ cm.

Find: (a) A (b) T (c) f.

(a) The initial stretch is the maximum displacement. So $A = 5.0$ cm.

(b) $T = 2\pi\sqrt{\frac{m}{k}} = 2\pi\sqrt{\frac{4.0\text{ kg}}{10\text{ N/m}}} = 4.0$ s.

(c) $f = \frac{1}{T} = \frac{1}{4.0\text{ s}} = 0.25$ Hz.

Example 13.4 The pendulum of a grandfather clock is 1.0 m long.

(a) What is its period on the Earth?

(b) What would its period be on the Moon where the acceleration due to gravity is 1.7 m/s^2?

Solution: Given: $L = 1.0$ m, $g_E = 9.8$ m/s^2, $g_M = 1.7$ m/s^2.

Find: (a) T_E (b) T_M.

(a) $T_E = 2\pi\sqrt{\frac{L}{g}} = 2\pi\sqrt{\frac{1.0\text{ m}}{9.8\text{ m/s}^2}} = 2.0$ s.

(b) $T_M = 2\pi\sqrt{\frac{1.0\text{ m}}{1.7\text{ m/s}^2}} = 4.8$ s.

Example 13.5 An object in simple harmonic motion obeys the following position vs. time equation: $y = (0.25 \text{ m}) \sin\left(\frac{\pi}{2}t\right)$. Find (a) the amplitude A, (b) the period T, (d) the maximum speed.

Solution: Given: $y = (0.25 \text{ m}) \sin\left(\frac{\pi}{2}t\right)$.

Find: (a) A (b) T (c) v_{max}.

(a) Comparing the equation of motion $y = (0.25 \text{ m}) \sin\left(\frac{\pi}{2}t\right)$ with the general form $y = A \sin(\omega t + \delta)$, we see that $A = 0.25$ m.

(b) Again, comparing the equation of motion with the general form, $\omega t = 2\pi f t = \frac{\pi}{2} t$.

So $f = 1/4 = 0.25$ Hz. Therefore $T = \frac{1}{f} = \frac{1}{0.25 \text{ Hz}} = 4.0$ s.

(c) From the equation of velocity as a function of time $v = A\omega \cos(\omega t + \delta)$, we know that the maximum speed is $A\omega = A(2\pi f) = (0.25 \text{ m})(2\pi)(1/4.0 \text{ Hz}) = 0.39$ m/s.

3. Wave Motion

Wave motion is the propagation of a disturbance or a process of energy and momentum transfer through a material. The medium though which the wave propagates is not transferred in wave motion. A periodic (as oposed to a pulse) wave can be characterized by the following quantities:

amplitude: the magnitude of the maximum displacement from the particles equilibrium positions,

wavelength: the distance between two successive crests (or troughs), see note below,

frequency: the number of wavelengths that passes by a given point in a second,

wave speed: the speed of the wave motion given by $v = \lambda f$.

Note: The wavelength is actually the distance between any two successive particles that are in phase (that is, at identical points on the wave form). Crests are used just for convenience.

Waves are divided into two types based on the direction of particle oscillation relative to the wave velocity. For a **transverse wave** (sometimes called a *shear wave*) the particle motion is *perpendicular* to the direction of the wave velocity. A wave on a rope is an example of a transverse wave. For a **longitudinal wave** (or sometimes called a *compressional wave*) the particle motion is *parallel* to the direction of the wave velocity. Sound waves are examples of longitudinal waves.

Example 13.6 The diagram shown is a "snapshot" of a wave at a given time. The frequency of the wave is 120 Hz.

(a) What is the amplitude?

(b) What is the wavelength?

(c) What is the wave speed?

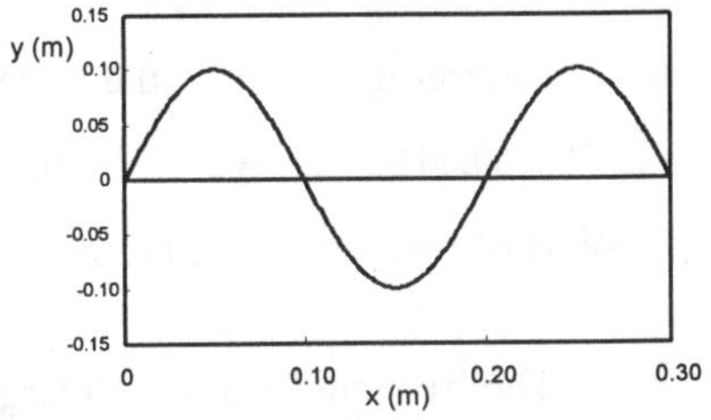

Solution: Given: $f = 120$ Hz and information in diagram.

Find: (a) A (b) λ (c) v.

(a) Amplitude is the maximum distance from the equilibrium position.

From the diagram, $A = 0.10$ m.

(b) Wavelength is the distance between successive crests.

From the diagram, $\lambda = 0.20$ m.

(c) $v = \lambda f = (0.20 \text{ m})(120 \text{ Hz}) = 24$ m/s.

4. Wave Phenomena

Interference occurs when waves meet or overlap. The **principle of superposition** states: *at any time, the combined waveform of two or more interfering waves is given by the sum of the displacements of the individual waves at each point in the medium.* If the amplitude of the combined wave is greater than that of any of the individual waves, we have what is called **constructive interference.** If the amplitude of the combined wave is smaller than that of any of the individual waves, it is called **destructive interference. Total constructive interference** occurs when two waves of the same frequency and amplitude are exactly in phase (the crest of one wave is aligned with the crest of the other) and **total destructive interference** takes place if two waves of the same frequency and amplitude are completely 180° out of phase (the crest of one wave is aligned with the trough of the other and vice versa).

Reflection occurs when a wave strikes an object or comes to a boundary of another medium and is at least partly bounced back. **Refraction** occurs when a wave crosses a boundary into another medium and the transmitted wave moves in a different direction. Generally, when a wave strikes the boundary, both reflection and refraction occur. **Diffraction** refers to the bending of waves around an edge of an object. A person in a room with an open door can hear sound from outside the room, which is partially due to diffraction (there may be some reflection).

5. Standing Waves and Resonance

Interfering waves of the same frequency and amplitude traveling in opposite directions in a rope or others can produce a **standing wave**. In this case, some points on the rope are always stationary and they are called **nodes**. The points of maximum amplitude, where constructive interference is greatest, are called **antinodes**.

The frequencies at which large-amplitude standing waves are produced are called **natural frequencies**, or **resonant frequencies**. The lowest frequency f_1 is called the **fundamental frequency**. In a stretched string fixed at both ends, all of the other natural frequencies are integral multiples of the fundamental frequency $f_n = n f_1$, where n is an integer. The set of frequencies $f_1, f_2 = 2 f_1, f_3 = 3 f_1$, and so on, is called a **harmonic series**: f_1 is the *first harmonic*, f_2 the *second harmonic*, and so on.

The speed of a wave in a stretched string is given by $v = \sqrt{\dfrac{F}{\mu}}$, where F is the tension in the string, and μ the linear mass density (mass per unit length, $\mu = m/L$). Therefore, the harmonic series *in a stretched string* is given by $f_n = \dfrac{n\,v}{2\,L} = \dfrac{n}{2\,L}\sqrt{\dfrac{F}{\mu}}$ (for $n = 1, 2, 3, \ldots$).

Resonance is a phenomena in which the vibrational amplitude of a system is greatly enhanced. It is caused by the match of the driving frequency of an external source to the natural frequency of the system.

Example 13.7 A 20-m long string has a mass of 0.010 kg. A 2.0-m segment of the string is fixed at both ends and when a tension of 20 N is applied to the string, three loops are produced. What is the frequency of the wave?

Solution: Three loops correspond to the 3rd harmonic.

Given: $L_o = 20$ m, $m = 0.010$ kg, $L = 2.0$ m, $F = 20$ N.

Find: f_3.

$$\mu = \frac{m}{L_o} = \frac{0.010\text{ kg}}{20\text{ m}} = 5.0\times10^{-4}\text{ kg/m}.$$

$$v = \sqrt{\frac{F}{\mu}} = \sqrt{\frac{20\text{ N}}{5.0\times10^{-4}\text{ kg/m}}} = 2.0\times10^{2}\text{ m/s}.$$

$$f_3 = 3f_1 = 3\,\frac{v}{\lambda} = 3\,\frac{v}{2\,L} = \frac{3(2.0\times10^{2}\text{ m/s})}{2(2.0\text{ m})} = 1.5\times10^{2}\text{ Hz}.$$

IV. Mathematical Summary

Hooke's Law	$\mathbf{F} = -k\,\mathbf{x}$ (13.1)	Relates spring force to displacement
Frequency and Period for SHM	$f = \frac{1}{T}$ (13.2)	Relates frequency and period for simpler harmonic motion
Total Energy of a Spring and Mass in SHM	$E = \frac{1}{2}kA^2 = \frac{1}{2}mv^2 + \frac{1}{2}kx^2$ (13.4–5)	Gives the total energy of a spring-mass system in simple harmonic motion
Velocity of Oscillating Mass on a Spring	$v = \pm\sqrt{\frac{k}{m}(A^2 - x^2)}$ (13.6)	Gives the velocity of an oscillating mass on a spring as a function of position
Maximum Speed of Oscillating Mass on a Spring	$v_{\text{max}} = \sqrt{\frac{k}{m}}\,A$ (13.7)	Gives the maximum speed of a oscillating mass on a spring
Period of a Mass Oscillating on a Spring	$T = 2\pi\sqrt{\frac{m}{k}}$ (13.11)	Computes the period of a mass on a spring in simple harmonic motion
Frequency of a Mass Oscillating on a Spring	$f = \frac{1}{2\pi}\sqrt{\frac{k}{m}}$ (13.12)	Computes the frequency of a mass on a spring in simple harmonic motion
Angular Frequency of a Mass Oscillating in a Spring	$\omega = 2\pi f = \sqrt{\frac{k}{m}}$ (13.13)	Computes the angular frequency of a mass on a spring in simple harmonic motion
Period of a Simple Pendulum	$T = 2\pi\sqrt{\frac{L}{g}}$ (13.14)	Computes the period of a simple pendulum at small-angle approximation
Displacement of a Mass in SHM	$y = A\sin(\omega t + \delta)$ (13.15) (with $\delta = 0$) $y = A\sin\omega t$ $= A\sin 2\pi f t = A\sin\frac{2\pi t}{T}$	Defines the equation of motion for a simple harmonic oscillator
Velocity of a Mass in SHM ($\delta = 0$)	$v = A\omega\cos\omega t$ (13.16)	Calculates the velocity of a simple harmonic oscillator as a function of time
Acceleration of a Mass in SHM ($\delta = 0$)	$a = -A\omega^2\sin\omega t$ $= -\omega^2 y$ (13.17)	Calculates the acceleration of a simple harmonic oscillator as a function of time
Wave Speed	$v = \frac{\lambda}{T} = \lambda f$ (13.18)	Relates wave speed with wavelength and frequency
Natural Frequencies in a Stretched String	$F_n = \frac{nv}{2L} = \frac{n}{2L}\sqrt{\frac{F}{\mu}}$ ($n = 1,2,3,\ldots$) (13.19–20)	Gives the natural frequencies in a stretched string as a harmonic series

V. Solutions of Selected Exercises and Paired Exercises

3. (a) Since $E = \frac{1}{2} k A^2 \propto A^2$, it is four times as large.

(b) Since $v_{\text{max}} = \sqrt{\frac{k}{m}}\, A \propto A$, it is twice as large.

10. From $\omega = \sqrt{\frac{k}{m}}$, we have $k = \omega^2 m$. So $E = \frac{1}{2} k A^2 = \frac{1}{2} m\, \omega^2 A^2$.

20. (a) $E = \frac{1}{2} k A^2 = \frac{1}{2}(80 \text{ N/m})(0.15 \text{ m})^2 = 0.90 \text{ J}$.

(b) No.

36. (a) Since $y = A \sin(\omega t + \delta)$, $\omega = 10$ rad/s.

So $T = \frac{1}{f} = \frac{2\pi}{\omega} = \frac{2\pi}{10 \text{ rad/s}} = 0.63 \text{ s}$.

(b) $\cos \omega t = \sin(\omega t + 90°)$. So the phase constant is $\delta = 90°$.

(c) $E = \frac{1}{2} m v_{\text{max}}^2 = \frac{1}{2} m (\omega A)^2 = \frac{1}{2}(0.025 \text{ kg})(10 \text{ rad/s})^2(0.080 \text{ m})^2 = 8.0\times10^{-3} \text{ J}$.

44. Since $v = \sqrt{\frac{k}{m}(A^2 - x^2)}$, $K = \frac{1}{2} m v^2 = \frac{1}{2} m \frac{k}{m}\left(A^2 - \frac{A^2}{4}\right) = \frac{3\,k A^2}{8}$.

50. (a) Since $T = 2\pi\sqrt{\frac{L}{g}}$ and the length is shorter, T is smaller.
So the clock runs faster or gains time.

(b) $\Delta T = 2\pi\sqrt{\frac{0.7500 \text{ m}}{9.80 \text{ m/s}^2}} - 2\pi\sqrt{\frac{0.7480 \text{ m}}{9.80 \text{ m/s}^2}} = 2.32\times10^{-3} \text{ s}$.

So in 24 hours = 86,400 s (or 43,200 periods), the time difference is $43{,}200(2.32\times10^{-3} \text{ s}) = 100 \text{ s}$ = 1.7 min.

(c) Yes. Because of linear expansion, the length depends on the temperature.

66. (a) $v = \sqrt{\frac{Y}{\rho}} = \sqrt{\frac{7.0\times10^{10} \text{ N/m}^2}{2.7\times10^3 \text{ kg/m}^3}} = 5091 \text{ m/s}$. $\lambda = \frac{v}{f} = \frac{5091 \text{ m/s}}{40 \text{ Hz}} = 1.3\times10^2 \text{ m}$.

(b) $v = \sqrt{\frac{11\times10^{10} \text{ N/m}^2}{8.9\times10^3 \text{ kg/m}^3}} = 3516 \text{ m/s}$. $\lambda = \frac{3516 \text{ m/s}}{40 \text{ Hz}} = 88 \text{ m}$.

79. (a) $v = \sqrt{\frac{F}{\mu}} = \sqrt{\frac{9.00 \text{ N}}{0.125 \text{ kg/m}}} = 8.49$ m/s.

(b) $f_n = \frac{nv}{2L} = n\frac{8.49 \text{ m/s}}{2(10.0 \text{ m})} = 0.425\,n$ Hz, $n = 1, 2, 3, \ldots.$

80. $v = \sqrt{\frac{F}{\mu}} = \sqrt{\frac{40 \text{ N}}{2.5\times 10^{-2} \text{ kg/m}}} = 40$ m/s. $f_n = \frac{nv}{2L} = \frac{40 \text{ m/s}}{2(2.0 \text{ m})}n = 10n$ Hz.

So the frequencies of the first four harmonics are 10 Hz, 20 Hz, 30 Hz, and 40 Hz.

93. $v = \sqrt{\frac{F}{\mu}} = \sqrt{\frac{550 \text{ N}}{(3.0\times 10^{-3} \text{ kg})/(0.60 \text{ m})}} = 332$ m/s.

$f_1 = \frac{v}{2L} = \frac{332 \text{ m/s}}{2(0.60 \text{ m})} = 2.8\times 10^2$ Hz.

$\lambda_1 = \frac{v}{f_1} = \frac{332 \text{ m/s}}{276 \text{ Hz}} = 1.2$ m, or $\lambda_1 = 2L = 1.2$ m.

101. The equation of motion is $x = A \sin(\omega t + \delta) = (16 \text{ cm}) \sin(\omega t + 90°) = (16 \text{ cm}) \cos \omega t$.

So $8.0 \text{ cm} = (16 \text{ cm}) \cos[\omega(0.50 \text{ s})]$, $\omega(0.50 \text{ s}) = \frac{2\pi}{T}(0.50 \text{ s}) = \cos^{-1} 0.50 = 1.05$ rad.

Therefore $T = \frac{2\pi (0.50 \text{ s})}{1.05 \text{ rad}} = 3.0$ s.

VI. Practice Quiz

1. A 2.0-kg mass, hung on a spring ($k = 18$ N/m), is displaced slightly from its equilibrium position and released. What is the frequency of vibration?

(a) zero (b) 1.5 Hz (c) 3.0 Hz (d) $\frac{1.5}{\pi}$ Hz (e) $\frac{3.0}{\pi}$ Hz

2. A mass attached to the free end of a spring executes simple harmonic motion according to the equation $y = (0.50 \text{ m}) \sin(18\pi t)$ where y is in meters and t in seconds. What is the period of the vibration?

(a) 0 (b) 9.0 s (c) 18 s (d) 1/18 s (e) 1/9 s

3. What is the wave speed if a wave has a frequency of 12 Hz and a wavelength of 3.0 m?

(a) 0.25 m/s (b) 4.0 m/s (c) 9.0 m/s (d) 15 m/s (e) 36 m/s

4. Resonance in a system, such as a string fixed at both ends or a bridge, occurs when

(a) it is oscillating in simple harmonic motion.

(b) it is oscillating in damped simple harmonic motion.

(c) the natural frequency of the system is the same as the external driving frequency.

(d) the natural frequency of the system is smaller than the external driving frequency.

(e) the natural frequency of the system is larger than the external driving frequency.

5. In seismology, the S wave is a transverse wave. As an S wave travels through the Earth, the relative motion between the S wave and the particles are

(a) first perpendicular, then parallel. (b) not at any particular angle to each other.

(c) perpendicular. (d) parallel. (e) antiparallel.

6. The mass of a simple pendulum is doubled. By what factor does the period change?

(a) 1 (b) 2 (c) 4 (d) 1/2 (e) 1/4

7. When a mass m oscillates on a spring, its period is T. A mass $2m$ is now placed on the spring and it again oscillates in SHM. What is its period?

(a) T (b) $T/2$ (c) $T/4$ (d) $1.4T$ (e) $2T$

8. A piano string of linear mass density 0.0050 kg/m is under a tension of 1350 N. What is the wave speed?

(a) 130 m/s (b) 260 m/s (c) 520 m/s (d) 1040 m/s (e) 2080 m/s

9. If a guitar string has a fundamental frequency of 500 Hz, which of the following frequencies can set the string into resonant vibration?

(a) 125 Hz (b) 250 Hz (c) 750 Hz (d) 1500 Hz (e) 1750 Hz

10. A mass on a spring oscillates in simple harmonic motion at a frequency of 4.0 Hz and an amplitude of 8.0 cm. If a timer is started when the mass is at the equilibrium position, what is its distance from its equilibrium position at $t = 3.0$ s?

(a) zero (b) 3.0 cm (c) 4.6 cm (d) 6.7 cm (e) 8.0 cm

Answers to Practice Quiz:

1.d 2.e 3.e 4.c 5.c 6.a 7.d 8.c 9.d 10.a

CHAPTER 14

Sound

I. Chapter Objectives

Upon completion of this chapter, you should be able to:

1. define sound and explain the sound frequency spectrum.
2. tell how the speed of sound differs in different media and describe the temperature dependence of the speed of sound in air.
3. define sound intensity and explain how it varies with distance from a point source, and calculate sound intensities on the decibel scale.
4. explain sound reflection, refraction, and diffraction, and distinguish between constructive and destructive interference.
5. describe and explain the Doppler effect, and give some examples of its occurrences and applications.
6. explain some of the sound characteristics of musical instruments in physical terms.

II. Key Terms

Upon completion of this chapter, you should be able to define and/or explain the following key terms:

sound waves
audible region
sound frequency spectrum
infrasonic region
ultrasonic region
intensity
decibel (dB)
sound intensity level
constructive interference
destructive interference
phase difference
path difference
beats
beat frequency
Doppler effect
sonic boom
March number
loudness
pitch
quality

The definitions and/or explanations of the most important key terms can be found in the following section: **III. Chapter Summary and Discussion.**

III. Chapter Summary and Discussion

1. Sound Waves

Sound waves are longitudinal waves (or compressional waves). The high- and low-density pressure regions (analogous to crests and troughs of transverse wave) of a sound wave are called *condensations* and *rarefactions*, respectively.

The **sound frequency spectrum** consists of three regions: frequencies lower than 20 Hz ($f < 20$ Hz) are in the **infrasonic region**, frequencies between about 20 Hz and 20 kHz (20 Hz $< f <$ 20 kHz) are in the **audible region**, and above 20 kHz ($f >$ 20 kHz) is the **ultrasonic region**.

2. The Speed of Sound

The **speed of sound** in a medium depends on the elasticity or the intermolecular interactions of the medium and the mass or density of its particles. Generally, the speeds of sound in solids and liquids are given by $v = \sqrt{\frac{Y}{\rho}}$ and $v = \sqrt{\frac{B}{\rho}}$, respectively, where Y is the Young's modulus, B is the bulk modulus, and ρ is the density.

For *normal environment temperatures*, the speed of sound in air increases by about 0.6 m/s for each Celsius degree above 0°C. A good approximation is given by $v = (331 + 0.6T_C)$ m/s, where T_C is the air temperature in degrees Celsius and 331 m/s is the speed of sound in air at 0°C. A useful general figure for the speed of sound in air is $\frac{1}{3}$ km/s (or $\frac{1}{5}$ mi/s).

Example 14.1 The speed of an ultrasonic sound of frequency 45 kHz in air is 342 m/s.

(a) What is the air temperature?

(b) What is the wavelength?

Solution: Given: $v = 342$ m/s, $f = 45$ kHz $= 45\times10^3$ Hz.

Find: (a) T_C (b) λ.

(a) From $v = (331 + 0.6T_C)$ m/s, we have $T_C = \frac{v - 331}{0.6} = \frac{342 - 331}{0.6} = 18°\text{C}$.

(b) From $v = \lambda f$, we have $\lambda = \frac{v}{f} = \frac{342 \text{ m/s}}{45\times10^3 \text{ Hz}} = 7.6\times10^{-3}$ m $= 7.6$ mm.

Example 14.2 Find the speed of sound in a steel rod.

Solution: Given: $Y_{\text{steel}} = 20\times10^{10}$ N/m^2, $\rho = 7.86\times10^3$ kg/m^3.

Find: v.

$$v = \sqrt{\frac{Y}{\rho}} = \sqrt{\frac{20\times10^{10}\ \text{N/m}^2}{7.86\times10^3\ \text{kg/m}^3}} = 5.0\times10^3\ \text{m/s}.$$

3. Sound Intensity

Sound **intensity** is the rate of the sound energy transfer or the energy transported per unit time across a unit area: $\text{intensity} = \dfrac{\text{energy/time}}{\text{area}} = \dfrac{\text{power}}{\text{area}}$. The SI unit of intensity is W/m^2.

The intensity of a point source of power P at a distance R from the source is given by $I = \dfrac{P}{A} = \dfrac{P}{4\pi R^2}$. Note *the intensity is inversely proportional to the square of the distance from the point source.*

Sound intensity is perceived by the ear as **loudness**. On the average, the human ear can detect sound waves (at 1 kHz) with an intensity as low as $I_o = 10^{-12}$ W/m^2 (the **threshold of hearing**). At an intensity of $I_p = 1.0$ W/m^2 (**threshold of pain**) the sound is uncomfortably loud and may be painful to the ear. The ratio between the two intensities is $\dfrac{I_p}{I_o} = \dfrac{1.0\ \text{W/m}^2}{10^{-12}\ \text{W/m}^2} = 10^{12}$, or 12 orders of magnitude. For a sound to be audible, it must have a frequency between 20 Hz and 20 kHz, *and* have a intensity greater than I_o.

The **sound intensity level**, or **decibel level** (β) of a sound of intensity I is defined as $\beta = 10 \log \dfrac{I}{I_o}$, where $I_o = 10^{-12}$ W/m^2 is the threshold of hearing. At the threshold of hearing, $\beta = 10 \log \dfrac{I_o}{I_o} = 10 \log 1 = 0$ dB and at the threshold of pain, $\beta = 10 \log \dfrac{1.0\ \text{W/m}^2}{10^{-12}\ \text{W/m}^2} = 10 \log 10^{12} = 10(12) = 120$ dB. The 12 orders of magnitude of sound intensity is only a difference of 120 dB − 0 dB = 120 dB.

Note: Intensity and intensity level are two different things. Intensity is a direct way to measure energy and is additive, that is, the sum of a 1.0 W/m^2 sound and a 2.0 W/m^2 sound is 1.0 W/m^2 + 2.0 W/m^2 = 3.0 W/m^2. Intensity level is based on a logarithmic scale and therefore is not additive, that is, the sum of a 10 dB sound and a 20 dB sound is *not* equal to 10 dB + 20 dB $\neq$ 30 dB.

Example 14.3 A barking dog delivers on average about 1.00 mW of power, which is assumed to be uniformly distributed in all directions.

(a) What is the sound intensity at a distance of 5.00 m from the dog?

(b) What is the intensity level at a distance of 5.00 m from the dog?

(c) What would be the intensity level if two identical dogs very close together are barking?

Solution: Given: $R = 5.00$ m, $P = 1.00 \text{ mW} = 1.00\times10^{-3}$ W.

Find: (a) I (b) β (c) β for two dogs.

$$\text{(a) } I = \frac{P}{4\pi R^2} = \frac{1.00\times10^{-3}\text{ W}}{4\pi(5.00\text{ m})^2} = 3.18\times10^{-6}\text{ W/m}^2.$$

$$\text{(b) } \beta = 10\log\frac{I}{I_o} = 10\log\frac{3.18\times10^{-6}\text{ W/m}^2}{10^{-12}\text{ W/m}^2} = 10\log 3.18\times10^{6} = 10(6.50) = 65.0\text{ dB}.$$

(c) If there are two identical dogs, the intensity doubles (the intensity level does not double as you will see). So the new intensity is $I = 2(3.18\times10^{-6}\text{ W/m}^2) = 6.36\times10^{-6}\text{ W/m}^2$.

Therefore $$\beta = 10\log\frac{6.36\times10^{-6}\text{ W/m}^2}{10^{-12}\text{ W/m}^2} = 10\log 6.36\times10^{6} = 10(68.0) = 68.0\text{ dB}.$$

Doubling the intensity only increases the intensity level by 68.0 dB − 65.0 dB = 3.0 dB! This is always true because $10\log 2I = 10(\log I + \log 2) = 10(\log I + 0.30) = 10\log I + 3.0$.

Example 14.4 The intensity level generated by 15 engines in a garage is 100 dB. What is the intensity level generated of one engine?

Solution: Intensity level is not additive (the intensity level of one engine is *not* 100/15 = 6.7 dB). We need to first find the intensity generated by 15 engines, then the intensity of one engine, and finally the intensity level of one engine.

Given: $\beta(15) = 100$ dB, $n = 15$.

Find: $\beta(1)$.

From $\beta = 10\log\frac{I}{I_o}$, we have $\frac{\beta}{10} = \log\frac{I}{I_o}$, so $\frac{I}{I_o} = 10^{\beta/10}$ (because if $y = \log x$, then $x = 10^y$).

Therefore $I(15) = I_o\,10^{\beta/10} = (10^{-12}\text{ W/m}^2)\,10^{100/10} = (10^{-12}\text{ W/m}^2)\,10^{10} = 10^{-2}\text{ W/m}^2$, and the intensity generated by one engine is $I(1) = \frac{I(15)}{15} = \frac{10^{-2}\text{ W/m}^2}{15} = 6.67\times10^{-4}\text{ W/m}^2$ (intensity is additive).

Hence $$\beta(1) = 10\log\frac{6.67\times10^{-4}\text{ W/m}^2}{10^{-12}\text{ W/m}^2} = 10\log 6.67\times10^{8} = 88\text{ dB}.$$ (Are you surprised?)

4. Sound Phenomena

Since sound is a wave, it has all the wave characteristics. It can be *reflected*, *refracted*, and *diffracted*: that is, bounce off objects or surfaces, have a change in direction due to a medium or density change, and be bent around corners or objects, respectively.

Sound waves *interfere* when they meet. There can be **constructive interference** or **destructive interference** depending on the **path difference** (ΔL) of the waves. If the path difference between two waves with the same frequency is zero or an integer (whole) number of wavelength, $\Delta L = n\,\lambda$ with $n = 0, 1, 2, \ldots$, constructive interference (two crests or two troughs coincide) occurs. Conversely, if the path difference between two waves with the same frequency is an odd number of half wavelength, $\Delta L = m\,\lambda/2$ with $m = 1, 3, 5, \ldots$, destructive interference (crest and trough coincide) takes place. The **phase difference** $\Delta\theta$ is related to path difference by the simple relationship: $\Delta\theta = \frac{2\pi}{\lambda}(\Delta L)$. In terms of the phase difference, the conditions for interference are

$\Delta\theta = n\,(2\pi)$ with $n = 0, 1, 2, \ldots$ for constructive interference,

$\Delta\theta = m\,(\pi)$ with $m = 1, 3, 5, \ldots$ for destructive interference.

Another interesting interference effect occurs when two tones of nearly the same frequency ($f_1 \approx f_2$) sound simultaneously. The ear senses pulsations in loudness known as **beats**. The **beat frequency** is equal to the difference between the two frequencies, or $f_b = |f_1 - f_2|$.

Example 14.5 A person stands between two loud speakers driven by the same source. Each speaker produces a tone with a frequency of 200 Hz on a day when the speed of sound is 330 m/s. The person is 1.65 m from one speaker and 4.95 m from the other speaker. What type of interference does the person sense?

Solution: To determine what type interference the person senses, we need to find the path difference as a function of wavelength.

Given: $f = 200$ Hz, $v = 330$ m/s, $L_1 = 1.65$ m, $L_2 = 4.96$ m.

Find: Interference type.

From $v = \lambda f$, we have $\lambda = \frac{v}{f} = \frac{330 \text{ m/s}}{200 \text{ Hz}} = 1.65$ m.

The path difference is $\Delta L = L_2 - L_1 = 4.95 \text{ m} - 1.65 \text{ m} = 3.30 \text{ m} = 2(1.65 \text{ m}) = 2\lambda$. So the path difference is an integer times the wavelength and therefore the interference is constructive.

Example 14.6 A music tuner uses a 550-Hz tuning fork to tune the frequency of a musical instrument. If the tuner hears a beat frequency of 2.0 Hz, what is the frequency of the instrument?

Solution: Given: $f_b = 2.0$ Hz, $f_1 = 550$ Hz.

Find: f_2.

From $f_b = |f_1 - f_2|$, we have $f_2 = |f_1 \pm f_b|$ (Note the absolute sign!).

So $f_2 = |550 \text{ Hz} \pm 2.0 \text{ Hz}| = 552$ Hz or 548 Hz.

There are two answers because we do not know which one (the tuning fork or the musical instrument) has a higher frequency.

5. The Doppler Effect

If there is relative motion between a sound source and an observer, the observer will detect a frequency which is different from the frequency of the source. This phenomena is called the **Doppler effect**. Generally, if the source and the observer are moving toward each other, the observed frequency is higher than the source frequency; and if the source and the observer are moving away from each other, the observed frequency is lower than the source frequency.

For a stationary observer, the observed frequency f_o by the observer due to a moving source of frequency f_s is given by $f_o = \frac{v}{v \pm v_s} f_s = \frac{1}{1 \pm \frac{v_s}{v}} f_s$, where v_s is the speed of source, v is the speed of sound, the minus sign corresponds to source moving toward stationary observer, and the plus sign corresponds to source moving away from stationary observer.

For a stationary source of frequency f_s, the observed frequency f_o by an observer due to the moving observer is given by $f_o = \frac{v \pm v_o}{v} f_s = \left(1 \pm \frac{v_o}{v}\right) f_s$, where v_o is the speed of observer, v is the speed of sound, the plus sign corresponds to observer moving toward the source, and the minus sign corresponds to observer moving away from the source.

Note: For source moving, use − for *toward* and use + for *away from* observer.

For observer moving, use − for *away from* and use + for *toward* the source. (Just opposite! Why?)

Objects traveling at supersonic (speed greater than the speed of sound) produce large pressure ridges or shock waves. This gives rise to the **sonic boom** heard when a supersonic aircraft passes. The ratio of the speed of the source v_s to the speed of sound v is called the **March number**: $M = \frac{v_s}{v}$ and $M > 1$ for supersonic speeds.

Example 14.7 The frequency of a train horn is 500 Hz. Assume the speed of sound in air is 340 m/s. What is the frequency heard by an observer if

(a) the observer is moving away from the stationary train with a speed of 30.0 m/s?

(b) the train is approaching the stationary observer with a speed of 30.0 m/s?

Solution: Given: $f_s = 500$ Hz, $v = 340$ m/s, (a) $v_o = 30.0$ m/s, (b) $v_s = 30.0$ m/s.

Find: (a) f_o (b) f_o.

(a) Since the observer is moving away from the source, we use the minus sign in $f_o = \frac{v \pm v_o}{v} f_s$.

$$f_o = \frac{v - v_o}{v} f_s = \frac{340 \text{ m/s} - 30.0 \text{ m/s}}{340 \text{ m/s}} (500 \text{ Hz}) = 456 \text{ Hz}.$$

(b) Since the train is approaching (toward), we use the minus sign in $f_o = \frac{v}{v \pm v_s} f_s$.

$$f_o = \frac{v}{v - v_s} f_s = \frac{340 \text{ m/s}}{340 \text{ m/s} - 30.0 \text{ m/s}} (500 \text{ Hz}) = 548 \text{ Hz}.$$

Example 14.8 The Concord airplane flies from the United States to Europe with a March number of 1.05 where the air temperature is 5.0°C. What is the speed of the plane?

Solution: Given: $M = 1.05$, $T_C = 5.0$°C.

Find: v_s.

The speed of sound in air is $v = (331 + 0.6T_C)$ m/s $= [331 + 0.6(5.0)]$ m/s $= 334$ m/s.

From $M = \frac{v_s}{v}$, we have $v_s = Mv = (1.05)(334 \text{ m/s}) = 351$ m/s.

---

6. Musical Instruments and Sound Characteristics

The secondary auditory effects of **loudness**, **pitch**, and **quality** are related to the physical wave properties of *intensity*, *frequency*, and *waveform* (harmonics), respectively.

Stringed musical instruments produce notes by setting up transverse standing waves in strings with different fundamental frequencies (see Chapter 13). Organ pipes and wind instruments produce notes by forming longitudinal standing waves in air columns.

For an open organ pipe (open at both ends), the harmonic series is given by $f_n = n\dfrac{v}{2L} = nf_1$ (for n = 1, 2, 3, . . .), where L is the length of the pipe. For a closed organ pipe (one end closed), the harmonic series is given by $f_m = m\dfrac{v}{4L} = mf_1$ (for m = 1, 3, 5, . . .). Note the missing even harmonics in a closed pipe.

Example 14.9 A 3.00-m long pipe is in a room where the temperature is 20°C.

(a) What is the frequency of the fundamental if the pipe is open?

(b) What is the frequency of the second harmonic if the pipe is open?

(c) What is the frequency of the fundamental if the pipe is closed?

(d) What is the frequency of the second harmonic if the pipe is closed?

Solution: Given: $L = 3.00$ m, $T_C = 20$°C.

Find: (a) f_1 (open) (b) f_2 (open) (c) f_1 (closed) (d) f_2 (closed).

The speed of sound in air is $v = (331 + 0.6T_C)$ m/s = [331 + 0.6(20)] m/s = 343 m/s.

(a) For an open pipe, $f_n = n\dfrac{v}{2L} = nf_1$ (for n = 1, 2, 3, . . .). So $f_1 = 1\dfrac{v}{2L} = \dfrac{343\text{ m/s}}{2(3.00\text{ m})} = 57.2\text{ Hz}$.

(b) $f_2 = 2f_1 = 2(57.2\text{ Hz}) = 114\text{ Hz}$.

(c) For a closed pipe, $f_n = m\dfrac{v}{4L} = mf_1$ (for m = 1, 3, 5, . . .). So $f_1 = 1\dfrac{v}{4L} = \dfrac{343\text{ m/s}}{4(3.00\text{ m})} = 28.6\text{ Hz}$.

(d) Even harmonics cannot exist in a closed pipe. There is no second harmonic.

IV. Mathematical Summary

Speed of Sound	$v = (331 + 0.6\,T_C)$ m/s (14.1)	Gives the speed of sound in air (in m/s)
Intensity of a Point Source	$I = \dfrac{P}{4\pi r^2}$ and $\dfrac{I_2}{I_1} = \left(\dfrac{R_1}{R_2}\right)^2$ (14.2–3)	Calculates the intensity of a point source as a function of distance from it
Intensity Level (in dB)	$\beta = 10\log\dfrac{I}{I_o}$ (12.4) where $I_o = 10^{-12}$ W/m^2	Calculates the intensity level from intensity

Phase Difference	$\Delta\theta = \frac{2\pi}{\lambda}(\Delta L)$ (14.5)	Computes the phase difference from path difference ΔL
Condition for Constructive Interference	$\Delta L = n\,\lambda\ (n = 0, 1, 2, 3, \ldots)$ (14.6)	Defines the condition for constructive interference
Condition for Destructive Interference	$\Delta L = m\,\lambda\ (m = 1, 3, 5, \ldots)$ (14.7)	Defines the condition for destructive interference
Beat Frequency	$f_b = \lvert f_1 - f_2 \rvert$ (14.9)	Calculates the beat frequency from two frequencies
Doppler Effect: Source Moving	$f_o = \frac{v}{v \pm v_s} f_s = \frac{1}{1 \pm \frac{v_s}{v}} f_s$ where v_s = speed of source v = speed of sound (14.12)	Relates observed frequency and source frequency − for source moving toward stationary observer + for source moving away from stationary observer
Doppler Effect: Observer Moving	$f_o = \frac{v \pm v_o}{v} f_s = \left(1 \pm \frac{v_o}{v}\right) f_s$ where v_o = speed of observer v = speed of sound (14.15)	Relates observed frequency and source frequency + for observer moving toward stationary source − for observer moving away from stationary source
Mach Number	$M = \frac{v_s}{v}$ (14.17)	Defines the Mach number of moving source
Natural Frequencies of Open Organ Pipe	$f_n = n \frac{v}{2L} = n f_1$ $(n = 1, 2, 3, \ldots)$ (14.18)	Gives the natural frequencies of an open (both ends open) organ pipe
Natural Frequencies of Closed Organ Pipe	$f_n = m \frac{v}{4L} = m f_1$ $(n = 1, 3, 5, \ldots)$ (14.19)	Gives the natural frequencies of a closed (one end open) organ pipe

V. Solutions of Selected Exercises and Paired Exercises

8. At the same temperature and pressure, water vapor has a smaller density than air. The speed of sound is related to the density of the medium. Generally, the less dense air (O_2 and N_2 molecules replaced by H_2O molecules), the slower sound travels. So the speed decreases with increasing humidity.

12. (a) From $v = (331 + 0.6T_C)$ m/s, we have $T_C = \dfrac{v - 331}{0.6}\ \text{°C} = \dfrac{346 - 331}{0.6} = 25\text{°C}$.

(b) $T_C = \dfrac{340 - 331}{0.6} = 15\text{°C}$.

16. (a) $\Delta t = \dfrac{d}{v_a} - \dfrac{d}{v_s} = \dfrac{300\ \text{m}}{343\ \text{m/s}} - \dfrac{300\ \text{m}}{4500\ \text{m/s}} = 0.81\ \text{s}$.

(b) 36 km/h = 10 m/s. So $\Delta t = \dfrac{300\ \text{m}}{343\ \text{m/s} + 10\ \text{m/s}} - \dfrac{300\ \text{m}}{4500\ \text{m/s}} = 0.78\ \text{s}$.

28. $I_2 = \dfrac{2}{3}I_1$. From $\dfrac{I_2}{I_1} = \dfrac{R_1^2}{R_2^2}$, we have $\dfrac{R_2}{R_1} = \sqrt{\dfrac{I_1}{I_2}} = \sqrt{\dfrac{3}{2}} = 1.2$ times.

30. (a) From $\beta = 10 \log \dfrac{I}{I_o}$, we have $\dfrac{I}{I_o} = 10^{\beta/10}$. So $I = 10^{\beta/10} I_o = 10^6\ (10^{-12}\ \text{W/m}^2) = 10^{-6}\ \text{W/m}^2$.

(b) $I = 10^{11}\ (10^{-12}\ \text{W/m}^2) = 10^{-1}\ \text{W/m}^2$.

47. For one bee, $I = I_o$. Assume it takes n bees.

From $\beta = 10 \log \dfrac{nI}{I_o} = 10 \log n$, we have $n = 10^{\beta/10} = 10^{4.0} = 10^4$ bees.

57. The beat frequency is $f_b = |f_2 - f_1| = 260\ \text{Hz} - 256\ \text{Hz} = 4\ \text{Hz}$.

For f_b to be zero, the two must have the same frequency.

60. The observer moves with a speed of 50 km/h = 13.9 m/s.

So $f_o = \dfrac{v + v_o}{v} f_s = \dfrac{331\ \text{m/s} + 13.9\ \text{m/s}}{331\ \text{m/s}}(800\ \text{Hz}) = 834\ \text{Hz}$.

66. The minimum separation for destructive interference corresponds to $\Delta L = \dfrac{\lambda}{2}$.

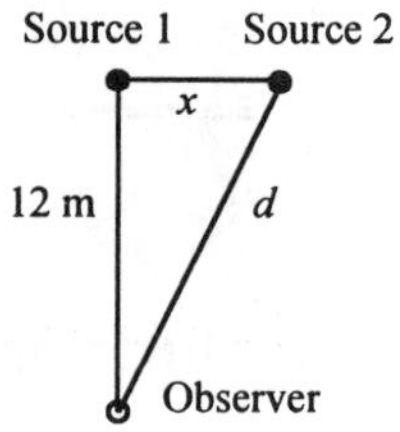

$\Delta L = d - 12\ \text{m} = \sqrt{x^2 + (12.0\ \text{m})^2} - 12.0\ \text{m}$

$= \dfrac{\lambda}{2} = \dfrac{v}{2f} = \dfrac{340\ \text{m/s}}{2(1000\ \text{Hz})} = 0.170\ \text{m}$.

So $x^2 + (12.0\ \text{m})^2 = (12.17\ \text{m})^2$. Solving for x, $x = 2.03$ m.

78. For the open pipe, $(f_1)_o = \dfrac{(1)\,v}{2L} = \dfrac{343\ \text{m/s}}{2(0.52\ \text{m})} = 330\ \text{Hz}$.

For the closed pipe, $(f_1)_c = \dfrac{(1)\,v}{4L} = \dfrac{(f_1)_o}{2} = 165\ \text{Hz}$.

81. If the observer were stationary, the frequency heard by the observer would have been $(f_o)_1 = \frac{v}{v \mp v_s} f_s$, which would be the frequency of the "source" to the moving observer. $(f_s)_2 = (f_o)_1$.

So $(f_o)_2 = f_o = \frac{v \pm v_o}{v}(f_s)_2 = \frac{v \pm v_o}{v}\frac{v}{v \mp v_s} f_s = \frac{v \pm v_o}{v \mp v_s} f_s$.

82. (a) $v_s = 90$ km/h $= -25$ m/s and $v_o = 65$ km/h $= 18.1$ m/s. Both the source and the observer are moving toward each other. From Exercise 14.81 and according to the sign conventions:

$$f_o = \frac{v + v_o}{v - v_s} f_s = \frac{354 \text{ m/s} + 18.1 \text{ m/s}}{354 \text{ m/s} - 25 \text{ m/s}}(500 \text{ Hz}) = 566 \text{ Hz}.$$

(b) Both the source and the observer are moving away from each other.

So $f_o = \frac{354 \text{ m/s} - 18.1 \text{ m/s}}{354 \text{ m/s} + 25 \text{ m/s}}(500 \text{ Hz}) = 443 \text{ Hz}$.

93. From $\beta = 10 \log \frac{I}{I_o}$, we have $\frac{I}{I_o} = 10^{\beta/10}$. So $\frac{I_1}{I_o} = 10^{6.00}$ and $\frac{I_2}{I_o} = 10^{6.50}$.

Therefore the intensity level heard by another person is

$$\beta_{tot} = 10 \log \frac{I_1 + I_2}{I_o} = 10 \log (10^{6.00} + 10^{6.50}) = 66.2 \text{ dB}.$$

101. Since $v = (331 + 0.6T_C)$ m/s, the percentage difference is

$$\frac{v_{20} - v_{10}}{v_{20}} = \frac{0.6(20 - 10)}{331 + 0.6(20)} = 0.017 = 1.7\%.$$

VI. Practice Quiz

1. An echo is heard 2.0 s from a cliff on a day the temperature is 15°C. Approximately how far is the cliff from the observer?

(a) 85 m (b) 170 m (c) 340 m (d) 680 m (e) 1360 m

2. The fundamental frequency in a closed pipe is 330 Hz. What is the frequency of the third harmonic?

(a) 110 Hz (b) 220 Hz (c) 330 Hz (d) 660 Hz (e) 990 Hz

3. The intensity of a point source at a distance d from the source is I. What is the intensity at a distance $2d$ from the source?

(a) $I/4$ (b) $I/2$ (c) I (d) $2I$ (e) $4I$

4. If the intensity level of a barking dog is 40 dB. What is the intensity level of two barking dogs together?

(a) 20 dB (b) 37 dB (c) 40 dB (d) 43 dB (e) 80 dB

5. Two tones have frequencies of 330 Hz and 332 Hz. What is the beat frequency?

(a) 0 Hz (b) 2 Hz (c) 331 Hz (d) 662 Hz (e) none of the above

6. The frequency of the third harmonic of an open pipe is 900 Hz. What is the length of the pipe? (Assume the speed of sound is 340 m/s.)

(a) 0.094 m (b) 0.189 m (c) 0.283 m (d) 0.567 m (e) 1.13 m

7. A sound source has a frequency of 1000 Hz. If a listener moves at a speed of 30.0 m/s away from the source, what is the frequency heard by the observer? (The speed of sound is 340 m/s.)

(a) 912 Hz (b) 919 Hz (c) 1000 Hz (d) 1090 Hz (e) 1097 Hz

8. Two loudspeakers are placed side by side and driven by the same frequency of 500 Hz. If the distance from a person to one of the speakers is 5.00 m and the observer detects little or no sound, what is the distance from the person to the other speaker? (The speed of sound is 340 m/s.)

(a) 7.72 m (b) 8.06 m (c) 8.40 m (d) 9.08 m (e) 5.0 m

9. The sound intensity level 5.0 m from a point source is 95 dB. At what distance will it be 75 dB?

(a) 25 m (b) 50 m (c) 75 m (d) 225 m (e) 500 m

10. What is the frequency heard by a stationary observer when a train approaches with a speed of 30.0 m/s. The frequency of the train horn is 600 Hz. (The speed of sound is 340 m/s.)

(a) 547 Hz (b) 551 Hz (c) 600 Hz (d) 653 Hz (e) 658 Hz

Answers to Practice Quiz:

1. c 2. e 3. a 4. d 5. b 6. d 7. a 8. b 9. b 10. e

CHAPTER 15

Electric Charge, Force, and Field

I. Chapter Objectives

Upon completion of this chapter, you should be able to:

1. distinguish between the two types of electric charge, state the charge-force law that operates between charged objects, and understand and use the law of charge conservation.
2. distinguish between conductors and insulators, explain the operation of the electroscope, and distinguish among charging by contact, induction, and polarization.
3. state Coulomb's law and use it to calculate the electric force between two charged particles, and on one charge due to several other charges.
4. understand how the electric field vector is defined, plot electric field lines for simple charge distributions, and calculate electric fields due to several point charges.
5. describe the electric field near the surface and the interior of a good conductor, determine where the highest concentration of excess charge accumulates on a charges conductor, and sketch the electric field line pattern outside a charged conductor.

*6. state the physical basis if Gauss's law and use the law to make qualitative predictions.

II. Key Terms

Upon completion of this chapter, you should be able to define and/or explain the following key terms:

electric charge
law of charges or charge-force law
coulomb (C)
net charge
conservation of charge
electrostatic charging
conductors
insulators
semiconductors
charging by friction
charging by contact
charging by induction
polarization
Coulomb's law
electric field
electric lines of force
electric dipole
***Gaussian surface**
***Gauss's law**

The definitions and/or explanations of the most important key terms can be found in the following section: **III. Chapter Summary and Discussion.**

III. Chapter Summary and Discussion

1. Electric Charge

Electric charge is the property of an object that determines its electrical behavior: the electric force it can exert and the electric force it can experience. There are two types of charges, which are distinguished as positive (+) or negative (−). A positive charge is arbitrarily associated with the proton, and a negative charge with the electron. The electronic charge (fundamental unit of charge) is $|e| = 1.60\times10^{-19}$ C, where the SI unit of charge is the coulomb (C). Both electron and proton carry this charge with opposite signs.

The directions of the electric forces on the charges of mutual interaction are given by the **law of charges** or the **charge-force law**: like charges repel and unlike charges attract.

An object with a **net charge** means that it has an *excess* of either positive or negative charges. The law of **conservation of charge** states that the net charge of an isolated system remains constant.

Example 15.1 A piece of plastic has a net charge of $+2.00\ \mu$C.

(a) Are there more protons or electrons?

(b) What happens if an identically charged piece of plastic is placed near the first one?

(c) How many fundamental units of charge does one piece of the plastic contain?

Solution: Given: $q = +2.00\ \mu\text{C} = 2.00\times10^{-6}$ C. Find: (c) n.

(a) Since the plastic has a positive charge, it must have more protons than electrons or the piece of plastic has lost some electrons.

(b) From the charge-force law, like charges repel. The two pieces of plastic therefore repel each other.

(c) From $q = ne$, we have $n = \dfrac{q}{e} = \dfrac{2.00\times10^{-6}\ \text{C}}{1.60\times10^{-19}\ \text{C}} = 1.25\times10^{13}$.

So the piece of plastic has 1.25×10^{13} excess protons or most likely it has lost 1.25×10^{13} electrons, since protons are in the atomic nucleus and stay fixed.

---

2. Electrostatic Charging

Electrostatic charging can be done by the following four processes. In **charging by friction**, certain insulators are rubbed with different materials and both the insulators and the other materials acquire opposite charges. In **charging by contact**, a charged object makes a contact with a uncharged object and some of the charge on the charged object transfers to the uncharged object. In **charging by induction**, a charged object is brought near (not touching) a uncharged object, the uncharged object is then "grounded", and the uncharged object acquires an opposite charge than the charged object. In **charging by polarization**, the positive and negative charges are simply separated or realigned within the object (the net charge of the object is still zero).

3. Electric Force

The magnitude of the electric force between two point charges q_1 and q_2 is given by **Coulomb's law**: $F = \dfrac{k\, q_1\, q_2}{r^2}$, where $k \approx 9.00\times 10^9$ N·m²/C² is called the Coulomb constant and r is the distance between the two point charges. Like the universal law of gravity ($F = \dfrac{G\, m_1\, m_2}{r^2}$), Coulomb's law is also an *inverse square law* because the force is proportional to $\dfrac{1}{r^2}$.

Note: Since force is a vector, you have to use vector addition to calculate the net force on a charge if there is more than one force on it (from several other charges).

Example 15.2 Two charges are separated by a distance d and exert mutual attractive forces of F_1 on each other. What are the mutual attractive forces if the charges are separated by a distance of $d/3$?

Solution: From Coulomb's law $F = \dfrac{k\, q_1\, q_2}{r^2}$, we have $\dfrac{F_2}{F_1} = \dfrac{r_1^2}{r_2^2}$ (since k, q_1, and q_2 are constants here).

So $F_2 = \dfrac{r_1^2}{r_2^2} F_1 = \dfrac{(d)^2}{(d/3)^2} F_1 = 9\, F_1$. The force increases by 9 times when the distance is reduced to 1/3 of its original value. This is an important feature of the inverse square law.

Example 15.3 A +3.0 C charge is at the origin and a +9.0 C charge is at $x = 4.0$ m. Where can a third charge q_3 be placed on the x axis so the net force on it is zero?

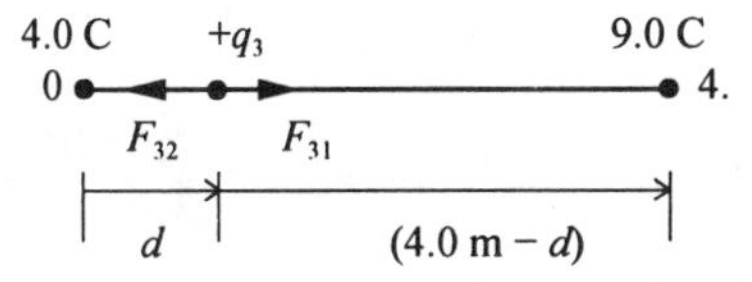

Solution: By the charge-force law, the third charge has to be placed in *between* the two charges. The third charge, however, can be either positive or negative. (Why?) Assume the third charge is placed at d from the +4.0 C charge, then it is (4.0 m − d) from the +9.0 C charge. For the force on q_3 be zero, the forces by q_1 and q_2 on q_3 must equal each other in magnitude.

Given: $q_1 = 4.0$ C, $q_2 = 9.0$ C, $r_1 = d$, $r_2 = (4.0\text{ m} - d)$, $F_1 = F_2$.

Find: d.

From Coulomb's law, we have $F_{31} = \dfrac{k\,q_1\,q_3}{d^2}$ and $F_{32} = \dfrac{k\,q_2\,q_3}{(4.0\text{ m} - d)^2}$.

Equating F_{31} and F_{32} gives $\dfrac{k\,(4.0\text{ C})\,q_3}{d^2} = \dfrac{k\,(9.0\text{ C})\,q_3}{(4.0\text{ m} - d)^2}$.

Simplifying and taking the square root on both sides $\dfrac{2.0}{d} = \dfrac{3.0}{(4.0\text{ m} - d)}$.

Cross multiplying, we arrive at $2.0(4.0\text{ m} - d) = 3.0d$, or $5.0d = 8.0$ m.

Solving for d, $d = \dfrac{8.0\text{ m}}{5.0} = 1.6$ m from the +4.0 C charge.

Example 15.4 Consider three point charges located at the corners of a triangle as shown in the figure. If $q_1 = 6.0$ nC, $q_2 = -1.0$ nC, and $q_3 = 5.0$ nC, what is the net force on the 5.0 nC charge.

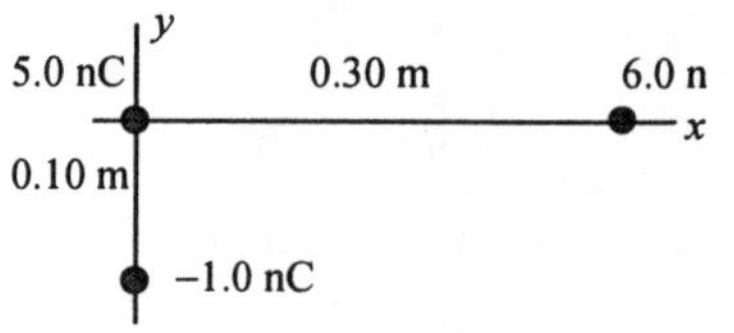

Solution: By the charge-force law, the force on the 5.0 nC charge by the 6.0 nC charge is to the left and the force by the −3.0 nC charge is down.

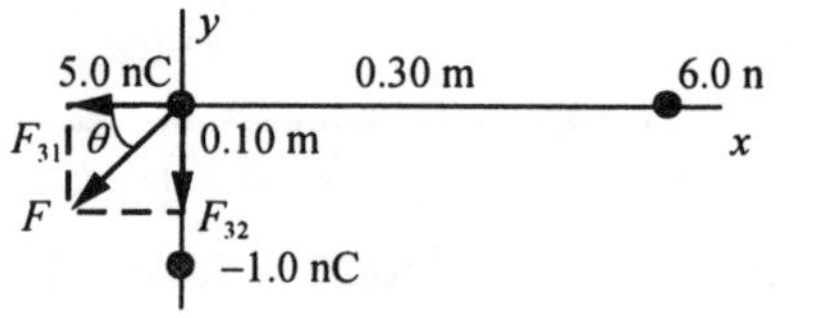

Given: $q_1 = 6.0\text{ nC} = 6.0\times10^{-9}$ C, $q_2 = -1.0\text{ nC} = -1.0\times10^{-9}$ C, $q_3 = 5.0\text{ nC} = 5.0\times10^{-9}$ C,

$r_1 = 0.30$ m, $r_2 = 0.10$ m.

Find: **F**.

From Coulomb's law, we have $F_{31} = \dfrac{(9.00\times10^9\text{ N}\cdot\text{m}^2/\text{C}^2)(6.0\times10^{-9}\text{ C})(5.0\times10^{-9}\text{ C})}{(0.30\text{ m})^2} = 3.0\times10^{-6}$ N,

and $F_{32} = \dfrac{(9.00\times10^9\text{ N}\cdot\text{m}^2/\text{C}^2)(1.0\times10^{-9}\text{ C})(5.0\times10^{-9}\text{ C})}{(0.10\text{ m})^2} = 4.5\times10^{-6}$ N (the negative sign of the –1.0-nC charge does not have to be included here because the direction of the force is already determined by the charge-force law).

So $F = \sqrt{F_{31}^2 + F_{32}^2} = \sqrt{(3.0\times10^{-6}\text{ N})^2 + (4.5\times10^{-6}\text{ N})^2} = 5.4\times10^{-6}$ N,

and $\theta = \tan^{-1}\dfrac{F_{32}}{F_{31}} = \tan^{-1}\dfrac{4.5\times10^{-6}\text{ N}}{3.0\times10^{-6}\text{ N}} = 56°$ below the $-x$ axis.

---

4. Electric Field

The **electric field** is a *vector field* that describes how nearby charges modify the space around them. It is defined as the electric force per unit positive charge: $\mathbf{F} = \dfrac{\mathbf{F}_{\text{on } q_o}}{q_o}$, where q_o is a positive test charge. The direction of the electric field at any point is in the direction of the force experienced by the positive test charge. The SI unit of the electric field is N/C. The magnitude of the electric field due to a point charge is $E = \dfrac{k\,q}{r^2}$.

Note: Since electric field is a vector, you have to use vector addition to calculate the net field at a point if there are more than one charge contributing to the fields.

Electric lines of force are the imaginary lines formed by connecting electric field vectors. Their closeness and direction indicate the magnitude and direction of the electric field at any point. In general, the electric lines of force originate from positive charges (or infinity if there is no positive charge) and terminates at negative charges (or infinity if there is no negative charge). The number of lines leaving or entering a charge is proportional to the magnitude of that charge.

Note: Since the direction of electric lines of force indicates the direction of the electric field, no two electric lines of force can cross. (Why?)

Example 15.5 A 5.0-C charge is 10 m from a small test charge of 1.0 nC.

(a) What is the electric field at the 1.0 nC charge?

(b) What is the direction of the electric field at the 1.0 nC charge?

(c) What is the magnitude of the force experienced by the 1.0 nC charge?

Solution: Given: $q = 5.0$ C, $q_o = 1.0$ nC $= 1.0\times10^{-9}$ C, $r = 10$ m.

Find: (a) E (b) direction of **E** (c) F.

(a) $E = \dfrac{k\,q}{r^2} = \dfrac{(9.00\times10^9\ \text{N·m}^2/\text{C}^2)(5.0\ \text{C})}{(10\ \text{m})^2} = 4.5\times10^8$ N/C.

(b) Since the 5.0-C charge is a positive charge, the direction of the electric field at the 1.0-nC charge points radially outward from the +5.0-C charge.

(c) From the definition of electric field $E = \dfrac{F}{q_o}$, we have $F = q_o\,E = (1.0\times10^{-9}\ \text{C})(4.5\times10^8\ \text{N/C}) = 0.45$ N.

Example 15.6 A 5.0-μC charge is placed at the 0 cm mark of a meterstick and a -4.0-μC charge is placed at the 50 cm mark.

(a) What is the electric field at the 30 cm mark?

(b) At what point along a line connecting the two charges is the electric field zero?

Solution: Given: $q_1 = 5.0\ \mu\text{C} = 5.0 \times 10^{-6}$ C, $q_2 = -4.0\ \mu\text{C} = 4.0 \times 10^{-6}$ C,

$r_1 = 0.30$ m, $r_2 = (0.50\text{ m} - 0.30\text{ m}) = 0.20$ m.

Find: (a) E (b) x.

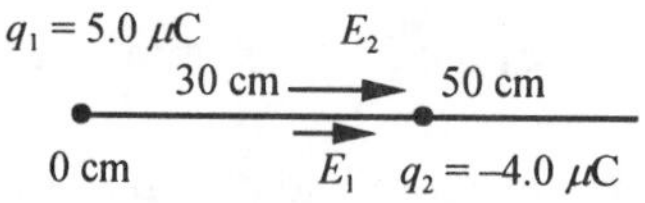

(a) The electric fields due to the two point charges at the 30 cm mark are in the same direction (why?).

From $E = \frac{kq}{r^2}$, we have $E_1 = \frac{(9.00 \times 10^9\ \text{N·m}^2/\text{C}^2)(5.0 \times 10^{-6}\ \text{C})}{(0.30\ \text{m})^2} = 5.0 \times 10^5$ N/C

and $E_2 = \frac{(9.00 \times 10^9\ \text{N·m}^2/\text{C}^2)(4.0 \times 10^{-6}\ \text{C})}{(0.20\ \text{m})^2} = 9.0 \times 10^5$ N/C (the negative sign of the -4.0-μC charge does not have to be included here because the direction of the electric field is already determined).

$E = E_1 + E_2 = 5.0 \times 10^5\ \text{N/C} + 9.0 \times 10^5\ \text{N/C} = 1.4 \times 10^6$ N/C. The direction is to the right.

(b) Since the direction of the fields between 0 and 50 cm marks are in the same direction, there is no point between them where the electric field is zero. We have to look outside the 0 and 50 cm marks along the meterstick. It is only possible to have zero electric field at a point to the right of the $-4.0\ \mu$C charge along the meterstick (why is it impossible to have zero electric field at a point to the left of the 5.0 μC charge along the meterstick). Assume the point is a distance x from the 0 cm mark, and then it is ($x - 0.50$ m) from the 50 cm mark. The directions of the electric fields are opposite.

$E_1 = \frac{k q_1}{x^2}$ and $E_2 = \frac{k q_2}{(x - 0.50\ \text{m})^2}$.

Equating E_1 and E_2 gives $\frac{k\,(5.0 \times 10^{-6}\ \text{C})}{x^2} = \frac{k\,(4.0 \times 10^{-6}\ \text{C})}{(x - 0.50\ \text{m})^2}$.

Simplifying and taking the square root on both sides $\frac{2.24}{x} = \frac{2.0}{x - 0.50\ \text{m}}$.

Cross multiplying, we arrive at $(2.24)(x - 0.50\ \text{m}) = 2.0x$, or $0.24x = 1.12$ m.

Solving for x, $x = 4.7$ m from the 0 cm mark.

Example 15.7 Four charges occupy the corners of a square as shown. Each charge has a magnitude of 3.0 μC and the side of the square is 0.50 m. Find the electric field at the center of the square shown.

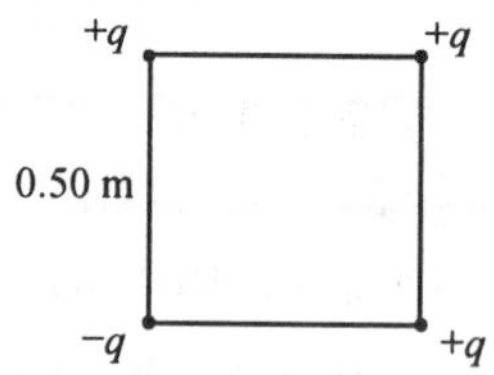

Solution: From symmetry, the fields due to the charges at the upper-left (1) and lower-right (4) corners cancel. The fields by the charges at the upper-right (2) and lower-left (3) are in the same direction (pointing toward the $-q$ charge). So the net field is simply the addition of these two fields.

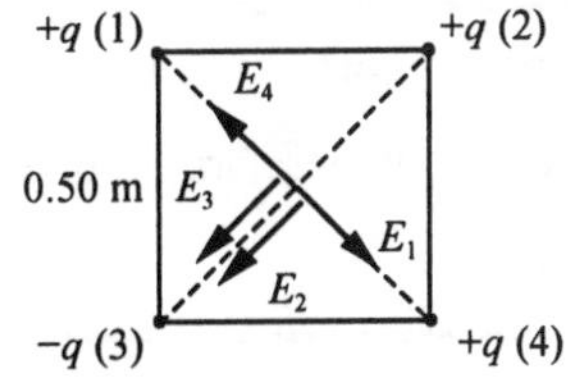

Given: $q = 3.0\ \mu\text{C}$, $L = 0.50$ m.

Find: **E**.

The length of the diagonal is $d = \sqrt{(0.50\text{ m})^2 + (0.50\text{ m})^2} = 0.707$ m. So the distance from the charges to the center of the square is $\frac{d}{2} = \frac{0.707\text{ m}}{2} = 0.354$ m.

From $E = \frac{kq}{r^2}$, we have $E_2 = E_3 = \frac{(9.00\times10^9\text{ N·m}^2/\text{C}^2)(3.0\times10^{-6}\text{ C})}{(0.354\text{ m})^2} = 2.15\times10^{15}$ N/C.

Therefore $E = E_2 + E_3 = 2(2.15\times10^{15}\text{ N/C}) = 4.3\times10^{15}$ N/C, toward the negative charge.

5. Conductor and Electric Fields

The electric fields associated with charged conductors in electrostatic equilibrium (isolated or insulated) have the following interesting properties:

(1) The electric field is zero everywhere inside a charged conductor.

(2) Any excess charge on an isolated conductor resides entirely on the surface of the conductor.

(3) The electric field at the outer surface of a charged conductor is perpendicular to the surface.

(4) Excess charge tends to accumulate at sharp points, or locations of greatest curvature, on charged conductors, and the places of highest charge accumulation are where the electric field from the conductor is the largest.

*6. Gauss's Law for Electric Fields: A Qualitative Approach

A qualitative statement of **Gauss's law** is that the net number of electric field lines passing through an imaginary surface is proportional to the amount of net charge enclosed within the surface. The surface is called a **Gaussian surface**. In general, by counting the number of lines entering or leaving a Gaussian surface, we can determine the sign and the relative magnitude of the charge enclosed within that surface. If there are a net number of lines leaving the surface (we label them "+"), the net charge enclosed is positive. If there are a net number of lines entering the surface (we label them "–"), the net charge enclosed is negative.

Example 15.8 A Gaussian surface encloses an object with a net charge of +2.0 C and there are 6 lines leaving the surface. Some charge is added to the object and now there are 12 lines *entering* the surface. How much charge is added?

Solution: Since there are 6 lines when there is +2.0 C, therefore +1.0 C is equivalent to 3 lines. After charge is added, there are 12 lines entering. So the net charge is now $-\dfrac{12 \text{ lines}}{3 \text{ lines/coulomb}} = -4.0 \text{ C}$.

Therefore, the charge added is $\Delta Q = Q_f - Q_i = -4.0 \text{ C} - 2.0 \text{ C} = -6.0 \text{ C}$.

IV. Mathematical Summary

Quantization of Electric Charge	$q = ne$ $\lvert e \rvert = 1.60\times10^{-19}$ C	(15.1)	Relates charge and fundamental unit of charge
Coulomb's Law	$F = \dfrac{k\, q_1\, q_2}{r^2}$ $k \approx 9.00\times10^9 \text{ N·m}^2/\text{C}^2$	(15.2)	Computes the magnitude of the electric force between two point charges
Electric Field	$\mathbf{E} = \dfrac{\mathbf{F}_{\text{on } q_o}}{q_o}$	(15.3)	Defines the electric field vector from the force vector on a charge q_o
Electric Field due to a Point Charge	$E = \dfrac{k\, q}{r^2}$	(15.4)	Calculates the magnitude of the electric field due to a point charge

V. Solutions of Selected Exercises and Paired Exercises

6. (a) -8.0×10^{-10} C according to charge conservation.

(b) $n = \dfrac{q}{e} = \dfrac{8.0\times10^{-10} \text{ C}}{1.6\times10^{-19} \text{ C}} = 5.0\times10^{9}$ electrons.

14. If you bring a positively charged object near the electroscope, the induction process will charge the electroscope with negative charges. You can prove the charges are negative by bringing the positively charged object near the leaves and seeing if the leaves fall.

18. (a) Since $F = \frac{k\,q_1\,q_2}{r^2} \propto \frac{1}{r^2}$, F is 4 times larger if r is halved.

(b) F is 1/9 as large if r is tripled.

24. From $F = \frac{k\,q_1\,q_2}{r^2}$, we have $\frac{F_2}{F_1} = \frac{r_1^2}{r_2^2}$.

So $r_2 = \sqrt{\frac{F_1}{F_2}}\, r_1 = \sqrt{\frac{1}{10}}$ (30 cm) = 9.5 cm.

32. $F_2 = F_3 = \frac{k\,q_1\,q_2}{r^2} = \frac{(9.0\times10^9\ \text{N·m}^2/\text{C}^2)(4.0\times10^{-6}\ \text{C})^2}{(0.20\ \text{m})^2} = 3.6\ \text{N}.$

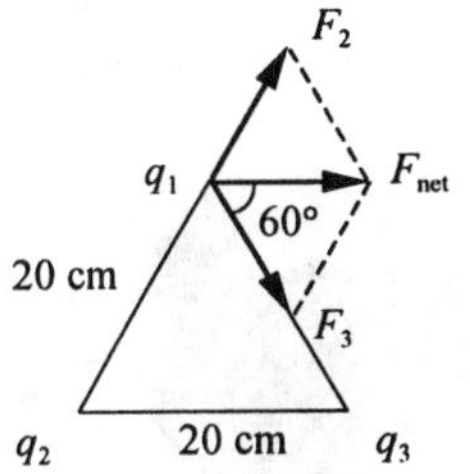

According to symmetry, the net force on q_1 points in the $+x$ direction.

$F_{net} = F_x = F_2 \cos 60° + F_3 \cos 60°$

$= 2(3.6\ \text{N}) \cos 60° = 3.6$ N in the $+x$ direction.

44. $E = \frac{F}{q} = \frac{6.4\times10^{-14}\ \text{N}}{1.6\times10^{-19}\ \text{C}} = 4.0\times10^5\ \text{N/m}.$

52. Due to symmetry, E_1 and E_2 cancel out.

So the net electric field is the electric field by q_3.

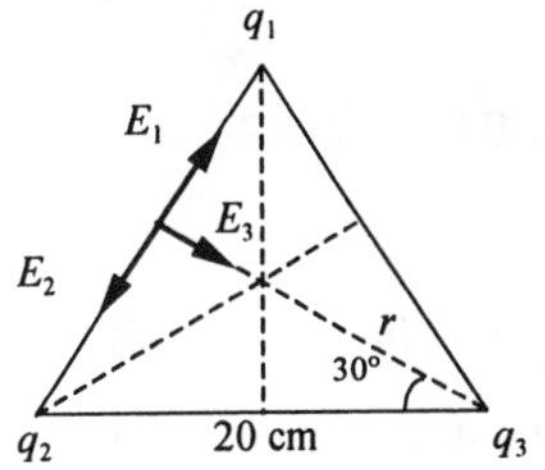

$E = E_3 = \frac{k\,q_3}{r^2} = \frac{(9.0\times10^9\ \text{N·m}^2/\text{C}^2)(4.0\times10^{-6}\ \mu\text{C})}{[(0.20\ \text{m})\cos 30°]^2}$

$= 1.2\times10^6$ N/C toward the $-4.0\ \mu$C charge.

60. (a) since there is no electric field inside a conductor in electrostatic equilibrium.

70. Since the number of lines is proportional to charge, the net charges are equal and opposite in sign.

75. (a) $F = q\,E = (-1.6\times10^{-19}\ \text{C})(2.5\times10^3\ \text{N/C}) = -4.0\times10^{-16}\ \text{N} = 4.0\times10^{-16}$ N in $-y$ direction.

(b) First find the acceleration. $a = \frac{F}{m}$, so $v^2 = v_o^2 + 2ax = \frac{2F\,x}{m}$.

Therefore $K = \frac{1}{2}mv^2 = F\,x = (4.0\times10^{-16}\ \text{N})(0.10\ \text{m}) = 4.0\times10^{-17}\ \text{J}.$

80. From $F = \frac{k\,q_1\,q_2}{r^2}$, we have $q_2 = \frac{F\,r^2}{k\,q_1} = \frac{(1.8\ \text{N})(0.30\ \text{m})^2}{(9.0\times10^9\ \text{N·m}^2/\text{C}^2)(6.0\times10^{-6}\ \text{C})} = 3.0\times10^{-6}\ \text{C} = 3.0\ \mu\text{C}.$

So the charge could be $+3.0\ \mu$C on the $-y$ axis or $-3.0\ \mu$C on the $+y$ axis from the charge-force law.

83. $Q_{net} = -10\times10^{-6}\text{ C} - 10\times10^{-6}\text{ C} + 5.0\times10^{-6}\text{ C} + 5.0\times10^{-6}\text{ C} = -10\times10^{-6}\text{ C}.$

So the net lines are negative.

88. Assume the electric field is in the direction shown by arrow 1. By symmetry argument, it should also be in the direction shown by arrow 3. Since it can have only one direction, it has to be in the direction shown by arrow 2, i.e., the field lines are perpendicular to the wire.

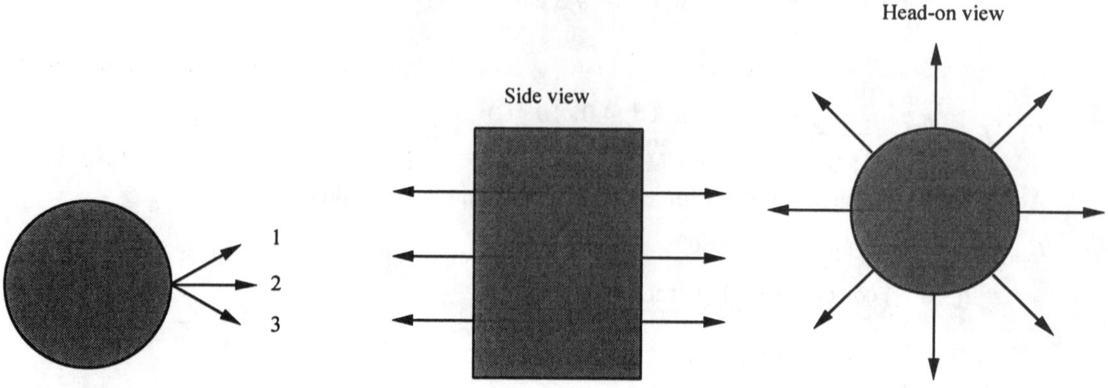

VI. Practice Quiz

1. Two point charges, initially 2.0 cm apart, experience a 1.0-N force. If they are moved to a new separation of 8.0 cm, what is the electric force between them?

   (a) 4.0 N (b) 16 N (c) 1/4 N (d) 1/16 N (e) none of the above

2. A negatively charged rod is brought near one end of an uncharged metal bar. The end of the metal bar farthest from the charged rod will be which of the following?

   (a) positive (b) negative (c) neutral (d) attracted (e) repelled

3. Two point charges each have a value of 3.0 C and are separated by a distance of 4.0 m. What is the electric field at a point midway between the two charges?

   (a) zero (b) 9.0×10^{7} N/C (c) 18.0×10^{7} N/C (d) 9.0×10^{7} N/C (e) 4.5×10^{7} N/C

4. Two point charges, separated by 1.5 cm, have charge values of +2.0 and −4.0 μC, respectively. What is the magnitude of the electric force between them?

   (a) 160 N (b) 320 N (c) 3.6×10^{-8} N (d) 8.0×10^{-12} N (e) 3.1×10^{-3} N

5. Which of the following is not a vector?

(a) electric force (b) electric field (c) electric charge (d) electric line of force (e) acceleration

6. Three point charges are located at the following positions: $q_1 = 2.00\ \mu C$ at $x = 1.00$ m; $q_2 = 3.00\ \mu C$ at $x = 0$; $q_3 = -5.00\ \mu C$ at $x = -1.00$ m. What is the magnitude of the force on the 3.00 μC charge?

(a) 5.40×10^{-2} N (b) 0.135 N (c) 8.10×10^{-2} N (d) 0.189 N (e) 9.0×10^{-2} N

7. A particle with a charge of 4.0 μC has a mass of 5.0×10^{-3} kg. What electric field directed upward will exactly balance the weight of the ball?

(a) 4.1×10^{2} N/C (b) 8.2×10^{2} N/C (c) 1.2×10^{4} N/C (d) 5.1×10^{6}N/C (e) 2.0×10^{-7} N/C

8. If a conductor is in electrostatic equilibrium near an electric charge

(a) the total charge on the conductor must be zero.

(b) any charge on the conductor must be uniformly distributed.

(c) the force between the conductor and the charge must be zero.

(d) the total electric field of the conductor must be zero.

(e) the electric field on the surface of the conductor is perpendicular to the surface.

9. Two point charges of +3.0 μC and −7.0 μC are placed at $x = 0$ and $x = 0.20$ m, respectively. What is the magnitude of the electric field at the point midway between them ($x = 0.10$ m)?

(a) 9.0×10^{5} N/C (b) 1.8×10^{6} N/C (c) 3.6×10^{6} N/C (d) 4.5×10^{6} N/C (e) 9.0×10^{6} N/C

10. What is the unit of the Coulomb constant k, which appears in Coulomb's law?

(a) N·m/C (b) N/C (c) $N^2\cdot m/C^2$ (d) $N\cdot m^2/C^2$ (e) N/C^2

Answers to Practice Quiz:

1. d 2. b 3. a 4. b 5. c 6. d 7. c 8. e 9. e 10. d

CHAPTER 16

Electric Potential, Energy, and Capacitance

I. Chapter Objectives

Upon completion of this chapter, you should be able to:

1. understand the concept of electric potential difference (voltage) and its relationship to electric potential energy, and calculate electric potential differences and electric potential energy in simple cases.
2. explain what is meant by an equipotential surface, sketch equipotential surfaces for simple charge configurations, and explain the relationship between equipotential surfaces and electric fields.
3. define capacitance and explain what it means physically, calculate the charge, electric field, and energy stored for parallel plate capacitors.
4. understand what a dielectric is and how it affects the physical properties of a capacitor.
5. find the equivalent capacitance of capacitors connected in series and in parallel, calculate the charges, voltages, and energy storage of individual capacitors in series and parallel configurations, and analyze capacitor networks that include both series and parallel arrangements.

II. Key Terms

Upon completion of this chapter, you should be able to define and/or explain the following key terms:

electric potential energy	**capacitor**
electric potential difference	**farad (F)**
volt (V)	**dielectric**
voltage	**dielectric constant (*K*)**
electric potential	**dielectric permittivity**
equipotential surfaces	**equivalent series capacitance**
electron volt (eV)	**equivalent parallel capacitance**

The definitions and/or explanations of the most important key terms can be found in the following section: **III. Chapter Summary and Discussion.**

III. Chapter Summary and Discussion

1. Electric Potential Energy and Potential Difference

The **electric potential difference (voltage)** between two points is the work per unit positive charge done by an external force in moving charge between these two points, or the change in electric potential energy per unit positive charge: $\Delta V = V_B - V_A = \frac{W}{q_o} = \frac{\Delta U}{q_o}$, where q_o is the positive test charge. The SI unit of electric potential difference is joule/coulomb (J/C) or volt (V).

Note: Potential difference and potential energy difference are *different*. Since potential difference is defined per unit charge, it does not depend on the amount of charge moved, whereas potential energy difference does.

In a uniform electric field E, the potential difference in moving a charge through a straight line distance d (in a direction opposite to the electric field, or from negative plate to positive plate in parallel plates) is $\Delta V = Ed$.

Example 16.1 A proton, initially at rest, is accelerated through an electric potential difference of 500 V.

(a) What is the kinetic energy of the proton?

(b) What is the speed of the proton?

Solution: Given: $\Delta V = 500$ V, $q_o = 1.60\times10^{-19}$ C, $m = 1.67\times10^{-27}$ kg.

Find: (a) K (b) v.

(a) When a positively-charged proton is accelerated through a potential difference, work is done on the proton. The proton gains kinetic energy while loosing potential energy. By the conservation of energy, $K = |\Delta U|$.

From $\Delta V = \frac{\Delta U}{q_o}$, we can calculate the kinetic energy gained by the proton.

$K = \Delta U = q_o\,\Delta V = (1.60\times10^{-19}\text{ C})(500\text{ V}) = 8.00\times10^{-17}$ J.

(b) From $K = \frac{1}{2}mv^2$, we have $v = \sqrt{\frac{2K}{m}} = \sqrt{\frac{2(8.00\times10^{-17}\text{ J})}{1.67\times10^{-27}\text{ kg}}} = 3.10\times10^5$ m/s.

Example 16.2 A 12-V battery maintains the electric potential difference between two parallel metal plates separated by 0.10 m. What is the electric field between the plates?

Solution: Given: $\Delta V = 12$ V, $d = 0.10$ m.

Find: E.

From $\Delta V = Ed$, we have $E = \frac{\Delta V}{d} = \frac{12\text{ V}}{0.10\text{ m}} = 1.2\times10^2$ V/m.

The electric potential difference due to a point charge is given by $\Delta V = \frac{kq}{r_B} - \frac{kq}{r_A}$, where r_A and r_B are the distances from the point charge to the two points A and B. If we define the potential at infinity as zero, then the **electric potential** at a single point due to a point charge is $V = \frac{kq}{r}$. The potential changes according to the following general rules:

- *Electric potential increases for locations closer to positive charges or farther from negative charges.*
- *Electric potential decreases for locations farther from positive charges or nearer to negative charges.*

Note: Unlike electric field, electric potential is a scalar quantity. When adding potentials due to point charges, you only need to add them algebraically (including the + or − signs). Also, the potential is proportional to $\frac{1}{r}$ and the magnitude of the electric field is proportional to $\frac{1}{r^2}$.

For a configuration of multiple point charges, the total potential energy U_t is $U_t = U_{12} + U_{23} + U_{13} + \ldots$, where $U_{ij} = \frac{k\,q_i\,q_j}{r_{ij}}$ is the potential energy due to two point charges i and j.

Example 16.3 A 5.0-nC charge is at (0,0) and a −2.0-nC charge is at (3.0 m, 0 m). If the potential is taken to be zero at infinity,

(a) what is the electric potential at point P (0, 4.0) m?

(b) what is the potential energy of a 1.0-nC charge at point P?

(c) what is the work required to bring a 1.0-nC charge from infinity to point P?

(d) what is the total potential energy of the three charge system?

Solution: Given: $q_1 = 5.0\text{ nC} = 5.0\times10^{-9}\text{ C}$, $q_2 = -2.0\text{ nC} = -2.0\times10^{-9}\text{ C}$, $q_3 = 1.0\text{ nC} = 1.0\times10^{-9}\text{ C}$,

$(x_1, y_1) = (0,0)$, $(x_2, y_2) = (3.0,0)$ m, $(x_P, y_P) = (0, 4.0)$ m.

Find: (a) V (b) U (c) W (d) U_t.

$r_{23} = \sqrt{(3.0\text{ m})^2 + (4.0\text{ m})^2} = 5.0\text{ m}.$

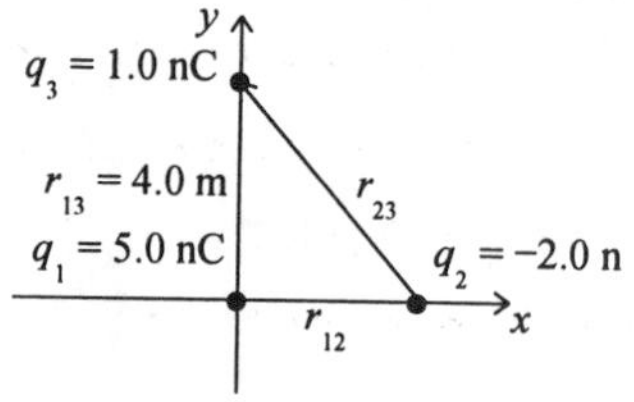

(a) From $V = \frac{kq}{r}$, the total potential, we have

$$V_t = V_1 + V_2 = \frac{(9.00\times10^9\text{ N}\cdot\text{m}^2/\text{C}^2)(5.0\times10^{-9}\text{ C})}{3.0\text{ m}}$$

$$+ \frac{(9.00\times10^9\text{ N}\cdot\text{m}^2/\text{C}^2)(-2.0\times10^{-9}\text{ C})}{5.0\text{ m}} = 11\text{ V}.$$

(b) From $V = \frac{U}{q_o}$, and so we have $U = q_3 V = (1.0\times10^{-9}\text{ C})(11\text{ V}) = 1.1\times10^{-8}\text{ J}$.

(c) From conservation of energy, the work required to bring the 1.0 nC charge from infinity is equal to its potential energy, $W = U = 1.1\times10^{-8}\text{ J}$.

(d) $U_t = U_{12} + U_{23} + U_{13} = \frac{k\,q_1 q_2}{r_{12}} + \frac{k\,q_2 q_3}{r_{23}} + \frac{k\,q_1 q_3}{r_{13}}$

$$= \frac{(9.00\times10^9\text{ N}\cdot\text{m}^2/\text{C}^2)(5.0\times10^{-9}\text{ C})(-2.0\times10^{-9}\text{ C})}{3.0\text{ m}}$$

$$+ \frac{(9.00\times10^9\text{ N}\cdot\text{m}^2/\text{C}^2)(-2.0\times10^{-9}\text{ C})(1.0\times10^{-9}\text{ C})}{5.0\text{ m}}$$

$$+ \frac{(9.00\times10^9\text{ N}\cdot\text{m}^2/\text{C}^2)(5.0\times10^{-9}\text{ C})(1.0\times10^{-9}\text{ C})}{4.0\text{ m}} = -2.2\times10^{-8}\text{ J}.$$

2. Equipotential Surfaces and the Electric Field

Equipotential surfaces (equipotentials) are surfaces on which a charge has a constant electric potential energy. The electric field is perpendicular to such surfaces at all points. It takes no work to move a charge along an equipotential. The electric field is related to the electric potential difference as follows: $E = -\left(\frac{\Delta V}{\Delta x}\right)_{max}$, where the minus sign means that **E** is in the direction opposite that in which V increases most rapidly, or in the direction V decreases most rapidly. The units of electric field are volts per meter (V/m), which has the same dimensions as N/C introduced in Chapter 15.

An **electron volt** (eV) is the kinetic energy gained by an electron accelerated from rest though a potential difference of 1 V. One eV is equivalent to 1.60×10^{-19} J.

Example 16.4 Two parallel plates, separated by 0.20 m, are connected to a 12-V battery. An electron is released from rest at a location 0.10 m from the negative plate. When the electron arrives at a distance of 0.050 m from the positive plate,

(a) what is the potential difference between the initial and final points?

(b) how much kinetic energy, in electron volts and joules, does the electron gain?

(c) what is the speed of the electron?

Solution: Given: $\Delta V = 12$ V, $d = 0.20$ m, $\Delta x = 0.20\text{ m} - 0.10\text{ m} - 0.050\text{ m} = 0.050\text{ m}$.

Find: (a) ΔV_e (b) K (c) v.

(a) The uniform electric field between the parallel plates is $E = \frac{\Delta V}{d} = \frac{12 \text{ V}}{0.20 \text{ m}} = 60 \text{ V/m}$.

From $E = -\left(\frac{\Delta V}{\Delta x}\right)_{\max}$, we have $\Delta V_e = E\,\Delta x = (60 \text{ V/m})(0.050 \text{ m}) = 3.0 \text{ V}$.

(b) From energy conservation, $K = |\Delta U| = |q\Delta V| = (1 \text{ e})(3.0 \text{ V}) = 3.0 \text{ eV} = (3.0 \text{ eV})\frac{1.60\times10^{-19} \text{ J}}{\text{eV}}$

$= 4.8\times10^{-19}$ J.

(c) From $K = \frac{1}{2}m\,v^2$, $v = \sqrt{\frac{2\,K}{m}} = \sqrt{\frac{2(4.8\times10^{-19} \text{ J})}{9.11\times10^{-31} \text{ kg}}} = 1.0\times10^6$ m/s.

3. Capacitance

A **capacitor** consists of two conductors. It stores charge, and therefore electric energy in the form of an electric field. **Capacitance** (C) is a quantitative measure of how effective a capacitor is in storing charge and is defined as $C = \frac{Q}{V}$, where Q is the charge on each conductor (the two conductors have equal but opposite charge) and V is the potential difference between the two conductors. The SI unit of capacitance is farad (F).

Note: From now on the symbol V used here stands for potential *difference* for convenience. It has the same meaning as ΔV used earlier.

A common capacitor is the parallel-plate capacitor. It consists of two parallel metal plates of area A and separated by a distance d. The capacitance of a parallel-plate capacitor is given by $C = \frac{\varepsilon_o A}{d}$, where ε_o is the *permittivity of free space* (vacuum) and is equal to 8.85×10^{-12} C²/N·m². ε_o is a fundamental constant and is related to Coulomb's constant by $k = \frac{1}{4\pi\varepsilon_o} = 9.00\times10^9$ N·m²/C².

The energy stored in a capacitor of capacitance C and potential difference V (therefore with a charge $Q = C\,V$) is given by $U = \frac{1}{2}Q\,V = \frac{Q^2}{2\,C} = \frac{1}{2}C\,V^2$. The form $\frac{1}{2}C\,V^2$ is usually the most practical, since the capacitance and the applied voltage are often known or can be measured.

Example 16.5 A parallel-plate capacitor consists of plates of area 1.5×10^{-4} m² and separated by 1.0 mm. The capacitor is connected to a 12-V battery.

(a) What is the capacitance?

(b) What is the charge on the plates?

(c) How much energy is stored in the capacitor?

(d) What is the electric field between the plates?

Solution: Given: $A = 1.5\times10^{-4}\ \text{m}^2$, $d = 1.0\ \text{mm} = 1.0\times10^{-3}\ \text{m}$, $V = 12\ \text{V}$.

Find: (a) C (b) Q (c) U (d) E.

(a) $C = \dfrac{\varepsilon_o A}{d} = \dfrac{(8.85\times10^{-12}\ \text{C}^2/\text{N}\cdot\text{m}^2)(1.5\times10^{-4}\ \text{m}^2)}{1.0\times10^{-3}\ \text{m}} = 1.3\times10^{-12}\ \text{F} = 1.3\ \text{pF}.$

(b) $Q = C\,V = (1.3\times10^{-12}\ \text{F})(12\ \text{V}) = 1.6\times10^{-11}\ \text{C}.$

(c) $U = \frac{1}{2}C\,V^2 = \frac{1}{2}(1.3\times10^{-12}\ \text{F})(12\ \text{V})^2 = 9.4\times10^{-11}\ \text{J}.$

(d) $E = \dfrac{\Delta V}{d} = \dfrac{V}{d} = \dfrac{12\ \text{V}}{1.0\times10^{-3}\ \text{m}} = 1.2\times10^4\ \text{V/m}.$

---

4. Dielectrics

A **dielectric** is any nonconducting material (insulator) capable of being partially polarized when placed in an electric field. When inserted between the plates of a capacitor, a dielectric raises the capacitance by a factor K, the **dielectric constant** (always greater than 1), which is defined as $K = \dfrac{C}{C_o}$, where C_o is the capacitance without the dielectric (or vacuum is between the plates). As a result, the energy of the capacitor also increases if the potential difference is maintained constant across the capacitor, $U = \frac{1}{2}CV^2 = \frac{1}{2}(K\,C_o)\,V^2 = K\,U_o$. The product $K\varepsilon_o$ is called the **dielectric permittivity**, ε (that is, $\varepsilon = K\varepsilon_o$).

For an isolated capacitor (battery disconnected) with charge Q_o and voltage V_o, the insertion of a dielectric decreases the voltage to V, while the charge remains constant. Therefore, capacitance increases to $C = \dfrac{Q_o}{V}$ $(V < V_o)$. For a capacitor connected to a battery of voltage V_o, the insertion of a dielectric increases the charge Q, while the voltage remains constant. Therefore, capacitance increases to $C = \dfrac{Q}{V_o}$ $(Q > Q_o)$.

Example 16.6 Repeat Example 16.5 if the capacitor is filled with Teflon. (The capacitor is always connected to the battery.)

Solution: Additional given: $K = 2.1$ (from Table 16.1).

(a) $C = K\,C_o = (2.1)(1.3\times10^{-12}\ \text{F}) = 2.7\times10^{-12}\ \text{F} = 2.7\ \text{pF}$. Note C increases.

(b) $Q = CV = (2.7\times10^{-12}\ \text{F})(12\ \text{V}) = 3.2\times10^{-11}\ \text{C}$. Note Q increases.

(c) $U = \frac{1}{2}CV^2 = \frac{1}{2}(2.7\times10^{-12}\text{ F})(12\text{ V})^2 = 1.9\times10^{-10}\text{ J}$. Note U increases.

(d) $E = \dfrac{\Delta V}{d} = \dfrac{V}{d} = \dfrac{12\text{ V}}{1.0\times10^{-3}\text{ m}} = 1.2\times10^4\text{ V/m}$. Why is E the same?

5. Capacitors in Series and Parallel

The **equivalent series capacitance** (C_s) is always less than that of the smallest capacitor of the series combination. It is given by $\dfrac{1}{C_s} = \dfrac{1}{C_1} + \dfrac{1}{C_2} + \dfrac{1}{C_3} + \ldots$. In a series combination, the charge on the equivalent capacitor is the same as that of each individual capacitor ($Q_s = Q_1 = Q_2 = Q_3 = \ldots$) and the voltage across the equivalent capacitor is equal to the sum of the voltages on each capacitor ($V_s = V_1 + V_2 + V_3 + \ldots$).

The **equivalent parallel capacitance** (C_p) is always larger than that of the largest capacitor of the parallel combination. It is given by $C_p = C_1 + C_2 + C_3 + \ldots$. In a parallel combination, the charge on the equivalent capacitor is equal to the sum of that of each capacitor ($Q_p = Q_1 + Q_2 + Q_3 + \ldots$) and the voltage across the equivalent capacitor is the same as that of each individual capacitor ($V_p = V_1 = V_2 = V_3 = \ldots$).

Note: If there are only two capacitors in a series combination, the equivalent series capacitance can be conveniently expressed as $C_s = \dfrac{1}{1/C_s} = \dfrac{1}{1/C_1 + 1/C_2} = \dfrac{C_1 C_2}{C_1 + C_2}$. If the two capacitors have the same capacitance $C_1 = C_2 = C$, then $C_s = \dfrac{CC}{C+C} = \dfrac{C}{2}$.

Example 16.7 For the capacitor network shown, $C_1 = 6.00\ \mu\text{F}$, $C_2 = 8.00\ \mu\text{F}$, $C_3 = 14.0\ \mu\text{F}$.

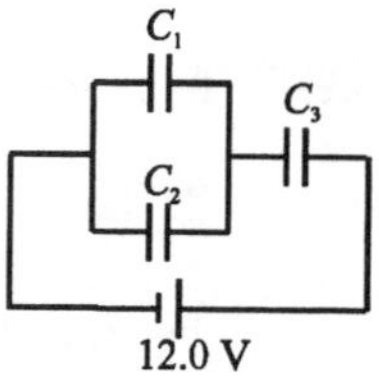

(a) What is the equivalent capacitance?

(b) What is the charge of each capacitor?

(a) What is the voltage across each capacitor?

Solution: Given: $C_1 = 6.00\ \mu\text{F}$, $C_2 = 8.00\ \mu\text{F}$, $C_3 = 14.0\ \mu\text{F}$, $V = 12.0$ V.

Find: (a) C_{eq} (b) Q of each capacitor (c) V across each capacitor.

When working with capacitor combinations, always start with the combination you recognize. In this Example, C_1 and C_3 are *not* in series, neither are C_2 and C_3. However, C_1 and C_2 are in parallel. So we start from this combination.

$C_p = C_1 + C_2 = 6.00\ \mu\text{F} + 8.00\ \mu\text{F} = 14.0\ \mu\text{F}$, and the circuit reduces to Figure 1.

Now we can see that C_p and C_3 are in series.

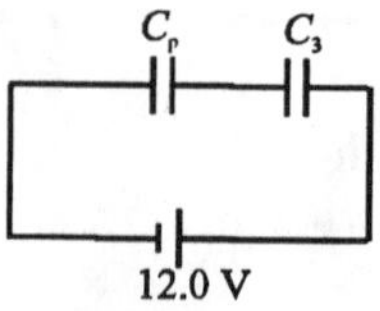

Figure 1

$$\frac{1}{C_s} = \frac{1}{C_p} + \frac{1}{C_3} = \frac{1}{14.0\ \mu\text{F}} + \frac{1}{14.0\ \mu\text{F}} = \frac{2}{14.0\ \mu\text{F}} + \frac{1}{7.00\ \mu\text{F}}.$$

So $C_s = 7.00\ \mu\text{F}$,

or $C_s = \frac{C_p C_3}{C_p + C_3} = \frac{(14.0\ \mu\text{F})(14.0\ \mu\text{F})}{14.0\ \mu\text{F} + 14.0\ \mu\text{F}} = 7.00\ \mu\text{F}.$

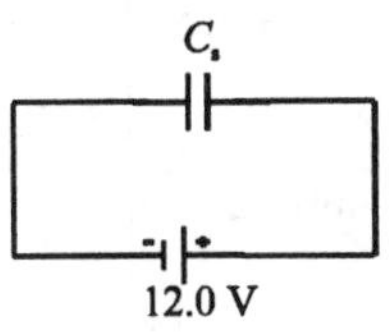

Figure 2

The circuit becomes Figure 2.

So the equivalent capacitance of the circuit is 7.00 μF.

(b) From Figure 2, the charge on the series capacitor (equivalent capacitor in this Example) is $Q_s = C_s V = (7.00\ \mu\text{F})(12.0\ \text{V}) = 84.0\ \mu\text{C}$. Since series capacitors have the same charge, C_p and C_3 will have the same charge as C_s. So $C_3 = 84.0\ \mu\text{C}$ and $C_p = 84.0\ \mu\text{C}$.

$V_3 = \frac{Q_3}{C_3} = \frac{84.0\ \mu\text{C}}{14.0\ \mu\text{F}} = 6.00\ \text{V}$. Since the voltage on the series equivalent capacitor is the sum of that of each individual capacitor, $V_s = V_p + V_3$, so $V_p = V_s - V_3 = 12.0\ \text{V} - 6.00\ \text{V} = 6.00\ \text{V}$.

Parallel capacitors have the same voltage, so $V_1 = V_2 = V_p = 6.00\ \text{V}$.

Therefore $Q_1 = C_1 V_1 = (6.00\ \mu\text{F})(6.00\ \text{V}) = 36.0\ \mu\text{C}$ and $Q_2 = C_2 V_2 = (8.00\ \mu\text{F})(6.00\ \text{V}) = 48.0\ \mu\text{C}$.

Hence, the answers are

(b) $Q_1 = 36.0\ \mu\text{C}$, $Q_2 = 48.0\ \mu\text{C}$, and $Q_3 = 84.0\ \mu\text{C}$.

(c) $V_1 = V_2 = V_3 = 6.00\ \text{V}$.

IV. Mathematical Summary

Electric Potential Difference (voltage)	$\Delta V = V_B - V_A = \frac{\Delta U}{q_o}$ (16.1)	Defines the electric potential difference (voltage) in terms of electric potential energy difference
Electric Potential Difference due to a Point charge	$\Delta V = \frac{kq}{r_B} - \frac{kq}{r_A}$ (16.3)	Computes the electric potential difference due to a point charge
Electric Potential Energy of a Configuration of Point Charges	$U_t = U_{12} + U_{23} + U_{13} + \ldots$ (16.6)	Calculates the electric potential energy of a configuration of point charges
Relationship between Potential and Electric Field	$E = -\left(\frac{\Delta V}{\Delta x}\right)_{max}$ (16.7)	Defines electric field in terms of electric potential gradient (difference)

Capacitance	$Q = CV$ or $C = \frac{Q}{V}$ (16.8)	Defines the capacitance of a capacitor
Capacitance of a Parallel-Plate Capacitor	$C = \frac{\varepsilon_o A}{d}$ (16.9) $\varepsilon_o = 8.85\times10^{-12}$ C²/N·m²	Computes the capacitance of a parallel-plate capacitor
Energy in a Charged Capacitor	$U = \frac{1}{2}QV = \frac{Q^2}{2C} = \frac{1}{2}CV^2$ (16.10)	Calculates the energy stored in a capacitor
Dielectric Effect on Capacitance	$C = KC_o$ (16.12)	Calculates the effect of dielectric on capacitance
Equivalent Capacitance for Capacitors in Series	$\frac{1}{C_s} = \frac{1}{C_1} + \frac{1}{C_2} + \frac{1}{C_3} + \ldots$ (16.14)	Computes the equivalent capacitance for capacitors connected in series
Equivalent Capacitance for Capacitors in Parallel	$C_p = C_1 + C_2 + C_3 + \ldots$ (16.15)	Computes the equivalent capacitance for capacitors connected in parallel

V. Solutions of Selected Exercises and Paired Exercises

10. (a) From $W_{ext} = q\,\Delta V$, we have $q = \frac{W_{ext}}{\Delta V} = \frac{1.6\times10^{-5}\text{ J}}{6\text{ V}} = 2.7\times10^{-6}\text{ C} = 2.7\ \mu\text{C}$.

(b) For a negative charge, it moves from positive to negative since work is done by an external source or the electric field does negative work.

16. (a) $\Delta V = \frac{kq}{r_B} - \frac{kq}{r_A} = \frac{(9.0\times10^9\text{ N·m}^2/\text{C}^2)(-5.5\times10^{-6}\text{ C})}{0.20\text{ m}} - \frac{(9.0\times10^9\text{ N·m}^2/\text{C}^2)(-5.5\times10^{-6}\text{ C})}{0.10\text{ m}}$

$= 2.5\times10^5$ V.

(b) Since ΔV is positive, the more distance point (20 cm) has a higher potential.

20. Field does work. $W_{elec} = -U_t = -\frac{k q_1 q_2}{r} = -\frac{(9.0\times10^9\text{ N·m}^2/\text{C}^2)(-1.4\times10^{-6}\text{ C})^2}{4.0\times10^{-3}\text{ m}} = -4.4$ J.

24. Electric force is a conservative force; the change in potential energy depends only on the initial and final points.

$W = \Delta U = q\,\Delta V = q\,(E\,\Delta x) = (1.6\times10^{-19}\text{ C})(15\text{ V/m})(0.25\text{ m})$

$= 6.0\times10^{-19}$ J.

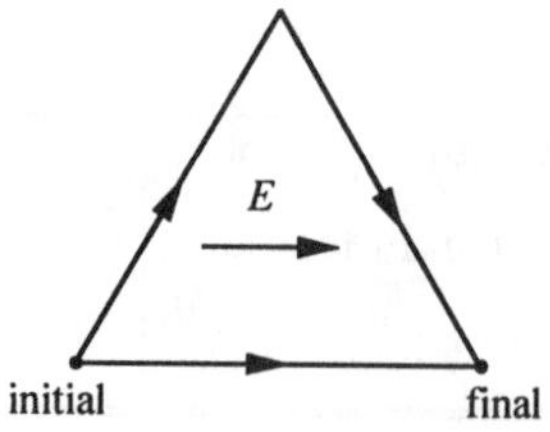

32. The distance from q_2 and q_3 to the point is $r = \sqrt{(0.10\text{ m})^2 + (0.05\text{ m})^2} = 0.112\text{ m}$.

$$V = \Sigma \frac{kq}{r} = \frac{(9.0\times10^9\text{ N·m}^2/\text{C}^2)(-10\times10^{-6}\text{ C})}{0.05\text{ m}} + \frac{(9.0\times10^9\text{ N·m}^2/\text{C}^2)(-10\times10^{-6}\text{ C})}{0.112\text{ m}}$$

$$+\frac{(9.0\times10^9\text{ N·m}^2/\text{C}^2)(5.0\times10^{-6}\text{ C})}{0.112\text{ m}} + \frac{(9.0\times10^9\text{ N·m}^2/\text{C}^2)(5.0\times10^{-6}\text{ C})}{0.05\text{ m}} = -1.3\times10^6\text{ V}.$$

47. An electron volt is $e\,\Delta V = q\,\Delta V = \Delta U$. It is also the kinetic energy gained by an electron when it goes through a potential difference of 1 V. So it is a unit of energy. A GeV is larger than a MeV by 1000 times.

52. $d = \dfrac{7.0\times10^3\text{ V}}{10\times10^3\text{ V/m}} = 0.70\text{ m} = 70\text{ cm}.$

58. (a) $W = \Delta K = K - K_o = K = -\Delta U = -q\,\Delta V = e\,\Delta V = e\,(100\times10^6\text{ V}) = 1.00\times10^8\text{ eV}.$

(b) $(100\times10^6\text{ eV}) \times \dfrac{1.60\times10^{-19}\text{ J}}{1\text{ eV}} = 1.60\times10^{-11}\text{ J}.$

62. (a) $\Delta V = \dfrac{\Delta U}{q} = \dfrac{\Delta U}{e} = \dfrac{3.5\text{ keV}}{\text{e}} = 3.5\text{ kV}.$

(b) 4.1 MV.

(c) 6.0 V.

(d) $(8.0\times10^{-13}\text{ J}) \times \dfrac{1\text{ eV}}{1.6\times10^{-19}\text{ J}} = 5.0\times10^6\text{ eV} = 5.0\text{ MeV}.$

So $\Delta V = 5.0\text{ MV}.$

74. (a) Charge remains constant and so it is the same.

(b) $U = \frac{1}{2}\frac{Q^2}{C}$. Now C is only half as large because d doubles.

So the energy stored doubles to 5.0×10^{-8} J.

82. $Q = CV = KC_o\,V = 2.6(50\times10^{-12}\text{ F})(24\text{ V}) = 3.1\times10^{-9}\text{ C},$

$U = \frac{1}{2}CV^2 = \frac{1}{2}KC_o\,V^2 = \frac{1}{2}2.6(50\times10^{-12}\text{ C})(24\text{ V})^2 = 3.7\times10^{-8}\text{ J}.$

91. The charge is the same for both capacitors. From $U_t = \frac{1}{2}C_s V^2$, we have

$C_s = \dfrac{2U_t}{V^2} = \dfrac{2(173\ \mu\text{J})}{(12\text{ V})^2} = 2.40\ \mu\text{F}.$ Also $\dfrac{1}{C_s} = \dfrac{1}{C_1} + \dfrac{1}{C_2}$

so $C_2 = \dfrac{C_1 C_s}{C_1 - C_s} = \dfrac{(4.0\ \mu\text{C})(2.40\ \mu\text{F})}{4.0\ \mu\text{F} - 2.40\ \mu\text{F}} = 6.0\ \mu\text{F}.$

101. Electric field lines are pointing toward the wire.

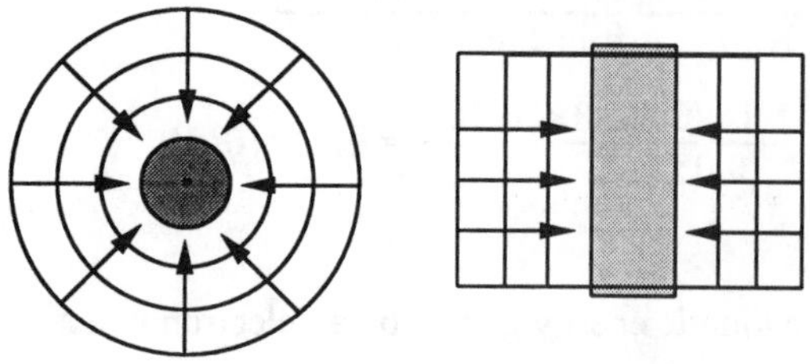

107. (a) The electric force must balance the gravitational force.

From $|F| = q\,E = q\dfrac{\Delta V}{\Delta x} = mg,$ we have

$$\Delta V = \frac{mg\Delta x}{q} = \frac{(9.11\times10^{-31}\text{ kg})(9.80\text{ m/s}^2)(0.015\text{ m})}{1.6\times10^{-19}\text{ C}} = 8.4\times10^{-13}\text{ V}.$$

(b) $\Delta V = \dfrac{(1.67\times10^{-27}\text{ kg})(9.80\text{ m/s}^2)(0.015\text{ m})}{1.6\times10^{-19}\text{ C}} = 1.5\times10^{-9}\text{ V}.$

(c) In (a), top is +; in (b) top is −.

VI. Practice Quiz

1. For an electron moving in a direction opposite to the electric field,

(a) its potential energy increases and its potential increases.

(b) its potential energy decreases and its potential increases.

(c) its potential energy increases and its potential decreases.

(d) its potential energy decreases and its potential decreases.

(e) both the potential energy and potential remain constants.

2. If a 12-V battery is connected between two parallel plates separated by 0.60 cm, what is the magnitude of the electric field between the plates?

(a) 2.0×10^3 V/m (b) 7.2×10^{-2} V/m (c) 0.50×10^{-3} V/m (d) 0.75×10^6 V/m (e) 0.50×10^6 V/m

3. A proton moves 0.10 m along the direction of an electric field line of magnitude 3.0 V/m. What is the magnitude of the change in potential energy of the proton?

(a) 4.8×10^{-20} J (b) 3.2×10^{-20} J (c) 1.6×10^{-20} J (d) 8.0×10^{-21} J (e) zero

4. If three capacitors of 1.0, 1.5, and 2.0 μF each are connected in series, what is the equivalent capacitance?

(a) 4.5 μF (b) 4.0 μF (c) 2.17 μF (d) 0.46 μF (e) 0.67 μF

5. If a 10-μF capacitor is charged so that it stores 2.0×10^{-3} J of electric potential energy, what is the voltage across it?

(a) 2.0 V (b) 5.0 V (c) 10 V (d) 15 V (e) 20 V

6. A parallel plate capacitor has a capacitance C. The area of the plates is doubled and the distance between the plates is halved. What is the new capacitance?

(a) $C/4$ (b) $C/2$ (c) C (d) $2C$ (e) $4C$

7. Two 3.00-μC charges are located on the ends of a meterstick. Find the electrical potential for the center of the meterstick.

(a) zero (b) 2.70×10^4 V (c) 5.40×10^4 V (d) 1.08×10^5 V (e) 2.16×10^5 V

8. A uniform electric field, with a magnitude of 500 V/m, is directed parallel to the positive x axis. If the potential at $x = 5.0$ m is 2500 V, what is the potential at $x = 2.0$ m?

(a) 500 V (b) 1000 V (c) 2000 V (d) 4000 V (e) 8000 V

9. What is the equivalent capacitance of the combination shown?

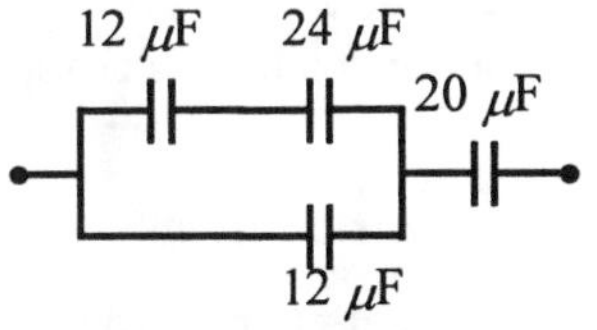

(a) 10 μF

(b) 25 μF

(c) 29 μF

(d) 40 μF

(e) 68 μF

10. If the combination in the previous problem is connected to a 12-V battery, what is the charge on the (lower) 12 μF capacitor?

(a) 18 μC (b) 36 μC (c) 72 μC (d) 120 μC (e) 240 μC

Answers to Practice Quiz:

1. b 2. a 3. a 4. d 5. e 6. e 7. d 8. d 9. a 10. c

CHAPTER 17

Electric Current and Resistance

I. Chapter Objectives

Upon completion of this chapter, you should be able to:

1. summarize the basic features of a battery and explain how a battery produces a direct current in a circuit.
2. define electric current, distinguish between electron flow and conventional current, and explain the concept of drift velocity.
3. define electrical resistance and explain what is meant by an ohmic resistor, summarize the factors that determine resistance, and calculate the effect of these factors in simple situations.
4. define electric power, calculate the power delivery of simple electric circuits, and explain joule heating and its significance.

II. Key Terms

Upon completion of this chapter, you should be able to define and/or explain the following key terms:

electric current	**kilowatt-hour (kWh)**
direct current (dc)	**anode**
ohm (Ω)	**drift velocity**
complete circuit	**temperature coefficient of resistivity**
battery	**superconductivity**
conventional current	**electric power**
resistivity	**electromotive force**
electric current (I)	**joule heat (I^2R) loss**
cathode	**terminal voltage**
ampere (A)	**resistance**
conductivity	**Ohm's law**

The definitions and/or explanations of the most important key terms can be found in the following section: **III. Chapter Summary and Discussion.**

III. Chapter Summary and Discussion

1. Batteries and Direct Current

A **battery** is a device that converts chemical potential energy into electrical energy. The potential difference across the two terminals, **anode** (+) and **cathode** (−), of a battery (or any dc power supply) when not connected to an external circuit is called the **electromotive force** (emf). The **terminal voltage** (operating voltage) across a battery or power supply is the voltage when it is connected to an external circuit. Terminal voltage is less than the emf because of internal resistance.

Electric current is the net rate at which charge flows past a given point. The direction of **conventional current** is that in which positive charge would move. In most materials (e.g., metals), the actual current is carried by electrons moving in the opposite direction. [In a battery circuit, the electrons can only flow in one direction, from negative to positive, the current is called **direct current** (dc).]

2. Current and Drift Velocity

A battery or some other voltage source connected to a continuous conducting path forms a **complete circuit**. In general, we say **electric current** is a flow of charge. Quantitatively, the electric current (I) in a wire is defined as the time rate of flow of net charge. If a net charge q passes a given point in a wire in time t at a constant rate of flow, the current is given by $I = \frac{q}{t}$. The SI unit of current is coulomb/s (C/s) or ampere (A).

Example 17.1 If 3.0×10^{15} electrons flow through a section of a wire of diameter 2.0 mm in 4.0 s, what is the electric current in the wire?

Solution: Given: $n = 3.0\times10^{15}$ electrons, $t = 4.0$ s, $d = 2.0$ mm.

Find: I.

The charge in 3.0×10^{15} electrons is $q = ne = (3.0\times10^{15}\text{ electrons})(1.60\times10^{-19}\text{ C/electron}) = 4.8\times10^{-4}$ C.

So $I = \frac{q}{t} = \frac{4.8\times10^{-4}\text{ C}}{4.0\text{ s}} = 1.2\times10^{-4}\text{ A} = 0.12$ mA.

The diameter of the wire is not needed in this Example since the current is simply the flow of charge.

The net electron flow is characterized by an average velocity called **drift velocity**, which is much smaller than the random velocities of the electrons themselves. The drift velocity is usually very slow (0.10 cm/s), but the electric field (which is what pushes the charges in the wire) travels at a speed close to the speed of light (on the order of 10^8 m/s). Hence the current starts "instantly" in circuits.

3. Resistance and Ohm's Law

The electrical **resistance** of a circuit element is defined as the potential difference (voltage) across the element divided by the resulting current, $R = \frac{V}{I}$. The SI unit of resistance is volt/ampere (V/A), or ohm (Ω). A resistor that has constant resistance is (at a given temperature) said to obey **Ohm's law**, or to be ohmic. Ohm's law is usually written as $V = IR$.

Example 17.2 A resistor of 20 Ω has a 1.5-mA current through it. What is the voltage on the resistor?

Solution: Given: $R = 20\ \Omega$, $I = 1.5\ \text{mA} = 1.5\times10^{-3}\ \text{A}$.

Find: V.

$V = IR = (1.5\times10^{-3}\ \text{A})(20\ \Omega) = 0.030\ \text{V}$.

The major factors affecting the resistance of a conductor of uniform cross-section are:

(1) the type of material or the intrinsic resistive properties,
(2) its length (L),
(3) its cross-sectional area (A),
(4) its temperature (T).

The resistive properties of a particular material are characterized by its **resistivity** (ρ), where for a conductor of uniform cross-section, $\rho = \frac{RA}{L}$. The SI unit of resistivity is Ω·m. This resistivity is somewhat temperature dependent, and over a small range of temperature change (ΔT), the resistivity changes according to $\rho = \rho_o (1 + \alpha\,\Delta T)$, where ρ_o is some reference resistivity at T_o and α is the **temperature coefficient of resistivity.** The reciprocal of resistivity is called the **conductivity,** $\sigma = \frac{1}{\rho}$. The unit of conductivity is $(\Omega\cdot\text{m})^{-1}$, or siemen (S).

For a conductor of uniform cross section, we may write the resistance as $R = \frac{\rho L}{A}$ and the variation of resistance with temperature as $R = R_o (1 + \alpha\,\Delta T)$. Most metallic materials have positive α's and their resistances increase with temperature. Some semiconductors have negative α's and their resistances decrease with increasing temperature.

Example 17.3 Calculate the current in a piece of 10.0-m long 22-gauge (the radius is 0.321 mm) nichrome wire if it is connected to a potential difference of 12.0 V. Assume the temperature is 20°C.

Solution: Given: $L = 10.0$ m, $r = 0.321$ mm $= 0.321\times10^{-3}$ m, $\rho = 100\times10^{-8}\ \Omega\cdot$m (Table 17.1), $V = 12.0$ V.

Find: I.

To find the current I, we first need to find the resistance R of the wire.

$$R = \frac{\rho L}{A} = \frac{\rho L}{\pi r^2} = \frac{(100\times10^{-8}\ \Omega\cdot\text{m})(10.0\ \text{m})}{\pi(0.321\times10^{-3}\ \text{m})^2} = 30.9\ \Omega.$$

From $V = IR$, we have $I = \frac{V}{R} = \frac{12.0\ \text{V}}{30.9\ \Omega} = 0.388$ A.

Example 17.4 A carbon resistor has a resistance of 18 Ω at a temperature 20°C. What is the resistance if it is heated up to a temperature of 120°C?

Solution: Given: $R_o = 18\ \Omega$, $T_o = 20°\text{C}$, $T = 120°\text{C}$, $\alpha = -5.0\times10^{-4}\ \text{C°}^{-1}$.

Find: R.

$$R = R_o(1 + \alpha\Delta T) = (18\ \Omega)[1 + (-5.0\times10^{-4}\ \text{C°}^{-1})(120°\text{C} - 20°\text{C})] = 17\ \Omega.$$

The resistance of carbon decreases with increasing temperature. When carbon is used with materials with positive temperature coefficient, we can make resistors which does not vary with temperature (why?).

4. Electric Power

Electric power is the time rate at which work is done or energy is transferred. The power delivered to a circuit element depends on its resistance and the potential difference across it. The SI unit of power is watt (W).

Power can be expressed as $P = IV = \frac{V^2}{R} = I^2 R.$

The thermal energy expended in a current-carrying resistor is sometimes referred to as **joule heat**, or $\boldsymbol{I^2 R}$ ("I squared R") **losses.**

Example 17.5 What is the nominal resistance of a 100-W household light bulb? The nominal line voltage of household electricity is 120 V.

Solution: Given: $P = 100$ W, $V = 120$ V.

Find: R.

From $P = \frac{V^2}{R}$, we have $R = \frac{V^2}{P} = \frac{(120\ \text{V})^2}{100\ \text{W}} = 144\ \Omega.$

Example 17.6 A 100-W driveway light bulb is on for 10 hours every day. Assuming the power company charges 10 cents for each kilowatt-hour of electricity used, estimate the annual cost to operate the bulb.

Solution: Given: $P = 100$ W, $t = 10$ h every day, cost = \$0.10 per kWh.

Find: total annual cost.

In one year, the light bulb is on for $t = (10 \text{ h/d})(365 \text{ d}) = 3.65\times10^3$ h.

The total electrical energy expended is

$E = Pt = (100 \text{ W})(3.65\times10^3 \text{ h}) = (0.10 \text{ kW})(3.65\times10^3 \text{ h}) = 3.65\times10^2$ kWh.

So the total cost is (\$0.10/kWh)(3.65×10^2 kWh) = \$36.50.

IV. Mathematical Summary

Electric Current	$I = \frac{q}{t}$ (17.1)	Defines electric current in terms of charge flow
Electrical Resistance	$R = \frac{V}{I}$ (17.2)	Defines electrical resistance (Ohm's law)
Resistivity	$\rho = \frac{RA}{L}$ (17.3)	Defines the resistivity of a material
Conductivity	$\sigma = \frac{1}{\rho}$ (17.4)	Defines the conductivity of a material in terms of its resistivity
Temperature Coefficient of Resistivity	$\rho = \rho_o(1 + \alpha\Delta T)$ (17.6) or $\Delta\rho = \rho_o\,\alpha\Delta T$ (where $\Delta\rho = \rho - \rho_o$)	Expresses the resistivity of a material as a function of temperature
Temperature-dependence of Resistance	$R = R(1 + \alpha\Delta T)$ (17.7) or $\Delta R = R_o\,\alpha\Delta T$ (where $\Delta R = R - R_o$)	Expresses the resistance of a material as a function of temperature
Electric Power	$P = IV = \frac{V^2}{R} = I^2R$ (17.8)	Computes the electric power delivery to a resistor

V. Solutions of Selected Exercises and Paired Exercises

10. (a) $\varepsilon = 6(1.5 \text{ V}) = 9.0$ V.

(b) 1.5 V.

16. (a) $q = It = (0.50\times10^{-3}\text{ A})(600\text{ s}) = 0.30\text{ C}.$

(b) Energy $=\Delta U = q\,\Delta V = (0.30\text{ C})(3.0\text{ V}) = 0.90\text{ J}.$

22. (a) $q = It = (9.5\times10^{-6}\text{ A})(1\text{ s}) = 9.5\times10^{-6}\text{ C} = (9.5\times10^{-6}\text{ C})\times\dfrac{1\,e}{1.6\times10^{-19}\text{ C}} = 5.9\times10^{13}$ protons.

So the answer is 5.9×10^{13} protons/s.

(b) $P = \dfrac{E}{t} = \dfrac{(5.9\times10^{13}\text{ protons})(20\times10^{6}\text{ eV/proton})(1.6\times10^{-19}\text{ J/eV})}{1\text{ s}} = 1.9\times10^{2}\text{ J/s}.$

32. $I = \dfrac{V}{R} = \dfrac{12\text{ V}}{150\ \Omega} = 8.0\times10^{-2}\text{ A}.$

42. $\rho = \rho_o(1 + \alpha\Delta T) = \rho_o[1 + (6.80\times10^{-3}\text{ C°}^{-1})(80\text{ C°})] = 1.544\,\rho_o.$

So the percentage variation is $\dfrac{\rho - \rho_o}{\rho_o} = \dfrac{\rho}{\rho_o} - 1 = 0.544 = 54\%$ increase.

50. At room temperature, $I_1 = \dfrac{V_1}{R_1} = \dfrac{12\text{ V}}{150\ \Omega} = 0.080\text{ A}.$

At 90°C, $R = R_o(1 + \alpha\Delta T) = (150\ \Omega)[1 + (-5.0\times10^{-4}\text{ C°}^{-1})(70\text{ C°})] = 144.8\ \Omega;$

$I_2 = \dfrac{12\text{ V}}{144.8\ \Omega} = 0.0829\text{ A}.$ So $\Delta I = 0.0829\text{ A} - 0.080\text{ A} = 0.0029\text{ A}.$

Therefore, the answer is 2.9 mA more at 90°C.

60. $P = \dfrac{V^2}{R} = \dfrac{(120\text{ V})^2}{10\times10^{3}\ \Omega} = 1.4\text{ W}.$

76. (a) $I = \dfrac{P}{V} = \dfrac{5.5\times10^{3}\text{ W}}{240\text{ V}} = 23\text{ A} > 20\text{ A}.$ So it should have a 30-A circuit breaker.

(b) The heat (energy) required is $Q = cm\,\Delta T = (4190\text{ J/kg·C°})(55\text{ gal})(3.785\text{ kg/gal})(60\text{ C°}) = 5.23\times10^{7}\text{ J}.$

The energy input is $E = \dfrac{5.23\times10^{7}\text{ J}}{0.85} = 6.16\times10^{7}\text{ J}.$

$t = \dfrac{E}{P} = \dfrac{6.16\times10^{7}\text{ J}}{5.5\times10^{3}\text{ W}} = 1.12\times10^{4}\text{ s} = 3.1\text{ h}.$

84. (a) From $R = \dfrac{V^2}{R} = \dfrac{\rho L}{A}$, we have $L = \dfrac{V^2 A}{P\rho} = \dfrac{(120\text{ V})^2(\pi)(2.0\times10^{-3}\text{ m})^2}{(500\text{ W})(100\times10^{-8}\ \Omega\text{·m})} = 3.6\times10^{2}\text{ m}.$

(b) Since $\dfrac{P_2}{P_1} = \dfrac{V_2^2}{V_1^2}$, $P_2 = \dfrac{V_2^2}{V_1^2}P_1 = \dfrac{(105\text{ V})^2}{(120\text{ V})^2}(500\text{ W}) = 3.8\times10^{2}\text{ W}.$

92. The volume of the wire is $V = \dfrac{m}{\rho} = AL,$

so $A = \dfrac{m}{\text{density} \cdot L} = \dfrac{0.010\text{ kg}}{(2.7\times10^3\text{ kg/m}^3)(0.50\text{ m})} = 7.41\times10^{-6}\text{ m}^2.$

$$I = \frac{V}{R} = \frac{V}{\rho L/A} = \frac{VA}{\rho L} = \frac{(10\text{ V})(7.41\times10^{-6}\text{ m}^2)}{(2.82\times10^{-8}\ \Omega\cdot\text{m})(0.50\text{ m})} = 5.3\times10^3\text{ A}.$$

103. (a) $R = \dfrac{V}{I} = \dfrac{40\text{ V}}{0.10\text{ A}} = 4.0\times10^2\ \Omega.$

(b) $P = VI = (40\text{ V})(0.10\text{ A}) = 4.0\text{ W}.$

(c) $E = Pt = (4.0\text{ W})(2.0)(60\text{ s}) = 4.8\times10^2\text{ J}.$

107. Since $P = VI$, a higher voltage lowers the current if P is kept constant. The resistance of the transmission lines is a fixed number. With lower current, the power loss on the lines is reduced.

$P_{\text{loss}} = I^2 R = \dfrac{P^2}{V^2} R \propto \dfrac{1}{V^2}$. So if V is raised by 10 times, P_{loss} will be reduced by a factor of 1/100.

VI. Practice Quiz

1. What is the voltage across a 5.0 Ω resistor if the current through it is 5.0 A?

(a) 100 V (b) 25 V (c) 4 V (d) 1 V (e) 0

2. If a 75 W light bulb operates at a voltage of 120 V, what is the current in the bulb?

(a) 1.1×10^{-4} A (b) 1.6 A (c) 9.0×10^3 A (d) 0.63 A (e) 9.0×10^3 A

3. A wire carries a steady current of 0.10 A over a period of 20 s. What total charge passes through the wire in this time interval?

(a) 200 C (b) 20 C (c) 2.0 C (d) 0.20 C (e) 0.005 C

4. A nichrome wire has a radius of 0.50 mm and a resistivity of $100\times10^{-8}\ \Omega\cdot\text{m}$. If the wire carries a current 0.50 A, what is the voltage across 1.0 m of the wire?

(a) 0.003 V (b) 0.32 V (c) 0.64 V (d) 1.6 V (e) 1.9 V

5. The length of a wire is doubled and the radius is doubled. By what factor does the resistance change?

(a) 4 times as large (b) twice as large (c) unchanged (d) half as large (e) quarter as large

6. During a power demand, the voltage output is reduced by 5.0%. By what percent is the power of a resistor affected?

(a) 2.5% less (b) 5.0% less (c) 10% less (d) 90% less (e) 95% less

7. A 12-V battery is connected to a 100-Ω resistor. How many electrons flow through the wire in 1.0 min?

(a) 1.5×10^{19} (b) 2.5×10^{19} (c) 3.5×10^{19} (d) 4.5×10^{19} (5.5×10^{19}

8. A 1500-W heater is connected to a 120-V source for 2.0 h. How much heat energy is produced?

(a) 1.5×10^{3} J (b) 3.0×10^{3} J (c) 1.1×10^{7} J (d) 1.8×10^{5} J (d) 9.0×10^{4} J

9. A 400-W computer (computer plus monitor) is turned on 8.0 hours per day. If electricity costs $0.10 per kWh, how much does it cost to run the computer annually?

(a) $116.80 (b) $1168.00 (c) $14.60 (d) $146.00 (e) $1460.00

10. A platinum wire is used to determine the melting point of indium. The resistance of the platinum wire is 2.000 Ω at 20°C and increases to 3.072 Ω as the indium starts to melt. What is the melting temperature of indium?

(a) 117°C (b) 137°C (c) 157°C (d) 351°C (e) 731°C

Answers to Practice Quiz:

1. b 2. d 3. c 4. c 5. d 6. c 7. d 8. c 9. a 10. c

CHAPTER 18

Basic Electric Circuits

I. Chapter Objectives

Upon completion of this chapter, you should be able to:

1. determine the equivalent resistance of resistors in series, parallel, and series-parallel combinations, and use equivalent resistance to solve simple circuits.
2. understand the physical principles that underlie Kirchhoff's circuit rules and apply these rules in the analyses of actual circuits.
3. understand the charging and discharging of a capacitor through a resistor and calculate the current and voltage at specific times during these processes.
4. understand how galvanometers are used as ammeters and voltmeters, how multirange versions of these devices are constructed, and how they are connected to measure current and voltage in real circuits.
5. understand how circuits are wired and the underlying principles that govern safety devices.

II. Key Terms

Upon completion of this chapter, you should be able to define and/or explain the following key terms:

equivalent series resistance	**ammeter**
equivalent parallel resistance	**voltmeter**
junction (node)	**galvanometer**
branch	**fuse**
Kirchhoff's first rule (junction rule)	**circuit breaker**
Kirchhoff's second rule (loop rule)	**grounded plug**
RC circuit	**polarized plug**
time constant	

The definitions and/or explanations of the most important key terms can be found in the following section: **III. Chapter Summary and Discussion.**

III. Chapter Summary and Discussion

1. Resistors in Series, Parallel, and Series-Parallel Combinations

When resistors are connected in *series* in a circuit, the **equivalent series resistance** (R_s) is given by $R_s = R_1 + R_2 + R_3 + \ldots = \Sigma_i R_i$, or the equivalent series resistance is simply the sum of the individual resistances. (Note: The equivalent series resistance is larger than that of the largest resistor in the series combination.) In a series circuit, the current is the same through all resistors, $I = I_1 = I_2 = I_3 = \ldots$, and the total voltage is the sum of the voltages of the individual resistors, that is, $V = V_1 + V_2 + V_3 + \ldots = \Sigma_i V_i = \Sigma_i I R_i$.

When resistors are connected in *parallel* in a circuit, the **equivalent parallel resistance** (R_p) is given by $\frac{1}{R_p} = \frac{1}{R_1} + \frac{1}{R_2} + \frac{1}{R_3} + \ldots = \Sigma_i \frac{1}{R_i}$, or the reciprocal of the equivalent parallel resistance is equal to the sum of the reciprocals of individual resistances. (Note:The equivalent parallel resistance is less than that of the smallest resistor in the parallel combination.) In a parallel circuit, the voltage is the same across all resistors, $V = V_1 = V_2 = V_3 = \ldots$, and the total current is the sum of the currents of the individual resistors, that is, $I = I_1 + I_2 + I_3 + \ldots = \Sigma_i I_i$. If there are only two resistors in parallel, $\frac{1}{R_p} = \frac{1}{R_1} + \frac{1}{R_2} = \frac{R_1 + R_2}{R_1 R_2}$, so $R_p = \frac{R_1 R_2}{R_1 + R_2}$. If the two resistors have equal resistance, $R_1 = R_2 = R$, then $R_p = \frac{R\,R}{R + R} = \frac{R}{2}$.

Note: The equivalent series capacitance is given by $\frac{1}{C_s} = \frac{1}{C_1} + \frac{1}{C_2} + \frac{1}{C_3} + \ldots = \Sigma_i \frac{1}{C_i}$; the equivalent parallel capacitance is equal to $C_p = C_1 + C_2 + C_3 + \ldots = \Sigma_i C_i$ (Chapter 16). The results for resistors are completely different.

Example 18.1 In the circuit shown, find:

(a) the equivalent resistance between points A and B,

(b) the current through the battery.

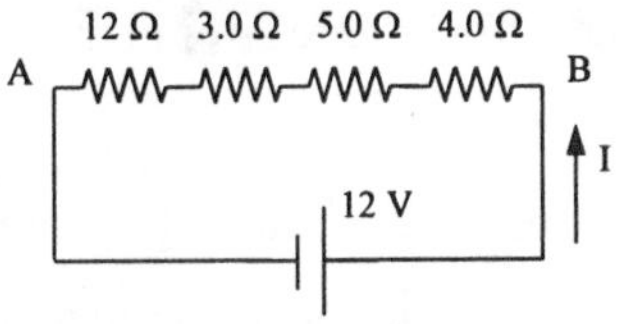

Solution: Given: $R_1 = 12\ \Omega$, $R_2 = 3.0\ \Omega$, $R_3 = 5.0\ \Omega$, $R_4 = 4.0\ \Omega$, $V = 12$ V.

Find: (a) R_s (b) I.

(a) All four resistors are in series combination, so

$R_s = R_1 + R_2 + R_3 + R_4 = 12\ \Omega + 3.0\ \Omega + 5.0\ \Omega + 4.0\ \Omega = 24\ \Omega$.

(b) $I = \frac{V}{R} = \frac{V}{R_s} = \frac{12\ \text{V}}{24\ \Omega} = 0.50$ A.

Example 18.2 In the circuit shown, find:

(a) the equivalent resistance between points A and B,

(b) the current through the battery.

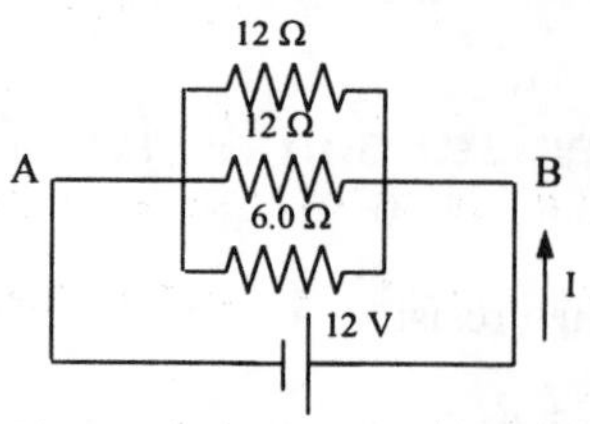

Solution: Given: $R_1 = 12\ \Omega$, $R_2 = 12\ \Omega$, $R_3 = 6.0\ \Omega$, $V = 12$ V.

Find: (a) R_p (b) I,

(a) All three resistors are in parallel combination.

$$\frac{1}{R_p} = \frac{1}{R_1} + \frac{1}{R_2} + \frac{1}{R_3} = \frac{1}{12\ \Omega} + \frac{1}{12\ \Omega} + \frac{1}{6.0\ \Omega} = \frac{1}{12\ \Omega\ 12\ \Omega} + \frac{1}{\ } \frac{2}{12\ \Omega} = \frac{4}{12\ \Omega} = \frac{1}{3.0\ \Omega}.$$

So $R_p = 3.0\ \Omega$.

Alternate method: Using $R_p = \dfrac{RR}{R+R} = \dfrac{R}{2}$ for the two equal resistors in parallel, we obtain the equivalent parallel resistance for the two 12 Ω resistors. $R_{p1} = \dfrac{12\ \Omega}{2} = 6.0\ \Omega$.

Now the two 6.0 Ω resistances are in parallel and the equivalent resistance is $R_p = \dfrac{6.0\ \Omega}{2} = 3.0\ \Omega$.

(b) From Ohm's law, $I = \dfrac{V}{R} = \dfrac{V}{R_p} = \dfrac{12\ \text{V}}{3.0\ \Omega} = 4.0$ A.

Resistors in a circuit may be connected in a variety of series-parallel combinations. The general procedure for analyzing circuits with different series-parallel combinations of resistors is to find the voltage across and the current through the various resistors as follows:

(1) Start from the resistor combination farthest from the voltage source, find the equivalent series and parallel resistances.

(2) Reduce the circuit until there is a single loop with one total equivalent resistance.

(3) Find the current delivered to the reduced circuit using $I = V/R$, where R is the total equivalent resistance.

(4) Expand the reduced circuit to find the currents and voltages for the resistors in each step.

Example 18.3 Find the equivalent resistance between points

(a) A and B,

(b) A and C,

(c) B and C.

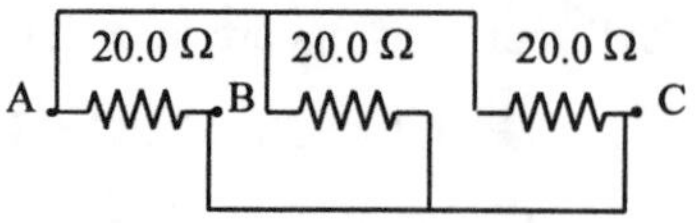

Solution: Given: $R_1 = R_2 = R_3 = R = 20.0\ \Omega$.

Find: (a) R_{AB} (b) R_{AC} (c) R_{BC}.

(a) All three resistors are in parallel between points A and B, that is, the ends of all three resistors are across points A and B. We redraw the circuit.

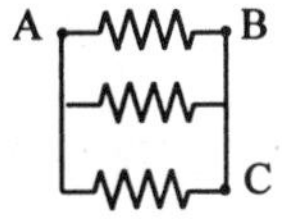

$$\frac{1}{R_{AB}} = \frac{1}{R_1} + \frac{1}{R_2} + \frac{1}{R_3} = \frac{3}{R},$$

or $$R_{AB} = \frac{R}{3} = \frac{20.0\ \Omega}{3} = 6.67\ \Omega.$$

(b) Since point C and point B are connected (they are the same point), $R_{AB} = R_{AC} = 6.67\ \Omega$.

(c) Again since points B and C are connected, there is no resistance between them (short circuit).

So $R_{BC} = 0$.

Example 18.4 In the circuit shown, find:

(a) the equivalent resistance between points A and B,

(b) the current through the battery,

(c) the current though the 4.0 Ω resistor.

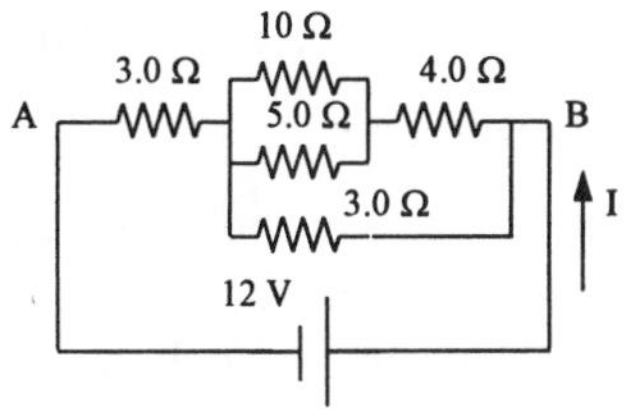

Solution: Here we have a variety of series-parallel combinations. We follow the general procedures outlined in the text and above.

(a) The 10 Ω and the 5.0 Ω are in parallel.

$$R_{p1} = \frac{(10\ \Omega)(5.0\ \Omega)}{10\ \Omega + 5.0\ \Omega} = 3.33\ \Omega.$$

The circuit reduces to Figure 1.

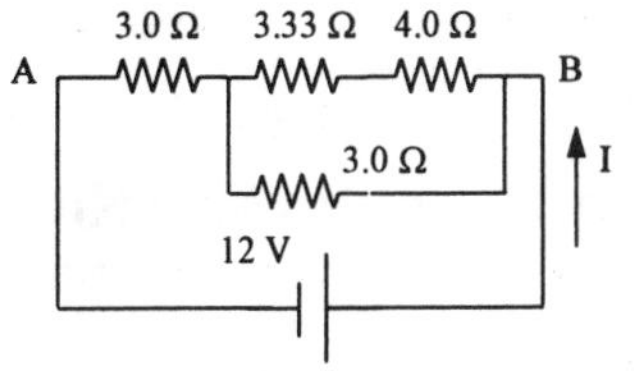

Figure 1

Now the 3.33 Ω and the 4.0 Ω are in series.

$R_{s1} = 3.33\ \Omega + 4.0\ \Omega = 7.33\ \Omega$.

The circuit reduces to Figure 2.

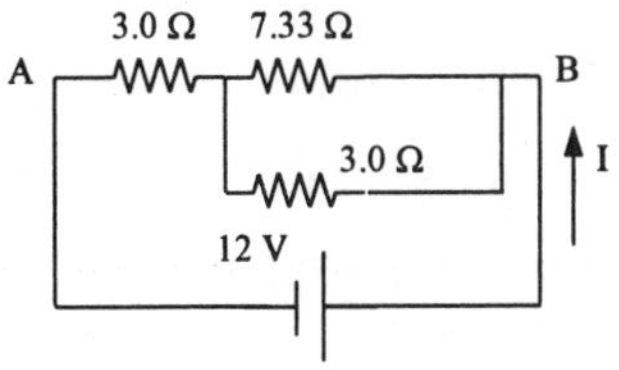

Figure 2

The 7.33 Ω and the 3.0 Ω are in parallel.

$$R_{p2} = \frac{(7.33\ \Omega)(3.0\ \Omega)}{7.33\ \Omega + 3.0\ \Omega} = 2.13\ \Omega.$$

The circuit reduces to Figure 3.

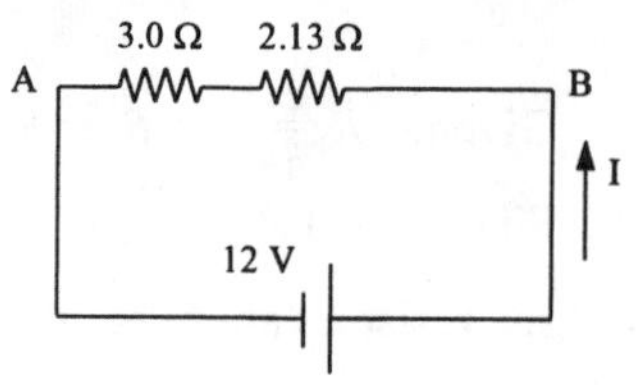

Figure 3

Finally, the 2.13 Ω and the 3.0 Ω are in series.

$R = R_{s2} = 2.13\ \Omega + 3.0\ \Omega = 5.13\ \Omega = 5.1\ \Omega$ (2 significant figures).

The circuit reduces to Figure 4.

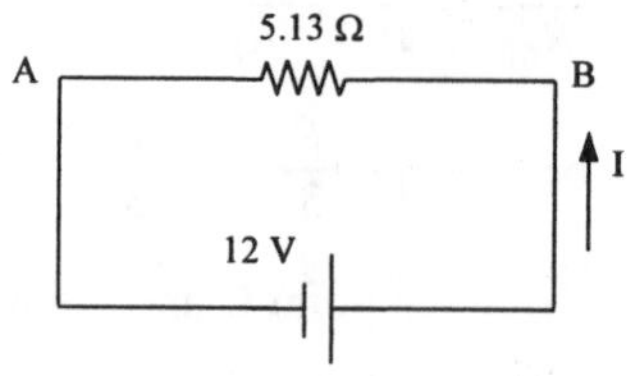

Figure 4

(b) From Ohm's law, $I = \frac{V}{R} = \frac{12\ \text{V}}{5.13\ \Omega} = 2.33\ \text{A} = 2.3\ \text{A}$ (2 significant figures).

(c) To find the current through the 4,.0 Ω resistor, we need to expand the combinations.

In Figure 3, the current through the 2.13 Ω and 3.0 Ω is the same as the total current, 2.33 A.

The voltage across the 2.13 Ω is then $V_{2.13} = (2.13\ \Omega)(2.33\ \text{A}) = 4.96\ \text{V}$.

In Figure 2, the voltage across the 7.33 Ω and 3.0 Ω is the same as that across the 2.13 Ω, 4.96 V.

So the current though the 7.33 Ω is $I_{7.33} = \frac{4.96\ \text{V}}{7.33\ \Omega} = 0.677\ \text{A}$.

In Figure 1, the current though the 3.33 Ω and the 4.0 Ω is the same as the current though the 7.33 Ω.

Therefore $I_{4.0} = 0.68\ \text{A}$ (2 significant figures).

In this Example, an extra digit is carried in intermediate results.

2. Multiloop Circuits and Kirchhoff's Rules

In a *multiloop circuit*, a **junction** or **node** is a point at which three or more connecting wires are joined together. A circuit path between two junctions is called a **branch** and may contain one or more circuit elements.

Kirchhoff's first rule, or **junction rule**, states that the algebraic sum of the currents at any junction is zero, $\Sigma_i I_i = 0$. This simply means that the sum of the currents going into a junction (taken as positive) is equal to the sum of the currents leaving the junction (taken as negative).

Note: You have to *assume* current directions at a particular junction because you generally cannot tell whether a particular current is directed into or out of a junction simply by looking at a multiloop circuit diagram. If these assumptions are wrong, you will find out later from the opposite sign in your mathematical results. A negative current will indicate the wrong choice of direction.

Kirchhoff's second rule, or **loop rule**, states that the algebraic sum of the potential differences (voltages) across all elements of any *closed* loop is zero, $\Sigma_i V_i = 0$. This means that the sum of the voltages rises equal to the sum of the voltage drops across the voltage sources and resistors around a closed loop, which must be true if energy is conserved. Voltage drops may be positive or negative, and we will use the following convention.

Sign convention for voltages across circuit elements in traversing a loop:

$+V$: when a battery is traversed from its negative terminal to positive terminal (voltage rises).

$-V$: when a battery is traversed from its positive terminal to negative terminal (voltage drops).

$-IR$: when a resistor is traversed in the direction of the assigned branch current, (voltage drop).

$+IR$: when a resistor is traversed in the direction opposite to the assigned branch current, (voltage rise).

The general steps in applying Kirchhoff's rules are as follows:

(1) Assign a current and current direction for each branch in the circuit. This is done most conveniently at junctions.

(2) Indicate the loops are the arbitrarily chosen directions in which they are to be traversed. Every branch *must* be in at least one loop.

(3) Apply Kirchhoff's first rule and write equations for the currents, one for each junction that gives different equation. (In general, this gives a set of equations that includes all branch currents.)

(4) Traverse the number of loops necessary to include all branches. In traversing a loop, apply Kirchhoff's second rule and write equations using the adopted sign convention.

(5) Solve the simultaneous equations for the currents.

These steps may seem complicated, but really straightforward, as the following Examples show.

Example 18.5 Find the currents in the branches of the circuit shown.

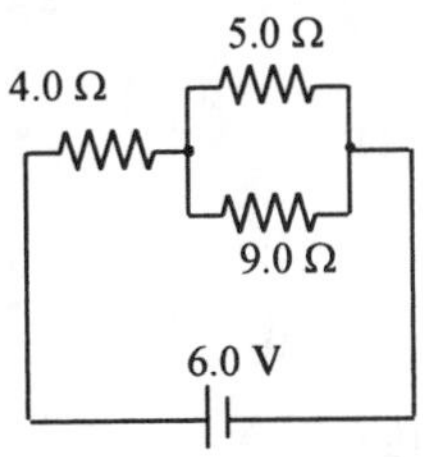

Solution: This is a simple circuit and can be solved with resistor combinations and Ohm's law. We use Kirchoff's rules here.

First, we assign junctions and currents.

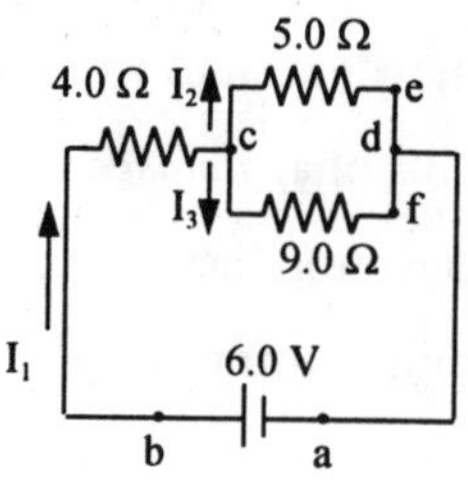

From the junction rule:

for junction c, $I_1 = I_2 + I_3$. Eq. (1)

From the loop rule:

for loop abceda, $+6.0\ \text{V} - I_1\,(4.0\ \Omega) - I_2\,(5.0\ \Omega) = 0,$

or $4.0I_1 + 5.0I_2 = 6.0;$ Eq. (2)

for loop abcfda, $+6.0\ \text{V} - I_1\,(4.0\ \Omega) - I_3\,(9.0\ \Omega) = 0,$

or $4.0I_1 + 9.0I_3 = 6.0.$ Eq. (3)

From Eq. (1), $I_3 = I_1 - I_2$. Substituting this into Eq. (3) to eliminate I_3 gives

$13I_1 - 9.0I_2 = 6.0.$ Eq. (4)

From Eq. (2), $I_1 = 1.5 - 1.25I_2$. Substituting this into Eq. (4) to eliminate I_1 yields $25.25I_2 = 13.5$.

Therefore, $I_2 = 0.535\ \text{A}$, $I_1 = 1.5 - 1.25I_2 = 1.5\ \text{A} - (1.25)(0.535\ \text{A}) = 0.832\ \text{A}$,

and $I_3 = I_1 - I_2 = 0.832\ \text{A} - 0.535\ \text{A} = 0.297\ \text{A}$.

Hence the currents are $I_1 = 0.53$ A, $I_2 = 0.83$ A, and $I_3 = 0.30$ A (two significant figures).

Example 18.6 Find the currents in the branches of the circuit shown.

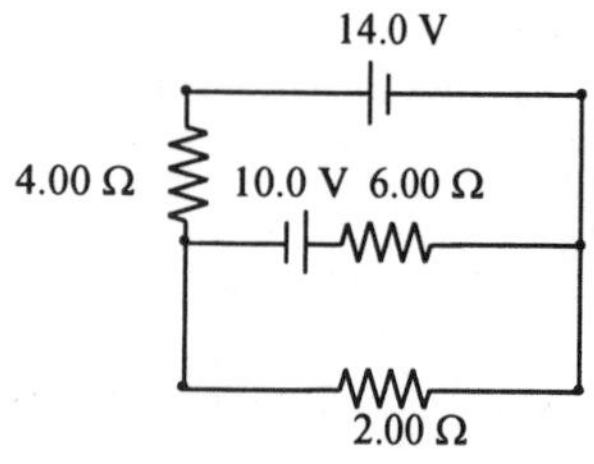

Solution: This is a circuit you cannot solve with resistor combinations and Ohm's law;,Kirchhoff's rules must be used.

First, we assign junctions and currents.

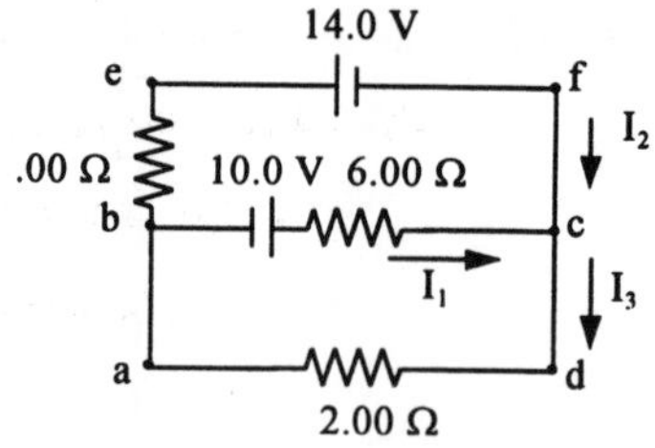

From the junction rule:

for junction c, $I_1 + I_2 = I_3$. Eq. (1)

From the junction rule:

for loop bcfeb, $+10.0\ \text{V} - (6.00\ \Omega)I_1 + 14.0\ \text{V} + (4.00\ \Omega)I_2 = 0,$

or $(6.00)I_1 - (4.00)I_2 = 24.0\ \text{V};$ Eq. (2)

for loop bcdab, $+10.0\ \text{V} - (6.00\ \Omega)I_1 - (2.00\ \Omega)I_3 = 0,$

or $(6.00)I_1 + (2.00)I_3 = 10.0\ \text{V}.$ Eq. (3)

Substituting Eq. (1) into Eq. (3) to eliminate I_3 gives $4.00I_1 + 2.00I_2 = 5.00.$ Eq. (4)

From Eq. (2), $I_2 = 1.50I_1 - 6.00$. Substituting this into Eq. (4) to eliminate I_2 yields $7.00I_1 = 17.0.$

Therefore, $I_1 = 2.43$ A, $I_2 = 1.50I_1 - 6.00 \text{ A} = 1.50(2.43 \text{ A}) - 6.00 \text{ A} = -2.36$ A,

and $I_3 = I_1 + I_2 = 2.43 \text{ A} + (-2.36 \text{ A}) = 0.0700$ A.

I_2 is negative, meaning our assumption (downward) is wrong. It should be upward.

Hence the currents are $I_1 = 2.43$ A, $I_2 = 2.36$ A, and $I_3 = 0.0700$ A (3 significant figures).

3. RC Circuits

An **RC circuit** consists of a resistor and a capacitor connected in series. The **time constant** ($\tau = RC$) for an RC circuit is a "characteristic time" by which we measure the capacitor's charge and discharge. In one time constant, a uncharged capacitor charges to 63% of its maximum value when charging and a fully charged capacitor discharges 63% of its charge when discharging (or decreases to 37% of its original value). In charging, the voltage (charge) on the capacitor increases; in discharging, the voltage (charge) on the capacitor decreases. The current in the circuit always decreases with time regardless of charging or discharging. (Why?) The voltages and currents as functions of time in a capacitor are given by

$V_C = V_o\left(1 - e^{-t/RC}\right)$ voltage across capacitor when charging

$V_C = V_o e^{-t/RC}$ voltage across capacitor when discharging

$I = I_o e^{-t/RC}$ current in the circuit when charging or discharging,

where V_o and I_o are the initial voltage across and current in the circuit, and $\tau = RC$ is the time constant of the circuit.

Example 18.7 In the circuit shown, $C = 5.0\text{-}\mu\text{F}$, $R = 4.0$ MΩ, and $V = 20$ V. The switch S is closed when the capacitor is initially uncharged.

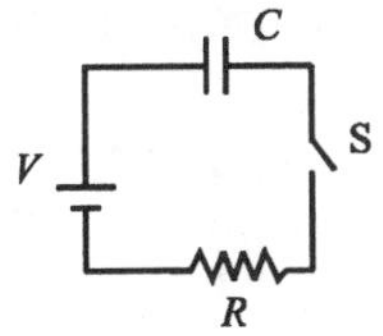

(a) Immediately after the switch is closed, find

(1) the voltage across the capacitor,

(2) the current through the resistor,

(b) What is the current in the circuit after 20 s?

(c) After the switch has been closed for a long period of time, find

(1) the current though the resistor,

(2) the charge and energy stored in the capacitor.

(d) What is the voltage across the capacitor 20 s after the switch opens once it is fully charged?

Solution: Given: $R = 4.0 \text{ M}\Omega = 4.0\times10^{6}\ \Omega$, $C = 5.0\ \mu\text{F} = 5.0\times10^{-6}$ F, $V = 20$ V.

(a) (1) Right after the switch is closed, the capacitor is not charged, so the voltage across the capacitor is zero ($V_C = 0$). Therefore, the voltage across the resistor is $V_R = V = 20$ V.

(2) $I_o = \dfrac{V}{R} = \dfrac{20\text{ V}}{4.0\times10^6\ \Omega} = 5.0\times10^{-6}\text{ A}.$

(b) The time constant $\tau = RC = (4.0\times10^6\ \Omega)(5.0\times10^{-6}\text{ F}) = 20\text{ s},$ so $\dfrac{t}{RC} = \dfrac{20\text{ s}}{20\text{ s}} = 1.$

So $I = I_o\, e^{-t/RC} = (5.0\times10^{-6}\text{ A})\, e^{-1} = 1.8\times10^{-7}\text{ A}.$ $(e^{-1} \approx 0.37)$

(c) (1) After a long period of time, the capacitor is fully charged and the voltage across it is $V_C = V = 20$ V, so the voltage across the resistor is zero. Therefore the current through the resistor is zero.

(2) The charge on the capacitor is $Q = CV_C = (5.0\times10^{-6}\text{ F})(20\text{ V}) = 1.0\times10^{-4}\text{ C}.$

The energy stored in the capacitor is $U = \frac{1}{2}\, CV^2 = \frac{1}{2}(5.0\times10^{-6}\text{ F})(20\text{ V})^2 = 1.0\times10^{-3}\text{ J}.$

(d) When the capacitor is fully charged, the initial voltage across it is $V_o = 20$ V. Now the capacitor starts discharging. The time constant of the circuit is still $\tau = RC = 20$ s.

$\dfrac{t}{RC} = \dfrac{20\text{ s}}{20\text{ s}} = 1.$ The voltage across the capacitor is $V_C = V_o\, e^{-t/RC} = (20\text{ V})\, e^{-1} = 7.4\text{ V}.$

4. Ammeters and Voltmeters

A **galvanometer** is a sensitive meter with an internal resistance r and whose needle deflection is proportional to the current though it. It allows a very small maximum current I_g to flow though it without burning the coil in the galvanometer.

An **ammeter** is a low resistance device to measure current; it consists of a galvanometer and a shunt resistor in parallel to it. An ammeter *must* be connected in series to the resistor whose current you want to measure. Since the ammeter has a low resistance, it does not appreciably affect the current measurement in the circuit. A shunt resistor (R_s) of proper value needs to be selected in constructing an ammeter. The equation which allows us to calculate R_s is $I_g = \dfrac{I R_s}{r + R_s}$, where I_g is the maximum current though the galvanometer, r the internal resistance of the galvanometer, R_s the shunt resistance, and I the current though the ammeter.

A **voltmeter** is a high resistance device used to measure voltage. It consists of a galvanometer and a multiplier resistor in series with it. A voltmeter *must* be connected in parallel with a circuit element whose voltage we wish to measure. Since the voltmeter has a high resistance, it does not affect the voltage across an element. A multiplier resistor of proper value must be selected in constructing a voltmeter. The equation which allows us to calculate R_m is $I_g = \dfrac{V}{r + R_m}$, where I_g is the maximum current though the galvanometer, r the internal resistance of the galvanometer, R_m the multiplier resistance, and V the voltage of the voltmeter.

Example 18.8 A galvanometer has an internal resistance 35 Ω and can safely carry 250 μA of current.

(a) What shunt resistor is needed to make an ammeter with a range of 0 to 10 A?

(b) What multiplier resistor is needed to make a voltmeter with a range of 0 to 100 V?

Solution: Given: $r = 35\ \Omega$, $I_g = 250\ \mu\text{A} = 250\times10^{-6}$ A, $I = 10$ A, $V = 100$ V.

Find: (a) R_s (b) R_m.

(a) From $I_g = \dfrac{IR_s}{r + R_s}$, $R_s = \dfrac{I_g r}{I - I_g} = \dfrac{(250\times10^{-6}\text{ A})(35\ \Omega)}{10\text{ A} - 250\times10^{-6}\text{ A}} = 8.8\times10^{-4}\ \Omega = 0.88\text{ m}\Omega$.

(b) From $I_g = \dfrac{V}{r + R_m}$, $R_m = \dfrac{V - I_g r}{I_g} = \dfrac{100\text{ V} - (250\times10^{-6}\text{ A})(35\ \Omega)}{250\times10^{-6}\text{ A}} = 4.0\times10^{5}\ \Omega = 400\text{ k}\Omega$.

5. Household Circuits and Electrical Safety

Fuses and **circuit breakers** are safety devices that open or break a circuit when the current exceeds a preset value. If the current in a circuit exceeds a preset (safe) value, this is usually an indication of some problem (short circuit) in the circuit.

A **grounded plug** (three-prong plug) uses a dedicated grounding wire to ground objects that may become conductors and thus dangerous. A **polarized plug** identifies the ground side of the line for use as a grounding safety feature.

The first precaution to take for personal safety is to avoid coming into contact with an electrical conductor that might cause a voltage across a human body or part of it, thus causing a current through the body which could be dangerous.

IV. Mathematical Summary

Equivalent Series Resistance	$R_s = R_1 + R_2 + R_3 + \ldots$ $= \Sigma_i R_i$ (18.2)	Computes the equivalent series resistance of resistors in series combination
Equivalent Parallel Resistance	$\dfrac{1}{R_p} = \dfrac{1}{R_1} + \dfrac{1}{R_2} + \dfrac{1}{R_3} + \ldots$ $= \Sigma_i \dfrac{1}{R_i}$ (18.3)	Computes the equivalent parallel resistance of resistors in parallel combination
Equivalent Parallel Resistance	$R_p = \dfrac{R_1 R_2}{R_1 + R_2}$ (18.3a)	Computes the equivalent parallel resistance of *two* resistors in parallel combination

Kirchhoff's Rules (1) junction rule (2) loop rule	(1) $\Sigma_i I_i = 0$ (2) $\Sigma_i V_i = 0$	(18.4) (18.5)	Applies the conservation of charge and energy in electric circuits
Time Constant for an RC Circuit	$\tau = RC$	(18.8)	Defines the time constant in a resistor-capacitor series circuit
Charging Voltage for an RC Circuit	$V_C = V_o\left(1 - e^{-t/RC}\right)$	(18.6)	Gives the voltage on a charging capacitor as a function of time for an RC circuit
Discharging Voltage for an RC Circuit	$V_C = V_o e^{-t/RC}$	(18.7)	Gives the voltage on a discharging capacitor as a function of time for an RC circuit
Current (Charging and Discharging) in an RC Circuit	$I = I_o e^{-t/RC}$	(18.8)	Gives the current as a function of time for an RC circuit (charging and discharging)
Ammeter Current	$I_g = \dfrac{I R_s}{r + R_s}$	(18.10)	Expresses the galvanometer current in terms of the shunt resistance in an ammeter
Voltmeter Current	$I_g = \dfrac{V}{r + R_m}$	(18.11)	Expresses the galvanometer current in terms of the multiplier resistance in a voltmeter

V. Solutions of Selected Exercises and Paired Exercises

9. Parallel: $\dfrac{1}{R_p} = \dfrac{1}{R_1} + \dfrac{1}{R_2} + \dfrac{1}{R_3} = 3\dfrac{1}{4.0\ \Omega} = \dfrac{3}{4.0\ \Omega}$,

so $R_p = 1.3\ \Omega$ for parallel.

Series: $R_s = R_1 + R_2 + R_3 = 3(4.0\ \Omega) = 12\ \Omega$ for series.

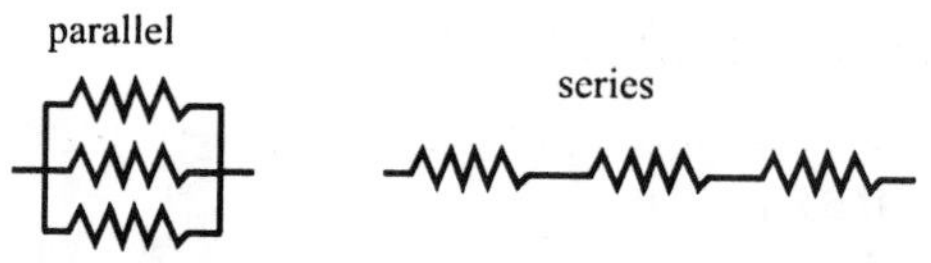

Two in series-parallel: $\dfrac{1}{R_p} = \dfrac{1}{R_s} + \dfrac{1}{R} = \dfrac{1}{R+R} + \dfrac{1}{R} = \dfrac{3}{2(4.0\ \Omega)}$,

so $R_p = 2.7\ \Omega$ for series-parallel.

Two in parallel-series: $R_s = R + R_p = R + \dfrac{R\,R}{R+R} = \dfrac{3R}{2} = \dfrac{3(4.0\ \Omega)}{2} = 6.0\ \Omega$ for parallel-series.

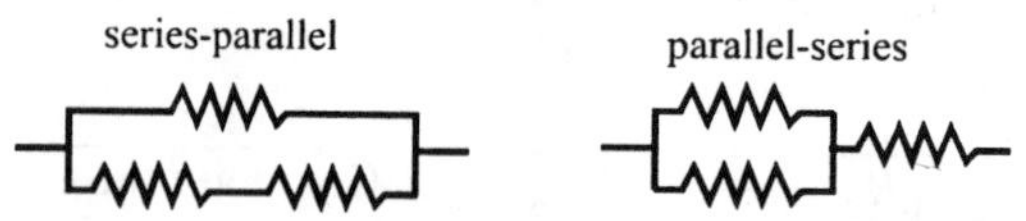

18. R_1 and R_2 are in series: $R_s = 2.0\ \Omega + 2.0\ \Omega = 4.0\ \Omega$.

R_s, R_3, and R_4 are in parallel: $\dfrac{1}{R_p} = \dfrac{1}{4.0\ \Omega} + \dfrac{1}{2.0\ \Omega} + \dfrac{1}{2.0\ \Omega} = \dfrac{5}{4.0\ \Omega}$,

so $R_p = \dfrac{4.0\ \Omega}{5} = 0.80\ \Omega$.

23. (a) Parallel consumes maximum power:

parallel

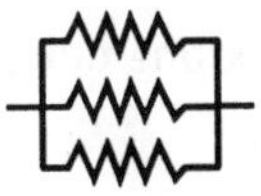

$\dfrac{1}{R_p} = 3\dfrac{1}{50\ \Omega} = \dfrac{3}{50\ \Omega}$, so $R_p = \dfrac{50\ \Omega}{3} = 16.7\ \Omega$.

So $P_{\text{max}} = \dfrac{V^2}{R_p} = \dfrac{(120\ \text{V})^2}{16.7\ \Omega} = 860\ \text{W}$.

series

(b) Series consumes minimum power:

$R_s = 3(50\ \Omega) = 150\ \Omega$. So $P_{\text{min}} = \dfrac{(120\ \text{V})^2}{150\ \Omega} = 96\ \text{W}$.

30. (a) $I_1 = \dfrac{V}{R_1} = \dfrac{6.0\ \text{V}}{6.0\ \Omega} = 1.0\ \text{A}$

R_2 and R_3 are in series. $R_s = 4.0\ \Omega + 6.0\ \Omega = 10\ \Omega$.

So $I_2 = I_3 = \dfrac{6.0\ \text{V}}{10\ \Omega} = 0.60\ \text{A}$, $I_4 = \dfrac{6.0\ \text{V}}{10\ \Omega} = 0.60\ \text{A}$.

(b) $P_1 = I_1^2 R_1 = (1.0\ \text{A})^2 (6.0\ \Omega) = 6.0\ \text{W}$, $P_2 = (0.60\ \text{A})^2 (4.0\ \Omega) = 1.4\ \text{W}$,

$P_3 = (0.60\ \text{A})^2 (6.0\ \Omega) = 2.2\ \text{W}$, $P_4 = (0.60\ \text{A})^2 (10\ \Omega) = 3.6\ \text{W}$.

(c) $P_{\text{sum}} = 6.0\ \text{W} + 1.44\ \text{W} + 2.16\ \text{W} + 3.6\ \text{W} = 13\ \text{W}$.

From Exercise 18.19, $P_{\text{total}} = \dfrac{(6.0\ \text{V})^2}{2.7\ \Omega} = 13\ \text{W}$. So $P_{\text{sum}} = P_{\text{tot}} = 13\ \text{W}$.

46. Around the left loop in a clockwise direction, $20\ \text{V} - I_1 (20\ \Omega) = 0$, so $I_1 = 1.0\ \text{A}$ down.

Around loop (whole circuit without R_1) in clockwise direction, $20\ \text{V} - I_2 (20\ \Omega) - I_2 (20\ \Omega) = 0$,

so $I_2 = 0.50\ \text{A}$ right and $I_3 = 0.50\ \text{A}$ down.

48. Around the left loop in clockwise direction, $10\ \text{V} - I_1 (4.0\ \Omega) + 5.0\ \text{V} = 0$, so $I_1 = 3.75\ \text{A}$.

Around the right loop in counterclockwise direction, $5.0\ \text{V} - I_3 (4.0\ \Omega) + 5.0\ \text{V} - I_2 (4.0\ \Omega) = 0$.

Also $I_2 = I_3$. So $I_2 = I_3 = 1.25\ \text{A}$.

Therefore, $I_1 = 3.75\ \text{A}$ up, $I_2 = 1.25\ \text{A}$ left, and $I_3 = 1.25\ \text{A}$ right.

55. (a) Just after the switch is closed, the capacitor is not charged.

So $V_C = 0$ and $V_R = V - V_C = V$.

(b) After one time constant, $V_C = 0.63V$ and $V_R = V - V_C = 0.37V$.

(c) After many time constants, the capacitor is fully charged.

So $V_C = V$ and $V_R = 0$.

64. (a) $V_C = V_o\left(1 - e^{-t/\tau}\right) = (24\text{ V})\left(1 - e^{-4}\right) = 23.56\text{ V}$.

So $Q = CV_C = (40\times10^{-6}\text{ F})(23.56\text{ V}) = 9.4\times10^{-4}\text{ C}$.

(b) After a long time, the capacitor is fully charged. $V_C = 24\text{ V}$ and $V_R = 0$.

70. From $I_g = \dfrac{I R_s}{r + R_s}$, we have $R_s = \dfrac{I_g r}{I - I_g} = \dfrac{(2000\times10^{-6}\text{ A})(100\ \Omega)}{3.0\text{ A} - 2000\times10^{-6}\text{ A}} = 0.0667\ \Omega = 67\text{ m}\Omega$.

76. (a) The current reading I is the total current through R and R_v.

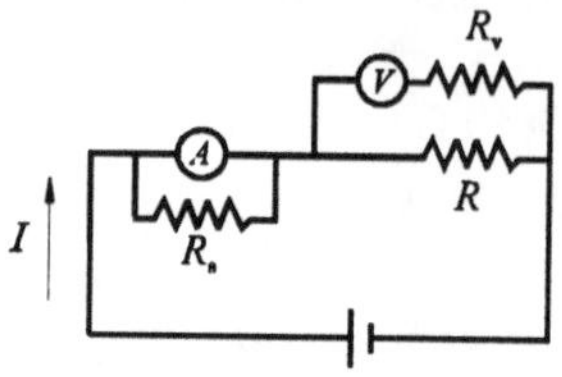

So the voltage reading is $V = I R_p = I\dfrac{R_v R}{R_v + R}$.

Therefore $R = \dfrac{V}{I - (V/R_v)}$.

(b) If R_v is ∞, $R = \dfrac{V}{I}$, i.e., the measurement is "perfect."

82. A conductor has very low resistance. The resistance of the wire between the feet is very very small, so the voltage between the feet is also small and so is the current through the bird.

91. (a) R_2, R_3, and R_4 are in parallel combination, $\dfrac{1}{R_p} = \dfrac{1}{25\ \Omega} + \dfrac{1}{50\ \Omega} + \dfrac{1}{25\ \Omega} = \dfrac{1}{10\ \Omega}$,

so $R_p = 10\ \Omega$.

So the total current is $I = I_1 = \dfrac{110\text{ V}}{100\ \Omega + 10\ \Omega} = 1.0\text{ A}$.

The voltage on R_p is $(1.0\text{ A})(10\ \Omega) = 10\text{ V}$.

Therefore $I_2 = I_4 = \dfrac{10\text{ V}}{25\ \Omega} = 0.40\text{ A}$, $I_3 = \dfrac{10\text{ V}}{50\ \Omega} = 0.20\text{ A}$.

(b) $P_1 = I^2 R_1 = (1.0\text{ A})^2(100\ \Omega) = 100\text{ W}$, $P_2 = P_4 = (0.40\text{ A})^2(25\ \Omega) = 4.0\text{ W}$,

$P_3 = (0.20\text{ A})^2(50\ \Omega) = 2.0\text{ W}$.

101. (a) Since they are in parallel, the voltage across them is 120 V.

(b) $P = VI$, so $I_{40} = \dfrac{P}{V} = \dfrac{40\text{ W}}{120\text{ V}} = 0.33\text{ A}$, $I_{60} = \dfrac{60\text{ W}}{120\text{ V}} = 0.50\text{ A}$ $I_{100} = \dfrac{100\text{ W}}{120\text{ V}} = 0.83\text{ A}$.

(c) Since the voltage on them is the nominal voltage of 120 V, the powers are 40 W, 60 W, and 100 W.

(d) $I_{total} = I_{40} + I_{60} + I_{100} = 0.333\text{ A} + 0.50\text{ A} + 0.833\text{ A} = 1.67\text{ A}$.

(e) $P_{\text{total}} = P_{40} + P_{60} + P_{100} = 40\text{ W} + 60\text{ W} + 100\text{ W} = 200\text{ W}.$

(f) The current, voltage, and power for the other two remain the same.

$I_{\text{total}} = 0.333\text{ A} + 0.50\text{ A} = 0.83\text{ A}$ and $P_{\text{total}} = 40\text{ W} + 60\text{ W} = 100\text{ W}.$

105. From $I_g = \dfrac{V}{r + R_m}$, we have $R_m = \dfrac{V}{I_g} - r.$

So $R_1 = \dfrac{10\text{ V}}{200\times10^{-6}\text{ A}} - 50\ \Omega = 4.995\times10^4\ \Omega = 49.95\text{ k}\Omega,$

$R_2 + R_1 = \dfrac{50\text{ V}}{200\times10^{-6}\text{ A}} - 50\ \Omega = 2.4995\times10^5\ \Omega = 249.95\text{ k}\Omega,$ so $R_2 = 200\text{ k}\Omega,$

$R_1 + R_2 + R_3 = \dfrac{250\text{ V}}{200\times10^{-6}\text{ A}} - 50\ \Omega = 1.24995\times10^6\ \Omega = 1.24995\text{ M}\Omega,$ so $R_3 = 1.0\text{ M}\Omega.$

106. From $I_g = \dfrac{I R_s}{r + R_s}$, we have $R_s = \dfrac{I_g r}{I - I_g}$. So $R_{s0.30} = \dfrac{(100\times10^{-6}\text{ A})(100\ \Omega)}{0.30\text{ A} - 100\times10^{-6}\text{ A}} = 33.3\text{ m}\Omega,$

$R_{s3.0} = \dfrac{(100\times10^{-6}\text{ A})(100\ \Omega)}{3.0\text{ A} - 100\times10^{-6}\text{ A}} = 3.33\text{ m}\Omega,$ and $R_{s30} = \dfrac{(100\times10^{-6}\text{ A})(100\ \Omega)}{30\text{ A} - 100\times10^{-6}\text{ A}} = 0.333\text{ m}\Omega.$

VI. Practice Quiz

1. Three resistors having values of 4.0 Ω, 6.0 Ω and 10.0 Ω, are connected in parallel. What is their equivalent resistance?

(a) 20 Ω (b) 7.3 Ω (c) 6.0 Ω (d) 4.0 Ω (e) 1.9 Ω

2. Three resistors having values of 4.0 Ω, 6.0 Ω and 10.0 Ω, are connected in parallel. If the circuit is connected in series to a 12.0 V battery and a 2.0 Ω resistor, what is the current through the 10 Ω resistor?

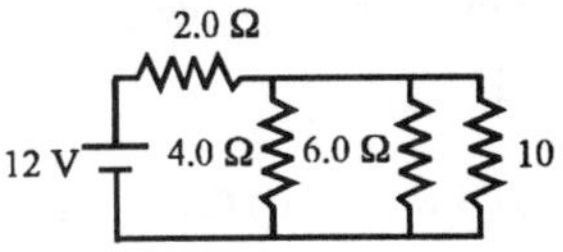

(a) 0.59 A (b) 1.0 A (c) 2.7 A (d) 11.2 A (e) 16.0 A

3. The following three appliances are connected to a 120 V house circuit: (1) toaster, 1200 W, (2) coffee pot, 650 W, and (3) microwave, 600 W. If all were operated at the same time what total current would they draw?

(a) 3.0 A (b) 4.0 A (c) 5.0 A (d) 10 A (e) 20 A

4. Find the current in the 15-Ω resistor.

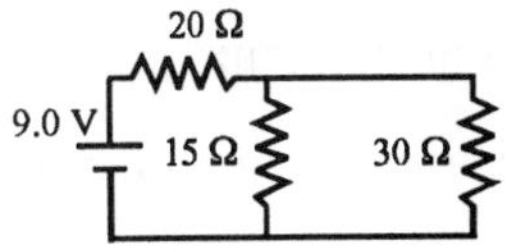

(a) 0.10 A (b) 0.13 A (c) 0.20 A (d) 0.26 A (e) 0.30 A

5. What is the maximum number of 100-W lightbulbs you can connect in parallel in a 120-W home circuit without tripping the 20-A circuit breaker?

(a) 11 (b) 17 (c) 23 (d) 27 (e) 29

6. What is Kirchhoff's junction rule in the figure?

(a) $I_2 = I_1 + I_3$

(b) $I_1 = I_2 + I_3$

(c) $I_3 = I_1 + I_2$,

(d) $I_1 + I_2 + I_3 = 0$

(e) none of the above

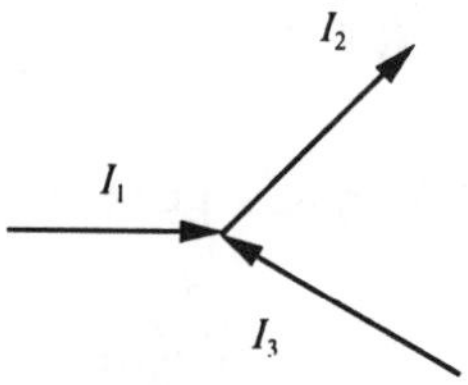

7. What is the current through the 2-Ω resistor in the circuit shown?

(a) 2 A

(b) 3 A

(c) 4 A

(d) 5 A

(e) 6 A

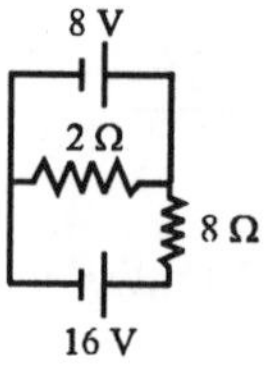

8. What is the equivalent resistance of the resistors in the figure?

(a) 0.443R (b) 0.75R (c) R (d) 2.5R (e) 4R

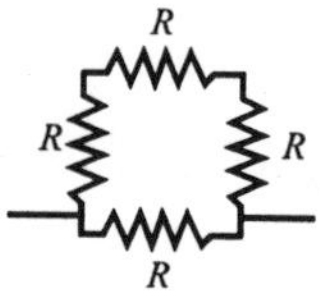

9. In the following simple circuit, G is a galvanometer and R is a resistor. What can we measure with this circuit?

(a) voltage (b) current (c) resistance (d) power (e) energy

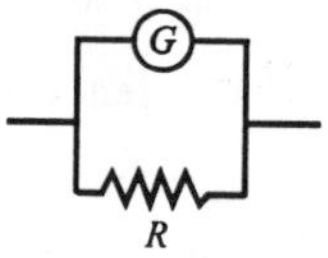

10. A voltage source of 10 V is connected to a series RC circuit with $R = 2.0\ \text{M}\Omega$ and $C = 3.0\ \mu\text{F}$. Initially the capacitor is uncharged. Find the time required for the current in the circuit to decrease to 5.0% of its original value.

(a) 3.0 s (b) 6.0 s (c) 9.0 s (d) 18 s (e) 36 s.

Answers to Practice Quiz:

1. e 2. a 3. e 4. c 5. c 6. a 7. c 8. b 9. b 10. d

CHAPTER 19

Magnetism

I. Chapter Objectives

Upon completion of this chapter, you should be able to:

1. state the force rule between magnetic poles and explain how the magnetic field is defined.
2. Determine the direction and magnitude of the magnetic force exerted by a magnetic field on a moving charged particle, and calculate the direction and magnitude of the magnetic field produced by an electric current.
3. how ferromagnetic materials enhance external magnetic fields, how "permanent" magnets are produced, and how "permanent" magnetism can be destroyed.
4. calculate the strength and direction of the magnetic force on a current-carrying wire and the torque on a current-carrying loop, and explain the concept of a magnetic moment for such a loop.
5. explain the operation of various instruments whose functions depend on electromagnetic interactions.

*6. state some of the general characteristics of the Earth's magnetic field, explain the theory about its possible source, and discuss some of the ways in which it affects the Earth's local environment.

II. Key Terms

Upon completion of this chapter, you should be able to define and/or explain the following key terms:

magnetic field
electromagnetism
tesla (T)
right-hand force rule
~~**right-hand force rule**~~
ferromagnetic materials
magnetic domains
magnetic permeability
Curie temperature
magnetic moment
dc motor
cathode ray tube (CRT)
mass spectrometer

The definitions and/or explanations of the most important key terms can be found in the following section: **III. Chapter Summary and Discussion**.

III. Chapter Summary and Discussion

1. Magnets and Magnetic Poles

Electricity and magnetism (electromagnetism) are manifestations of a single fundamental force or interaction, the electromagnetic force.

Magnets have two different poles or "centers" of force, which are designated as north and south poles. By the *law of poles*, like magnetic poles repel, and unlike magnetic poles attract. A magnet can create a **magnetic field**, similar to the electric field created by an electric charge.

The direction of a magnetic field (**B**) at any location is in the direction that the north pole of a compass would point if a compass were at that location. Hence, the magnetic field lines are directed away from a magnetic north pole and toward a magnetic south pole.

2. Electromagnetism, Magnetic Forces, and the Source of Magnetic Fields

A **magnetic field** can exert forces only on *moving* charges. The *magnitude* of a magnetic field can be defined in terms of the magnetic force (F) on a moving charge (q) by, $B = \dfrac{F}{qv\sin\theta}$, where v is the speed of the charge and θ is the angle between the velocity vector and the magnetic field vector. When **v** and **B** are perpendicular, $\theta = 90°$, $F = qvB\sin\theta = qvB\sin 90° = qvB$ (maximum force). The *direction* of the magnetic force on a charged particle is determined by the **force right-hand rule**: when the fingers of the right hand are pointed in the direction of **v** and then turned or curled toward the vector **B**, the extended thumb points in the direction of **F** for a positive charge. (For a negative charge, the force is in the opposite direction.) The SI unit of magnetic field is tesla (T).

Note: The magnetic force is always perpendicular to both the velocity vector and the magnetic field vector, or equivalently the magnetic force is perpendicular to a plane formed by the velocity and magnetic field vectors.

Example 19.1 An electron moves with a speed of 8.0×10^6 m/s along the $+x$-axis. It enters a region where there is a uniform magnetic field of 2.5 T, directed at an angle of 60° to the x axis and lying in the xy plane. Calculate the initial force and acceleration of the electron.

Solution: Given: $q = 1.60\times10^{-19}$ C, $m = 9.11\times10^{-31}$ kg, $v = 8.0\times10^6$ m/s, $B = 2.5$ T, $\theta = 60°$.

Find: **F** and **a**.

According to the right-hand rule, the direction of the magnetic force in the diagram is into the page ($-z$) for the electron because it has negative charge.

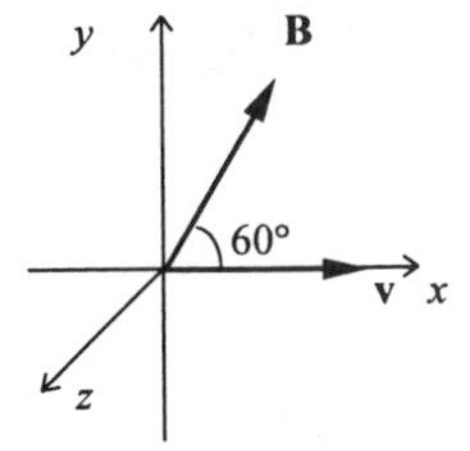

$F = qvB\sin\theta = (1.60\times10^{-19}\text{ C})(8.0\times10^{6}\text{ m/s})(2.5\text{ T})\sin 60° = 2.8\times10^{-12}\text{ N}$.

By Newton's second law, $a = \dfrac{F}{m} = \dfrac{2.8\times10^{-12}\text{ N}}{9.11\times10^{-31}\text{ kg}} = 3.1\times10^{18}\text{ m/s}^2$.

The directions of both the magnetic force and acceleration are in the $-z$ axis.

Example 19.2 A proton has a speed of 3.0×10^{6} m/s in a direction perpendicular to a uniform magnetic field, and the proton moves in a circle of radius 0.20 m. What is the magnitude of the magnetic field?

Solution: Given: $q = 1.60\times10^{-19}$ C, $m = 1.67\times10^{-27}$ kg, $v = 3.0\times10^{6}$ m/s, $r = 0.20$ m.

Find: B.

Since the magnetic force is always perpendicular to the velocity of the proton (right-hand rule), it is a centripetal force that causes the particle to move in a circular path. The centripetal force is provided by the magnetic force. By Newton's second law: $F_c = m\,a_c = \dfrac{m\,v^2}{r} = qvB$,

so $B = \dfrac{m\,v}{q\,r} = \dfrac{(1.67\times10^{-27}\text{ kg})(3.0\times10^{6}\text{ m/s})}{(1.60\times10^{-19}\text{ C})(0.20\text{ m})} = 0.16\text{ T}$.

Magnetic fields are produced by electric currents. The magnitude of the magnetic field around *a long, straight, current-carrying wire* is given by $B = \dfrac{\mu_o I}{2\pi d}$, where $\mu_o = 4\pi\times10^{-7}$ T·m/A is a constant called the *magnetic permeability of free space*, I the current in the wire, and d the perpendicular distance from the wire. The direction of the magnetic field is determined by the **right-hand source rule**: if a current-carrying wire is grasped with the right hand with the extended thumb pointing in the direction of the conventional current, the curled fingers indicate the circular sense of the magnetic field.

The magnitude of the magnetic field at the *center of a circular current-carrying wire loop* is given by $B = \dfrac{\mu_o I}{2r}$, where r is the radius of the loop. The direction of the magnetic field is determined by the right-hand source rule and is perpendicular to the plane of the loop at its center.

The magnitude of the magnetic field near the center of *a current-carrying solenoid* is given by $B = \dfrac{\mu_o N I}{L}$, where N is the number of turns (loops) in the solenoid and L is the length of the solenoid. The direction of the magnetic field is determined by the right-hand source rule as applied to one of the loops of the solenoid. The quantity $n = \frac{N}{L}$ is called the *linear turn density* (number of turns per unit length). The magnitude of the magnetic field can also be expressed in terms of n as $B = \mu_o nI$. The magnetic field near the center of a solenoid is independent of the radius of the solenoid and is approximately constant if you stay away from the ends.

Note: Since magnetic field is a vector, you must use vector addition to find the net field if there are contributions from two or more sources.

Example 19.3 At what distance from a long straight wire carrying a current of 5.0 A (typical household current) is the magnitude of the magnetic field due to the wire equal to the strength of the Earth's magnetic field of about 5.0×10^{-5} T?

Solution: Given: $B = 5.0\times10^{-5}$ T, $I = 5.0$ A.

Find: d.

From $B = \frac{\mu_o I}{2\pi d}$, we have $d = \frac{\mu_o I}{2\pi B} = \frac{(4\pi\times10^{-7}\text{ T·m/A})(5.0\text{ A})}{2\pi(5.0\times10^{-5}\text{ T})} = 2.0\times10^{-2}\text{ m} = 2.0\text{ cm}.$

Example 19.4 Calculate the magnitude of the magnetic field at the center of a solenoid which has 100 turns, is 0.10 m long and carries a current of 1.0 A.

Solution: Given: $N = 100$, $L = 0.10$ m, $I = 1.0$ A.

Find: B.

$$B = \frac{\mu_o NI}{L} = \frac{(4\pi\times10^{-7}\text{ T·m/A})(100)(1.0\text{ A})}{0.10\text{ m}} = 1.3\times10^{-3}\text{ T} = 1.3\text{ mT}.$$

Example 19.5 Two long parallel wires carry currents of 20 A and 5.0 A in opposite directions. The wires are separated by 0.20 m.

(a) What is the magnetic field midway between the two wires?

(b) At what point between the wires are the contributions from the two wires the same?

Solution: (a) According to the right-hand source rule, the magnetic fields between the wires due to the two wires are in the same direction (both are into the page). At midway between the two wires, the distances from either wire is 0.10 m.

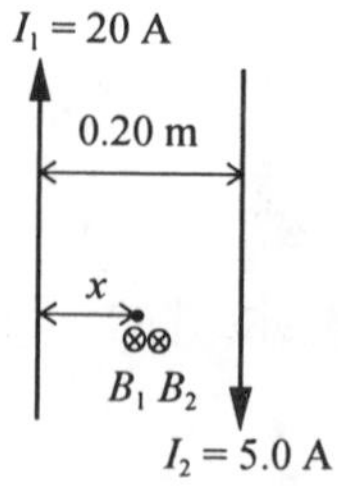

$$B_1 = \frac{\mu_o I_1}{2\pi d_1} = \frac{(4\pi\times10^{-7}\text{ T·m/A})(20\text{ A})}{2\pi(0.10\text{ m})} = 4.0\times10^{-5}\text{ T}.$$

$$B_2 = \frac{(4\pi\times10^{-7}\ \text{T·m/A})(5.0\ \text{A})}{2\pi(0.10\ \text{m})} = 1.0\times10^{-5}\ \text{T}.$$

So the net magnetic field is $B = B_1 + B_2 = 4.0\times10^{-5}\ \text{T} + 1.0\times10^{-5}\ \text{T} = 5.0\times10^{-5}\ \text{T}$, into the page.

(b) Assume the contributions from both wires are the same at a distance x from the 20-A wire. Then the distance from the 5.0 A wire is $(0.20\ \text{m} - x)$. $B_1 = B_2$, so $\dfrac{\mu_o I_1}{2\pi x} = \dfrac{\mu_o I_2}{(0.20\ \text{m} - x)}$.

Therefore, $I_1\,(0.20\ \text{m} - x) = I_2\, x$, or $(20\ \text{A})(0.20\ \text{m} - x) = (5.0\ \text{A})\, x$.

Solving for x, we have $x = 0.16$ m from the 20 A wire.

3. Magnetic Materials

Ferromagnetic materials are those in which the electrons spins "lock" together to create **magnetic domains** where the magnetic fields of individual electrons add constructively. The materials are easily magnetized. The magnetic permeability of these materials is many times that of free space, reflecting the fact that they enhance an external magnetic field. In the external magnetic field, the domains parallel to the field grow at the expense of other domains and the orientation of some domains may become more aligned with the field.

Common ferromagnetic materials are iron, nickel, and cobalt. Iron is commonly used in the cores of electromagnets. This type of iron is termed "soft" iron because the domains become unaligned and the iron unmagnetized when the external field is removed. "Hard" iron retains some magnetism after a field is removed, and this type of iron is used to make "permanent" magnets by heating the iron above the **Curie temperature** and cooling in a strong magnetic field. Above the Curie temperature, the magnetic domains become thermally disordered and the iron loses its "permanent" magnetic field.

The magnetic permeability is defined as $\mu = K_m\, \mu_o$, where K_m is the magnetic analog of the dielectric constant and is called the *relative permeability*. $K_m = 1$ for free space (vacuum).

Example 19.6 If an iron core is inserted into the solenoid in Example 19.4, what is the magnetic field near the center of the solenoid?

Solution: Given: $K_m = 1500$, $B_o = 1.3$ mT (from Example 19.4)

Find: B.

$B = \mu\, nI = K_m\, \mu_o\, n\, I = K_m\, B_o = (1500)(1.3\ \text{mT}) = 2.0\ \text{T}$.

The magnetic field increases by 1500 times!

4. Magnetic Forces on Current-Carrying Wires

A magnetic field can exert a force (F) on a current-carrying wire. The magnitude of the magnetic force is given by $F = ILB\sin\theta$, where L is the length of the wire, I the current in the wire, and θ the angle between the current and the magnetic field vector. The direction of the force is determined by the right-hand rule: when the fingers of the right hand are pointed in the direction of the conventional current and then turned or curled toward the vector **B**, the extended thumb points in the direction of **F**.

Forces exist between two parallel current-carrying wires. This is because the magnetic field produced by the current in one wire exerts a force on the other wire. The force per unit length can be calculated with $\frac{F}{L} = \frac{\mu_o I_1 I_2}{2\pi d}$, where d is the distance between the wires and I_1 and I_2 the currents in the wires, respectively. According to the right-hand rules (source plus force) and the law of poles, the forces are attractive if the currents are in the same direction and repulsive if the currents are in opposite directions (can you show why?).

A magnetic field can exert torque on a current-carrying loop. The magnitude of the torque is equal to $\tau = NIAB\sin\theta$, where N is the number of turns (loops), I the current in the loop, A the area enclosed by the loop, and θ the angle between the normal to the loop and the magnetic field vector. The quantity IA is often referred as the **magnetic moment**, $m = IA$. The magnitude of the torque can then be expressed as $\tau = NmB\sin\theta$.

Note: The angle θ in the torque equation is between the *normal to the plane of the loop* and the magnetic field, *not* the angle between the plane of the loop and the magnetic field.

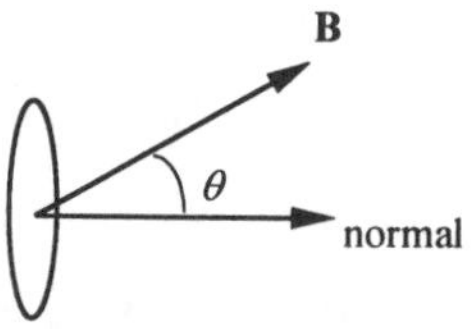

Example 19.7 A wire carries a current of 10 A in a direction of 30° with respect to the direction of a 0.30-T magnetic field. Find the magnitude of the magnetic force on a 0.50 m length of the wire.

Solution: Given: $I = 10$ A, $\theta = 30°$, $B = 0.30$ T, $L = 0.50$ m.

Find: F.

$$F = ILB\sin\theta = (10\text{ A})(0.50\text{ m})(0.30\text{ T})\sin 30° = 0.75\text{ N}.$$

Example 19.8 A circular loop of wire of radius 0.50 m is in a uniform 0.30-T magnetic field. The current in the loop is 2.0 A. Find the magnitude of the torque when

(a) the plane of the loop is parallel to the magnetic field,

(b) the plane of the loop is perpendicular to the magnetic field,

(c) the plane of the loop is at 30° to the magnetic field.

Solution: The angle in the torque equation is between the normal to the plane of the loop and the magnetic field, not between the plane of the loop and the magnetic field.

Given: $N = 1$, $r = 0.50$ m, $B = 0.30$ T, $I = 2.0$ A,

(a) $\theta = 90°$, (b) $\theta = 0°$, (c) $\theta = 90° - 30° = 60°$.

Find: τ in (a), (b), and (c).

(a) $\tau = NIAB\sin\theta = NA\pi r^2 B\sin\theta = (1)(2.0\text{ A})(\pi)(0.50\text{ m})^2(0.30\text{ T})\sin 90° = 0.47\text{ m·N}$.

(b) $\tau = (1)(2.0\text{ A})(\pi)(0.50\text{ m})^2(0.30\text{ T})\sin 0° = 0\text{ m·N}$.

(c) $\tau = (1)(2.0\text{ A})(\pi)(0.50\text{ m})^2(0.30\text{ T})\sin 60° = 0.41\text{ m·N}$.

Example 19.9 Two long, straight wires separated by a distance of 0.30 m carry currents in the same direction. If the current in one wire is 5.0 A and the current in the other is 8.0 A, find the magnitude and direction of the forces per unit length (per meter) between the two wires. What if the currents are in opposite directions?

Solution: Given: $I_1 = 5.0$ A, $I_2 = 8.0$ A, $d = 0.30$ m.

Find: F/L.

There is a force on each wire because the magnetic field produced by one wire can exert force on the other. According to Newton's third law, the two forces on two wires are equal in magnitude and opposite in direction. The field by wire 1 at wire 2 is $B_1 = \dfrac{\mu_o I_1}{2\pi d}$, so the magnitude of the force on wire 2 is $F_2 = I_2 L B_1 = \dfrac{\mu_o I_1 I_2}{2\pi d} L$.

Therefore the force per unit length is

$$\frac{F_2}{L} = \frac{\mu_o I_1 I_2}{2\pi d} = \frac{(4\pi\times10^{-7}\text{ T·m/A})(5.0\text{ A})(8.0\text{ A})}{2\pi(0.30\text{ m})} = 2.7\times10^{-5}\text{ N/m}.$$

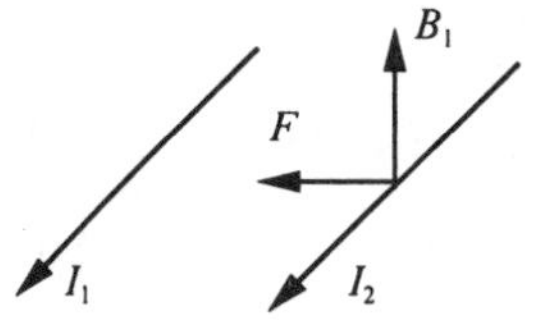

According to the right-hand rule, the magnetic field due to one wire exerts an attractive force on the other, and vice versa, so the forces are attractive. If the currents are opposite, the magnetic field due to one wire exerts a repulsive force on the other, and vice versa.

---

5. Applications of Electromagnetism

Applications of electromagnetic interactions include the galvanometer, the **dc motor**, the **cathode ray tube (CRT)**, the **mass spectrometer**, and the electronic balance. A galvanometer consists of a small coil. When there is a current in the coil, the magnetic field exerts a torque on the coil, and therefore deflects the coil. The deflection of the coil is directly proportional to the current through the galvanometer. A dc motor is a device that converts electrical energy into mechanical energy. A cathode ray tube (CRT) is a vacuum tube that is used in an oscilloscope, computer monitor, and television. A mass spectrometer is a device used to separate isotopes, or atoms of different masses. It uses the different radii of circular orbits in a magnetic field by particles of the same charge but different mass. In an electronic balance, the magnetic force on a current-carrying wire replaces the known force in a conventional balance.

In a mass spectrometer, a *velocity selector* is used to allow charged particles of certain velocity to enter the spectrometer. If a beam of charged particles are not deflected in the presence of a uniform electric field and a magnetic field mutually perpendicular, it must satisfy $v = \frac{E}{B_1} = \frac{V}{B_1 d}$, where B_1 is the magnetic field in the velocity selector, V the voltage across and d the separations of the plates in the velocity selector. The mass of the particle (m) depends on the radius of the circular orbit (R) in another magnetic field B_2: $m = \frac{qdB_1 B_2}{V} R$.

Example 19.10 In a mass spectrometer, a single-charged particle has a speed of 1.0×10^6 m/s and enters a uniform magnetic field of 0.20 T. The radius of the circular orbit is 0.020 m.

(a) What is the mass of the particle?

(b) What is the kinetic energy of the particle?

Solution: Given: $q = 1.6\times10^{-19}$ C, $v = 1.0\times10^6$ m/s, $B_2 = 0.20$ T, $R = 0.020$ m.

Find: (a) m (b) K.

(a) Since $v = \frac{E}{B_1} = \frac{V}{B_1 d}$,

$$m = \frac{qdB_1 B_2}{V} R = \frac{qB_2 R}{v} = \frac{(1.6\times10^{-19}\text{ C})(0.20\text{ T})(0.020\text{ m})}{1.0\times10^6\text{ m/s}} = 6.4\times10^{-28}\text{ kg}.$$

The particle is the helium nucleus.

(b) $K = \frac{1}{2} mv^2 = \frac{1}{2}(6.4\times10^{-28}\text{ kg})(1.0\times10^6\text{ m/s})^2 = 3.2\times10^{-16}\text{ J} = (2.0\text{ keV})$.

---

*6. The Earth's Magnetic Field

The Earth's magnetic field resembles that which would be produced by a large interior bar magnet (with the magnet's south pole near the Earth's north-geographical pole and magnet's north pole near the Earth's south-geographical pole). However, this is not possible because the Earth's interior temperature is above the Curie temperature of ferromagnetic materials. Scientists associate the Earth's magnetic field with motions (electric current) in its liquid outer core.

The magnetic north pole and the geographical south pole do not coincide (nor do the magnetic south pole and geographical north pole), and the magnetic poles "wander" or move about over periods of hundreds of thousands of years. The deviation between the north-south magnetic poles and the south-north geographical poles is called the *magnetic declination.* There is evidence that the Earth's magnetic poles have reversed periodically over geologic time.

Charged particles from the Sun and other cosmic rays are trapped in the Earth's magnetic field, and regions or concentrations of these charged particles are called *Van Allen belts.* It is believed that the recombination of ionized air molecules and electrons that have been ionized by articles from the lower belt give rise to the Aurora Borealis (northern lights) and Aurora Australis (southern lights).

IV. Mathematical Summary

Magnetic Force on a Moving Charge	$F = qvB\sin\theta$ (19.3) θ is the angle between **v** and **B**	Computes the magnitude of the magnetic force on a moving charge
Magnetic Field around a Long, Straight Current-Carrying Wire	$B = \frac{\mu_o I}{2\pi d}$ (19.4) $\mu_o = 4\pi \times 10^{-7}$ T·m/A	Calculates the magnitude of the magnetic field due to a long, straight current-carrying wire
Magnetic Field at the Center of a Circular Loop	$B = \frac{\mu_o I}{2r}$ (19.5)	Calculates the magnitude of the magnetic field at the center of a circular loop of current-carrying wire
Magnetic Field in a Solenoid ($n = N/L$)	$B = \frac{\mu_o NI}{L}$ (19.6) or $B = \mu_o nI$ (19.7)	Computes the magnitude of the magnetic field along the axis of a long solenoid
Magnetic Permeability	$\mu = K_m \mu_o$ (19.8)	Relates the magnetic permeability of a materiel and the permeability of free space
Force on a Straight Current-carrying Wire	$F = ILB\sin\theta$ (19.10)	Computes the magnitude of the magnetic force on a straight current-carrying wire
Torque on a Single Current-carrying Loop	$\tau = IAB\sin\theta$ (19.11)	Computes the magnitude of the magnetic torque on a single current-carrying loop
Magnetic Velocity Selector	$v = \frac{E}{B_1} = \frac{V}{B_1 d}$ (19.16)	Calculates the velocity of a charged particle in a magnetic velocity selector
Particle Mass (mass spectrometer)	$m = \frac{qdB_1B_2}{V}R$ (19.17)	Calculates the mass of a charged particle in a mass spectrometer

V. Solutions of Selected Exercises and Paired Exercises

9. If the charges are electrons (negative), an excess negative charge accumulates on the left side of the strip, and if positive, on the right side of the strip. The sign of the voltage drop across the strip would then indicate the type of charge.

10. $F = qvB \sin\theta = (0.25 \text{ C})(2.0\times10^2 \text{ m/s})(0.40 \text{ T}) \sin 90° = 20 \text{ N}.$

21. (a) According to the right-hand rule: (1) negative charge, (2) no charge, (3) positive charge.

(b) The radius of the circular orbit is proportional to the mass, so $m_3 > m_1$.

24. (a) The fields by the two wires are equal in magnitude and opposite in direction. So the net field is 0.

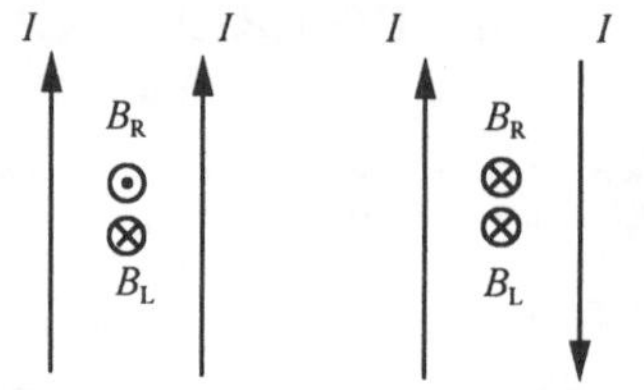

(b) The fields by the two wires are equal in magnitude and in the same direction. So the net field is

$$B = 2B_R = 2\frac{\mu_o I}{2\pi d} = \frac{\mu_o I}{\pi d} = \frac{(4\pi\times10^{-7} \text{ T·m/A})(4.0 \text{ A})}{\pi(0.25 \text{ m})} = 6.4\times10^{-6} \text{ T}.$$

28. (a) Magnetic force provides centripetal force. $F = qvB \sin\theta = m\frac{v^2}{R}$,

so $$R = \frac{mv}{qB\sin\theta} = \frac{(1.67\times10^{-27} \text{ kg})(4.0\times10^6 \text{ m/s})}{(1.6\times10^{-19} \text{ C})(3.0 \text{ T}) \sin 90°} = 1.4\times10^{-2} \text{ m}.$$

(b) Magnetic force must equal electric force for the proton to travel along a straight line.

From $qvB\sin\theta = qE$, we have $E = vB\sin\theta = (4.0\times10^6 \text{ m/s})(3.0 \text{ T}) = 1.2\times10^7 \text{ N/C}$.

32. From $B = \frac{\mu_o I}{2r}$, we have $I = \frac{2rB}{\mu_o} = \frac{2(0.10 \text{ m})(10^{-5} \text{ T})}{4\pi\times10^{-7} \text{ T·m/A}} = 1.6 \text{ A}$.

41. (a) From the result of Exercise 19.28(a), $R = \frac{mv}{qB\sin\theta} = \frac{mv}{qB}$.

The frequency is $f = \frac{1}{T} = \frac{1}{2\pi R/v} = \frac{v}{2\pi R} = \frac{qB}{2\pi m}$.

(b) $T = \frac{1}{f} = \frac{2\pi m}{qB}$, independent of m and v.

(c) $R = \frac{(9.11\times10^{-31} \text{ kg})(10^5 \text{ m/s})}{(1.6\times10^{-19} \text{ C})(10^{-4} \text{ T})} = 5.7\times10^{-3} \text{ m}$, $f = \frac{(1.6\times10^{-19} \text{ C})(10^{-4} \text{ T})}{2\pi(9.11\times10^{-31} \text{ kg})} = 2.8\times10^6 \text{ Hz}$.

58. $F = ILB\sin\theta = (5.0\text{ A})(1.0\text{ m})(0.30\text{ T})\sin 90° = 1.5\text{ N}$, $+y$ direction (taking $+z$ as upward).

64. From $F = ILB\sin\theta$, we have $I = \dfrac{F}{LB\sin\theta} = \dfrac{5.0\text{ N}}{(0.25\text{ m})(3.0\text{ T})\sin 90°} = 6.7\text{ A in } -z$.

72. (a) For the magnetic force to repel, the current must be in the opposite direction.

(b) Magnetic force cancels gravity. $F = mg$.

So $\dfrac{F}{L} = \dfrac{\mu_0 I_1 I_2}{2\pi d} = \dfrac{mg}{L}$. Therefore, for a length of 1 m, we have

$$I = I_1 = I_2 = \sqrt{\frac{2\pi\, d\, mg}{\mu_0}} = \sqrt{\frac{2\pi(0.020\text{ m})(1.5\times10^{-3}\text{ kg})(9.80\text{ m/s}^2)}{4\pi\times10^{-7}\text{ T}\cdot\text{m/A}}} = 38\text{ A}.$$

88. $v = \dfrac{V}{Bd} = \dfrac{E}{B} = \dfrac{1.0\times10^3\text{ V/m}}{0.10\text{ T}} = 1.0\times10^4\text{ m/s}$.

In circular motion magnetic force provides centripetal force. $F = qvV\sin\theta = m\dfrac{v^2}{R}$,

so $m = \dfrac{qB\sin\theta\, R}{v} = \dfrac{(1.6\times10^{-19}\text{ C})(0.10\text{ T})(0.012\text{ m})\sin 90°}{1.0\times10^4\text{ m/s}} = 1.9\times10^{-26}\text{ kg}$.

91. From $K = \frac{1}{2}mv^2$, we have $v = \sqrt{\dfrac{2K}{m}} = \sqrt{\dfrac{2(10\times10^3\text{ eV})(1.6\times10^{-19}\text{ J/eV})}{1.67\times10^{-27}\text{ kg}}} = 1.38\times10^6\text{ m/s}$.

In circular motion magnetic force provides centripetal force.

Since $F = qvB\sin\theta = m\dfrac{v^2}{R}$, $B = \dfrac{mv}{qR\sin\theta} = \dfrac{(1.67\times10^{-27}\text{ kg})(1.38\times10^6\text{ m/s})}{(1.6\times10^{-19}\text{ C})(0.50\text{ m})\sin 90°} = 2.9\times10^{-2}\text{ T}$.

98. $B = \mu_0\dfrac{N_1}{L}I_1 + \mu_0\dfrac{N_2}{L}I_2 = \dfrac{4\pi\times10^{-7}\text{ T}\cdot\text{m/A}}{0.10\text{ m}}[(3000)(5.0\text{ A}) + (2000)(10\text{ A})] = 0.44\text{ T}$.

101. $\tau = IAB\sin\theta = IL^2B\sin\theta$, where θ is the angle between the normal to the plane and the field.

So $L = \dfrac{\tau}{IB\sin\theta} = \dfrac{0.15\text{ N}\cdot\text{m}}{(10\text{ A})(0.500\text{ T})\sin 40°} = 0.22\text{ m}$.

103. $v = \dfrac{V}{Bd} = \dfrac{E}{B}$,

so $B = \dfrac{E}{v} = \dfrac{100\text{ V/m}}{2.0\times10^2\text{ m/s}}$

$= 0.50$ T, **B** is east if **v** is north and **E** is upward.

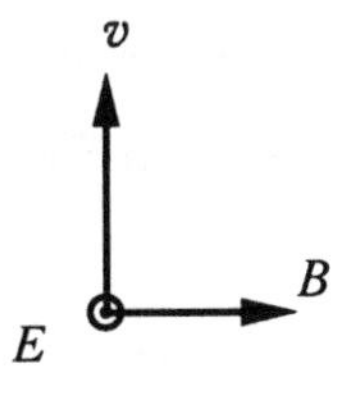

VI. Practice Quiz

1. An electron moves with a speed of 3.0×10^4 m/s perpendicular to a uniform magnetic field of 0.40 T. What is the magnitude of the force it experiences?

   (a) 4.8×10^{-14} N (b) 1.9×10^{-15} N (c) 2.2×10^{-24} N (d) 1.1×10^{-24} N (e) zero

2. A 2.0-m wire carrying a current of 0.60 A is oriented parallel to a uniform magnetic field of 0.50 T. What is the magnitude of the force it experiences?

   (a) 6.0 N (b) 0.30 N (c) 0.15 N (d) 0.60 N (e) zero

3. An electron moving along the $+x$ axis enters a region where there is a uniform magnetic field in the $+y$ direction. What is the direction of the magnetic force on the electron, if x axis is to the right, y axis points upward, then $+z$ is out of the page?

   (a) $+z$ direction (b) −z direction (c) $+y$ direction (d) $-y$ direction (e) $-x$ direction

4. A solenoid with 500 turns is 0.10 m long, and carries a current of 4.0 A. What strength of magnetic field will it have at its center?

   (a) 1.5×10^{-3} T (b) 3.1×10^{-3} T (c) 6.2×10^{-3} T (d) 13×10^{-3} T (e) 25×10^{-3} T

5. The direction of the force on a current-carrying wire in a magnetic field is described by which of the following?

   (a) perpendicular to the current (b) perpendicular to the magnetic field (c) parallel to the current (d) both (a) and (b) are valid (e) neither (a) nor (b) is valid

6. A circular loop carrying a current of 2.0 A is in a magnetic field of 3.5 T. The loop has an area of 0.12 m^2 and its plane is oriented at a 37° angle to the field. What is the magnitude of the magnetic torque on the loop?

   (a) 46 m·N (b) 0.67 m·N (c) 0.51 m·N (d) 0.10 m·N (e) zero

7. Two long parallel wires carry equal currents. The magnitude of the force between the wires is F. The current in each wire is now doubled. What is the magnitude of the new force between the two wires?

   (a) $4F$ (b) $2F$ (c) F (d) $F/2$ (e) $F/4$

8. A proton travels though a potential of 1.0 kV and then moves into a magnetic field of 0.040. What is the radius of the proton's resulting orbit?

   (a) 0.080 m (b) 0.11 m (c) 0.14 m (d) 0.17 m (e) 0.21 m

9. A thin copper rod 1.0 m long has a mass of 0.050 kg and is in a magnetic field of 0.10 T. What minimum current in the rod is needed in order for the magnetic force to cancel the weight of the rod?

(a) 0.58 A (b) 1.2 A (c) 2.5 A (d) 4.9 A (e) 9.8 A

10. A stationary proton is in a uniform magnetic field of 0.20 T. What is magnitude of magnetic force on the proton?

(a) 1.6×10^{-20} N (b) 1.6×10^{-21} N (c) 3.2×10^{-20} N (d) 3.2×10^{-21} N (e) zero

Answers to Practice Quiz:

1. b 2. e 3. b 4. e 5. d 6. b 7. a 8. b 9. d 10. e

CHAPTER 20

Electromagnetic Induction

I. Chapter Objectives

Upon completion of this chapter, you should be able to:

1. define magnetic flux and explain how induced emf's are created by changing magnetic flux, and calculate the magnitude and predict the polarity of an induced emf.
2. understand the operation of electrical generators and calculate the emf produced by an ac generator at any time in its operating cycle, and explain the origin of back emf and its effect on the behavior of motors.
3. explain transformer action in terms of Faraday's law, calculate the output of step-up and step-down transformers, and understand the importance of transformers in electric energy delivery systems.
4. explain the physical nature, origin, and means of propagation of electromagnetic waves, and describe the properties and uses of various types of electromagnetic waves.

II. Key Terms

Upon completion of this chapter, you should be able to define and/or explain the following key terms:

electromagnetic induction
magnetic flux
Faraday's law of induction
Lenz's law
alternating current
ac generator
dc generator
back emf
transformer
primary coil
secondary coil
step-up transformer
step-down transformer
eddy currents
electromagnetic waves (radiation)
Maxwell's equations
radiation pressure
power waves
radio and TV waves
microwaves
infrared radiation
visible light
ultraviolet radiation
X-rays
gamma rays

The definitions and/or explanations of the most important key terms can be found in the following section: **III. Chapter Summary and Discussion**.

III. Chapter Summary and Discussion

1. Induced Emf's: Faraday's Law and Lenz's Law

A measure of the number of magnetic field lines though a particular loop area is defined as the **magnetic flux**, $\Phi = BA\cos\theta$, where B is the magnetic field, A the area of the loop, and θ the angle between the magnetic field vector and the normal to the loop area. The SI unit of magnetic flux is weber (Wb), or T·m^2.

Note: The angle θ is the angle between the magnetic field and the *normal to the plane of the loop*, *not* the angle between the magnetic field and the plane of the loop.

Electromagnetic induction refers to the creation of induced emf's whenever the magnetic flux though a coil changes. The magnitude of induced emf in a conducting loop depends on the time rate change of magnetic flux though the loop, $\mathcal{E} = -N\frac{\Delta\Phi}{\Delta t}$, where N is the number of loops. This relationship is known as **Faraday's law of induction**. Since $\Phi = BA\cos\theta$, any change in B, A, and θ will result in an induced emf. Faraday's law can be written in detail as $\mathcal{E} = -N\left[\frac{\Delta B}{\Delta t}A\cos\theta + B\frac{\Delta A}{\Delta t}\cos\theta + BA\frac{\Delta(\cos\theta)}{\Delta t}\right]$. There are three different ways to produce induced emf.

When a conductor of length L moves perpendicular to a magnetic field B with a speed v, the magnitude of the induced emf is called *motional emf* and is given by $\mathcal{E} = BLv$.

The minus sign in Faraday's law of induction gives the polarity of the induced emf, which is found by **Lenz's law**: when a change in magnetic flux induces an emf in a conducting loop, the resulting current is in such a direction as to create a magnetic field that tends to oppose the change in flux. That is, an induced emf gives rise to a current whose magnetic field opposes the change in magnetic flux that produced it.

Example 20.1 A circular loop of radius 0.10 m is rotating in a uniform magnetic field of 0.20 T. Find the magnetic flux though the loop when the plane of the loop and the magnetic field vector

(a) are parallel,

(b) are perpendicular

(c) are at 30°.

Solution: The angle θ in the magnetic flux definition is the angle between the magnetic field and the *normal* to the plane of the loop, *not* the angle between the magnetic field and the plane of the loop.

Given: $r = 0.10$ m, $B = 0.20$ T, (a) $\theta = 90°$, (b) $\theta = 0°$, (c) $\theta = 90° - 30° = 60°$.

Find: Φ in (a), (b), and (c).

(a) $\Phi = BA\cos\theta = B\pi r^2 \cos\theta = (0.20\ \text{T})(\pi)(0.10\ \text{m})^2 \cos 90° = 0$.

(b) $\Phi = (0.20\ \text{T})(\pi)(0.10\ \text{m})^2 \cos 0° = 6.3\times10^{-3}\ \text{T}\cdot\text{m}^2$.

(c) $\Phi = (0.20\ \text{T})(\pi)(0.10\ \text{m})^2 \cos 60° = 3.1\times10^{-3}\ \text{T}\cdot\text{m}^2$.

Example 20.2 A coil is wrapped with 200 turns of wire on a square frame with sides 18 cm. A uniform magnetic field is applied perpendicular to the plane of the coil. If the field changes uniformly from 0 to 0.50 T in 8.0 s, find the average value of the magnitude of the induced emf.

Solution: For B, A, and θ in this Example, only B changes.

Given: $N = 200$, $d = 0.18$ m, $T_i = 0$, $T_f = 0.50$ T, $\Delta t = 8.0$ s, $\theta = 0°$.

Find: $\mathcal{E}$.

From Faraday's law of induction, the magnitude of the induced emf is

$$\mathcal{E} = N\frac{\Delta B}{\Delta t}A\cos\theta = (200)\frac{0.50\ \text{T} - 0}{8.0\ \text{s}}(0.18\ \text{m})^2\cos 0° = 0.41\ \text{V}.$$

Both $\dfrac{\Delta A}{\Delta t}$ and $\dfrac{\Delta(\cos\theta)}{\Delta t}$ are zero because A and θ do not change.

Example 20.3 A square coil of wire with 15 turns and an area of 0.40 m² is placed parallel to a magnetic field of 0.75 T. The coil is flipped so its plane is perpendicular to the magnetic field in 0.050 s. What is the magnitude of the average induced emf?

Solution: For B, A, and θ in this Example, only θ changes.

Given: $N = 15$, $A = 0.40\ \text{m}^2$, $\Delta t = 0.050$ s, $B = 0.75$ T, $\theta_i = 90°$, $\theta_f = 0°$.

Find: $\mathcal{E}$.

From Faraday's law of induction, the magnitude of the induced emf is

$$\mathcal{E} = NBA\frac{\Delta(\cos\theta)}{\Delta t} = (15)(0.75\ \text{T})(0.40\ \text{m}^2)\frac{\cos 0° - \cos 90°}{0.050\ \text{s}} = 90\ \text{V}.$$

Both $\dfrac{\Delta B}{\Delta t}$ and $\dfrac{\Delta A}{\Delta t}$ are zero because B and A do not change.

Example 20.4 An airplane with a wing span of 50 m flies horizontally to the Earth at a location where the downward component of the Earth's magnetic field is 6.0×10^{-5} T. Find the magnitude of the induced emf between the tips of the wing when the speed of the plane is 250 m/s.

Solution: Given: $L = 50\text{m}$, $B = 6.0 \times 10^{-5}$ T, $v = 250$ m/s.

Find: $\mathcal{E}$.

The magnitude of the motional emf is $\mathcal{E} = BLv = (6.0 \times 10^{-5}\ \text{T})(50\ \text{m})(250\ \text{m/s}) = 0.75$ V.

Example 20.5 Find the direction of the induced current in the figures shown.

(a)

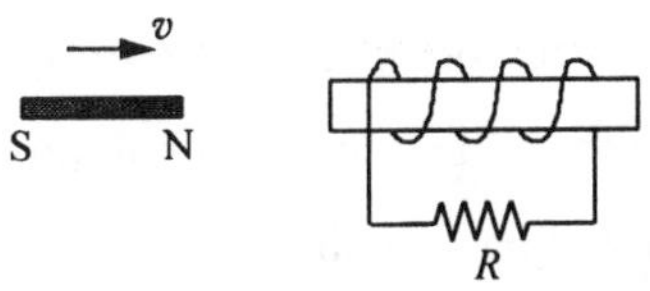

(b)

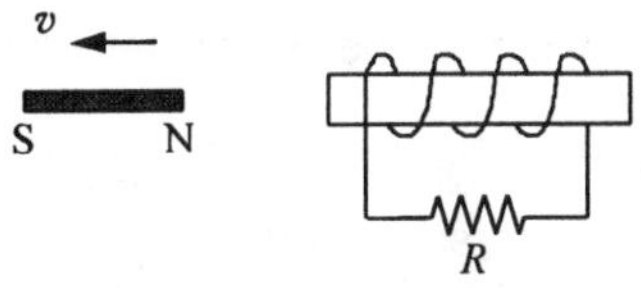

(c)

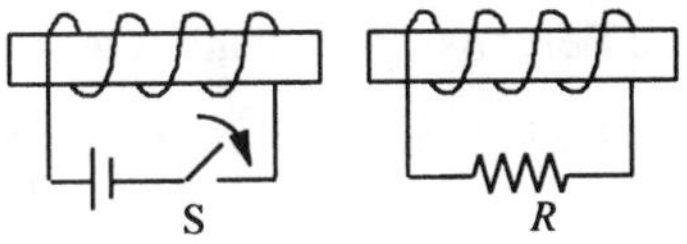

(d)

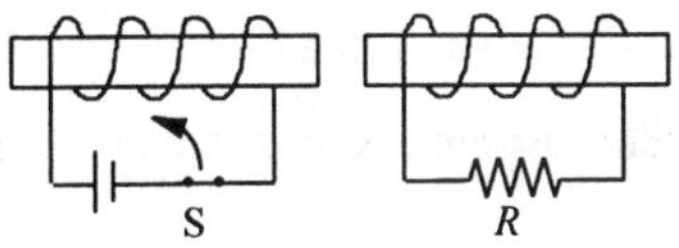

(e)

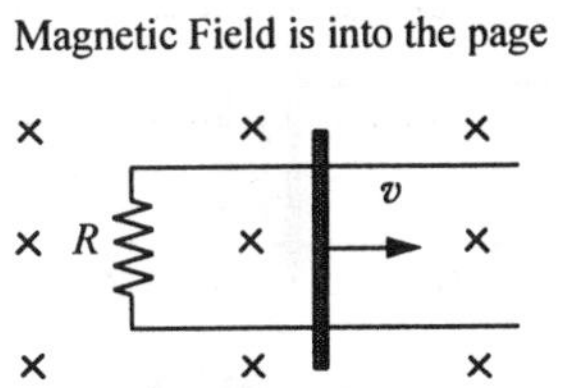

(f)

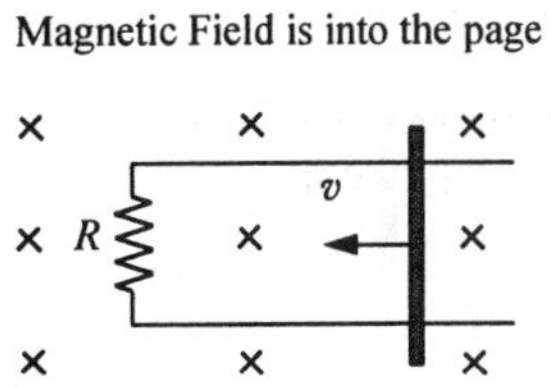

Solution: We know from Lenz's law that the magnetic field produced by the induced current will tend to oppose or resist the change in flux.

(a) As the north pole of the magnet approaches the coil, the field lines going into the coil increase. The induced current in the coil produces a magnetic field that opposes the increase, so the left end of the coil will be a north pole. By the right-hand source rule, the induced current is from right to left through the resistor.

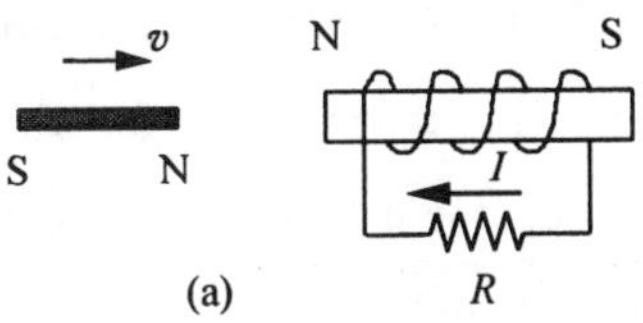

(a)

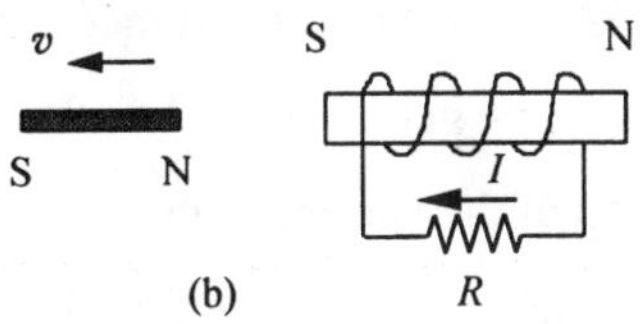

(b)

(b) As the north pole of the magnet moves away from the coil, the field lines going into the coil decrease. The induced current in the coil produces a magnetic field that opposes the decrease, so the left end of the coil will be a south pole. By the right-hand source rule, the induced current is from left to right through the resistor.

Now try to repeat parts (a) and (b) by flipping the magnet, that is, south pole near the coil.

(c) When the switch is closed, the magnetic field due to the coil on the left increases, so the field lines going though the coil on the right increase. The induced current in the coil on the right will produce a magnetic field to oppose that increase, so the left end of the coil on the right will be a north pole. Therefore, the current is from right to left.

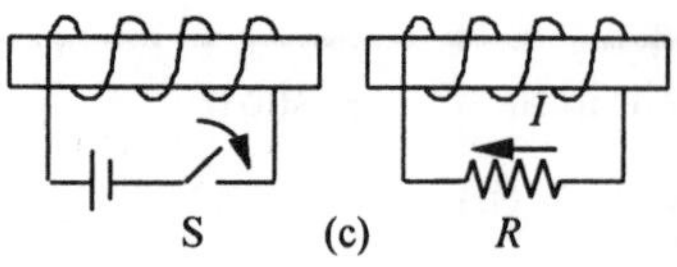

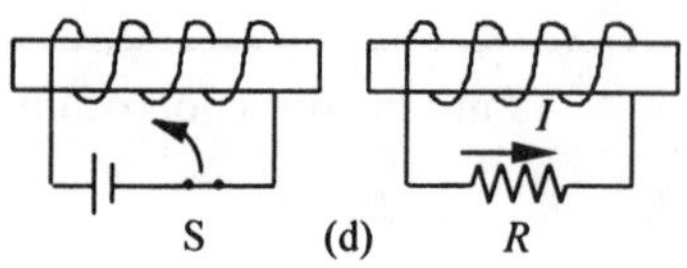

(d) When the switch is opened, the magnetic field due to the coil on the left decreases, so the field lines going though the coil on the right decrease. The induced current in the coil on the right will produce a magnetic field to oppose that decrease, so the left end of the coil on the right will be a south pole. Therefore, the current is from left to right.

(e) As the rod moves to the right, the area of the loop increases, so the magnetic flux through the area also increases. The induced current in the loop will produce a magnetic field to oppose that increase, so the magnetic field due to the induced current is out of the page. By the right-hand source rule, the current is from bottom to top through the rod.

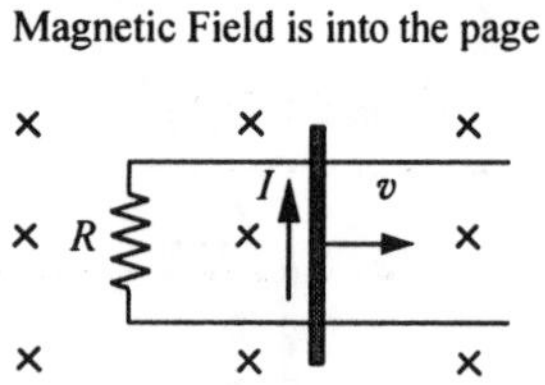

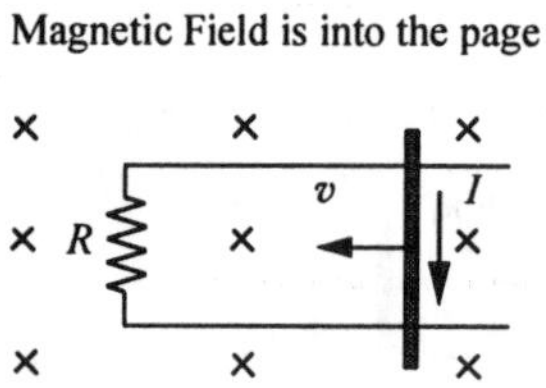

(f) As the rod moves to the left, the area of the loop decreases, so the magnetic flux through the area also decreases. The induced current in the loop will produce a magnetic field to oppose that decrease, so the magnetic field due to the induced current is onto the page. The current is from top to bottom through the rod.

2. Generators and Back Emf

A **generator** is a device to convert mechanical energy into electrical energy. In principle, it is the reverse of an electrical motor. An **ac generator** or alternator produces **alternating current** (ac) for which the polarity of the voltage and direction of the current change periodically. A **dc generator** produces direct current (dc). When coils are rotated in a magnetic field (or the field is rotated while the coils are kept fixed), an emf is induced in the coils and used to create an electric current. This creation of emf and current is due to the term $\frac{\Delta(\cos\theta)}{\Delta t}$ in Faraday's law of induction. The induced emf for an ac generator is expressed as $\mathcal{E} = \mathcal{E}_o \sin \omega t = \mathcal{E}_o \sin 2\pi f t$, where $\omega = 2\pi f$ is the angular frequency and $\mathcal{E}_o = NBA\omega$ the maximum emf (N is the number of turns of the coil, B the magnetic field, A the area of the coil).

A **back emf** is a reverse emf created by induction in motors when their armature is rotated in a magnetic field. The direction is opposite to the applied emf due to Lenz's law. The back emf partially cancels the voltage that drives the motor, so the current though the armature when the armature is rotating (operating) is smaller than when it is not (start-up). When a motor is in operating speed, the back emf is given by $\mathcal{E}_b = V - IR$, where V is the applied driving voltage, I the current through the armature, and R the resistance of the armature.

Example 20.6 An ac generator consists of 100 turns of wire of area 0.090 m^2 and total resistance 12 Ω. The loop rotates in a magnetic field of 0.50 T at a constant angular speed of 60 revolutions per second. Find the maximum induced emf and the maximum current.

Solution: Since the generator is rotating at 60 rev/s, the frequency is 60 cycles/s = 60 Hz.

Given: $N = 100$, $A = 0.090 \text{ m}^2$, $R = 12\ \Omega$, $B = 0.50$ T, $f = 60$ Hz.

Find: $\mathcal{E}_o$.

The angular frequency is equal to $\omega = 2\pi f = 2\pi(60 \text{ Hz}) = 120\pi$ rad/s.

So the maximum emf is $\mathcal{E}_o = NAB\omega = (100)(0.090 \text{ m}^2)(0.50 \text{ T})(120\pi \text{ rad/s}) = 1.7\times10^3 \text{ V} = 1.7$ kV.

Example 20.7 A dc motor of internal resistance 6.0 Ω is connected to a 24 V power supply. The operating current is 1.0 A.

(a) What is the start-up current?

(b) What is the back emf when the motor is running at full speed?

(c) What is the back emf when the motor is running at half speed?

Solution: Given: $R = 6.0\ \Omega$, $V = 24$ V, $I = 1.0$ A.

Find: (a) I (start-up) (b) $\mathcal{E}_b$.

(a) At start-up, there is no back emf since the motor is not rotating.

So the current is $I = \frac{V}{R} = \frac{24 \text{ V}}{6.0\ \Omega} = 4.0$ A.

(b) At full operating speed, the back emf is equal to $\mathcal{E}_b = V - IR = 24 \text{ V} - (1.0 \text{ A})(6.0\ \Omega) = 18$ V.

(c) From Faraday's law (better see the generator equation), the induced emf (back emf is induced emf after all) is directly proportional to the angular speed, so at half speed, the back emf is also halved to $\frac{18\text{ V}}{2} = 9.0\text{ V}$.

The current at this speed is $I = \frac{V - \mathcal{E}_b}{R} = \frac{24\text{ V} - 9.0\text{ V}}{6.0\ \Omega} = 2.5\text{ A}$.

---

3. Transformers and Power Transmission

A **transformer** is a device that changes the ac voltage supplied to it by means of induction. (Transformer cannot work with dc voltage.) Why? It consists of two coils, a **primary coil** (the coil on the input side) and a **secondary coil** (the coil on the output side) and usually an iron core. If the voltage on the secondary is larger than that on the primary, it is a **step-up transformer**. A **step-down transformer** has lower voltage on the secondary side. Whether a transformer is a step-up or step-down transformer is solely determined by the relative number of turns in the coils. A step-up transformer has more turns in the secondary, while a step-down transformer has less turns in the secondary. If the voltage is increased, the current is reduced, and vice versa, according to the conservation of energy. A real (in contrast to an ideal one) transformer has $I^2 R$ energy loss in the windings and another mechanism of energy loss caused by **Eddy currents** in the transformer core. The number of turns (N), voltage (V), and current (I) ratios are related by $\frac{I_p}{I_s} = \frac{V_s}{V_p} = \frac{N_s}{N_p}$, where the subscript p stands for primary and s for secondary.

In power transmission, the voltage is stepped up before transmission so as to decrease the current and the $I^2 R$ loss in the transmission lines (see Example below). The voltage is then stepped down before the power is supplied to businesses and homes.

Example 20.8 A generator produces 60 A of current at 120 V. The voltage is stepped up to 4500 V by a transformer and transmitted through a power line of total resistance 1.0 Ω. Find:

(a) the number of turns in the secondary coil if the primary coil has 200 turns,

(b) the power lost in the transmission line,

(c) the power lost in the transmission line if no transformer is used.

Solution: Given: $I_p = 60\text{ A}$, $V_p = 120\text{ V}$, $V_s = 4500\text{ V}$, $R = 1.0\ \Omega$, (a) $N_p = 200$.

Find: (a) N_s (b) P_{loss} (c) P_{loss} (no transformer).

(a) From $\frac{I_p}{I_s} = \frac{V_s}{V_p} = \frac{N_s}{N_p}$, we have $N_s = \frac{V_s}{V_p} N_p = \frac{4500\text{ V}}{120\text{ V}}(200) = 7500$ turns.

(b) $I_s = \frac{V_p}{V_s} I_p = \frac{120 \text{ V}}{4500 \text{ V}} (60 \text{ A}) = 1.6 \text{ A}.$

So the total power loss in the transmission lines is $P_{\text{loss}} = I^2 R = (1.6 \text{ A})^2 (1.0 \ \Omega) = 2.6 \text{ W}.$

The total power generated is $P_{\text{tot}} = (60 \text{ A})(120 \text{ V}) = 7200 \text{ W}.$

So the percentage loss is $\frac{2.6 \text{ W}}{7200 \text{ W}} = 0.036\%.$

(c) If no transformer is used, the current in transmission is 60 A.

The total power loss is $P_{\text{loss}} = (60 \text{ A})^2 (1.0 \ \Omega) = 3600 \text{ W}.$

So the percentage loss is $\frac{3600 \text{ W}}{7200 \text{ W}} = 50\%!$

4. Electromagnetic Waves

An **electromagnetic wave** (or **radiation**) consists of mutually perpendicular, time-varying electric and magnetic fields that propagate at a constant speed in vacuum (the speed of light, $c = 3.00 \times 10^8$ m/s). **Maxwell's equations** are a set of four equations that describe all magnetic and electric field phenomena. Radiation carries energy and momentum and thus can exert force and therefore **radiation pressure** (radiation force per unit area).

The different types of electromagnetic radiation differ in frequency f, and thus in wavelength λ, because $c = \lambda f$. The major types of electromagnetic radiation in increasing frequency (decreasing wavelength) are **power waves, radio and TV waves, microwaves, infrared radiation, visible light, ultraviolet radiation, X-rays,** and **gamma rays.**

Example 20.9 A radar receiver indicates that a transmitted pulse is returned as an echo 20 μs after transmission. How far away is the reflecting object?

Solution: Given: $\Delta t_{\text{tot}} = 20 \ \mu\text{s} = 20 \times 10^{-6}$ s, $c = 3.00 \times 10^8$ m/s.

Find: d.

Since the time is for the echo (forward and back), the time for one-way distance is $\Delta t_{1/2} = 10 \times 10^{-6}$ s.

So the distance is $d = c\Delta t_{1/2} = (3.00 \times 10^8 \text{ m/s})(10 \times 10^{-6} \text{ s}) = 3.0 \times 10^3 \text{ m} = 3.0 \text{ km}.$

Example 20.10 What are the wavelengths for the carrier waves in the following two radio stations?

(a) WBRN 1460 (b) WCKC 107

Solution: The call number of a radio station indicates the frequency of the carrier wave. For the AM band, the frequency is expressed in kilohertz (kHz) and for the FM band, the frequency is in megahertz (MHz).

Given: $c = 3.00 \times 10^8$ m/s, (a) f = 1460 kHz = 1460×10^3 Hz, (b) f = 107 MHz = 107×10^6 Hz.

Find: λ in (a) and (b).

(a) From $c = \lambda f$, we have $\lambda = \dfrac{c}{f} = \dfrac{3.00 \times 10^8 \text{ m/s}}{1460 \times 10^3 \text{ Hz}} = 205$ m.

(b) $\lambda = \dfrac{3.00 \times 10^8 \text{ m/s}}{107 \times 10^6 \text{ Hz}} = 2.80$ m.

The length of the antenna of a radio station is determined by the wavelength. By knowing the wavelength, we can construct the antenna to the right length.

IV. Mathematical Summary

Magnetic Flux	$\Phi = BA\cos\theta$ (20.1) θ is the angle between the normal to the plane of the loop and the magnetic field vector	Defines the magnetic flux through an area
Faraday's Law of Induction	$\mathcal{E} = -N\dfrac{\Delta\Phi}{\Delta t}$ (20.2) $= -N\left(\dfrac{\Delta B}{\Delta t}\right)(A\cos\theta)$ $-NB\left(\dfrac{\Delta A}{\Delta t}\right)(\cos\theta)$ $-NBA\left(\dfrac{\Delta(\cos\theta)}{\Delta t}\right)$ (20.3)	Calculates the induced emf as the change in magnetic flux divided by the change in time
Generator emf	$\mathcal{E} = \mathcal{E}_o \sin \omega t$ (20.4) $= \mathcal{E}_o \sin 2\pi f t$ (20.5) where $\mathcal{E}_o = NBA\,\omega$	Expresses the generated emf as a function of time
Back emf in a motor	$\mathcal{E}_b = V - IR$ (20.6)	Computes the back emf in an electric motor
Currents, Voltages, and Turn Ratios for a Transformer	$\dfrac{I_p}{I_s} = \dfrac{V_s}{V_p} = \dfrac{N_s}{N_p}$ (20.9)	Relates the current, voltage, and turn ratio in a transformer

V. Solutions of Selected Exercises and Paired Exercises

5. $\Phi = BA\cos\theta$, where θ is the angle between the field and the normal to the loop.

(a) $\Phi = BA\cos 90° = 0$.

(b) $\Phi = (0.25\text{ T})(0.030\text{ m}^2)\cos(90° - 37°) = 4.5\times10^{-3}\text{ T}\cdot\text{m}^2$.

(c) $\Phi = (0.25\text{ T})(0.030\text{ m}^2)\cos 0° = 7.5\times10^{-3}\text{ T}\cdot\text{m}^2$.

6. $$\varepsilon = -N\frac{\Delta\Phi}{\Delta t} = -N\frac{A\,\Delta B}{\Delta t} = -(1)\frac{(2.0\times10^{-3}\text{ m}^2)(0 - 6.0\text{ T})}{0.0045\text{ s}} = 2.7\text{ V}.$$

14. $\varepsilon = IR = (40\text{ A})(2.5\ \Omega) = 100\text{ V}$. Also $\varepsilon = -N\dfrac{\Delta\Phi}{\Delta t}$,

so $$\Delta t = -N\frac{\Delta\Phi}{\varepsilon} = -(1)\frac{-30\text{ T}\cdot\text{m}^2}{100\text{ V}} = 0.30\text{ s}.$$

22. $\varepsilon = IR = (3.0\text{ A})(0.10\ \Omega) = 0.30\text{ V}$. Also $\varepsilon = -N\dfrac{\Delta\Phi}{\Delta t} = -N\dfrac{A\,\Delta B}{\Delta t}$.

So $$\frac{\Delta B}{\Delta t} = -\frac{\varepsilon}{NA} = -\frac{0.30\text{ V}}{(1)(0.20\text{ m})(0.30\text{ m})} = -5.0\text{ T/s}.$$

38. $\varepsilon_o = NBA\omega = 2\pi NBAf = 2\pi(20)(0.800\text{ T})(\pi)(0.10\text{ m})^2(60\text{ Hz}) = 1.9\times10^2\text{ V}$.

This maximum value (positive or negative) is attained every half of a period.

$$t = \frac{T}{2} = \frac{1}{2f} = \frac{1}{2(60\text{ Hz})} = 1/120\text{ s}.$$

42. (a) From $\varepsilon_b = V - IR$, we have $I = \dfrac{V - \varepsilon_b}{R} = \dfrac{12\text{ V} - 10\text{ V}}{0.40\ \Omega} = 5.0\text{ A}$.

(b) When initially starting up, there is no back emf, so $I = \dfrac{12\text{ V}}{0.40\ \Omega} = 30\text{ A}$.

48. (a) Step-up since $N_s > N_p$.

(b) $\dfrac{I_p}{I_s} = \dfrac{V_s}{V_p} = \dfrac{N_s}{N_p} = \dfrac{450}{75} = 6$

(c) $\dfrac{V_p}{V_s} = \dfrac{N_p}{N_s} = \dfrac{75}{450} = 1/6$.

56. (a) From $\frac{I_p}{I_s} = \frac{V_s}{V_p} = \frac{N_s}{N_p}$, we have $N_s = \frac{V_s}{V_p} N_p = \frac{20\text{ V}}{120\text{ V}}(300) = 50$ turns.

(b) $I_p = \frac{50}{300}(0.50\text{ A}) = 8.3\times10^{-2}\text{ A}$.

61. (a) At the area substation, $\frac{N_s}{N_p} = \frac{V_s}{V_p} = \frac{100{,}000\text{ V}}{200{,}000\text{ V}} = 1/2$;

at the distributing station, $\frac{N_s}{N_p} = \frac{7200\text{ V}}{100{,}000\text{ V}} = 1/13.9$;

at the utility pole, $\frac{N_s}{N_p} = \frac{240\text{ V}}{7200\text{ V}} = 1/30$.

(b) At the area substation, $\frac{I_s}{I_p} = \frac{V_p}{V_s} = \frac{200{,}000\text{ V}}{100{,}000\text{ V}} = 2.0$;

at the distributing station, $\frac{I_s}{I_p} = \frac{100{,}000\text{ V}}{7200\text{ V}} = 13.9$;

at the utility pole, $\frac{I_s}{I_p} = \frac{7200\text{ V}}{240\text{ V}} = 30$;

(c) Overall, $\frac{I_s}{I_p} = \frac{200{,}000\text{ V}}{240\text{ V}} = 833$.

73. For AM: $L = \frac{\lambda}{4} = \frac{c/f}{4} = \frac{c}{4f} = \frac{3.00\times10^8\text{ m/s}}{4(0.53\times10^6\text{ Hz} + 1.7\times10^6\text{ Hz})/2} = 67\text{ m}$.

For FM, $L = \frac{3.00\times10^8\text{ m/s}}{4(88\times10^6\text{ Hz} + 108\times10^6\text{ Hz})/2} = 0.77\text{ m}$.

80. (a) $B = \frac{\mu_0 I}{2r} = \frac{(4\pi\times10^{-7}\text{ T·m/A})(2.50\text{ A})}{0.28\text{ m}} = 1.12\times10^{-5}\text{ T}$.

$\Phi = BA\cos\theta = (1.12\times10^{-5}\text{ T})(\pi)(0.14\text{ m})^2\cos 0° = 6.9\times10^{-7}\text{ T·m}^2$.

(b) $B' = \frac{(4\pi\times10^{-7}\text{ T·m/A})(0.50\text{ A})}{0.28\text{ m}} = 2.24\times10^{-6}\text{ T}$.

$\Phi' = (2.24\times10^{-6}\text{ T})(\pi)(0.14\text{ m})^2\cos 0° = 1.38\times10^{-7}\text{ T·m}^2$.

So $\mathcal{E} = -N\frac{\Delta\Phi}{\Delta t} = -(1)\frac{1.38\times10^{-7}\text{ T·m}^2 - 6.90\times10^{-7}\text{ T·m}^2}{0.35\text{ s}} = 1.6\ \mu\text{V}$.

86. $I_s = \frac{P_s}{V_s} = \frac{6.6\times10^3\text{ W}}{220\text{ V}} = 30\text{ A}$; $I_p = \frac{V_s}{V_p} I_s = \frac{220\text{ V}}{20{,}000\text{ V}}(30\text{ A}) = 0.33\text{ A}$.

90. $\lambda = 2(5.0\text{ cm}) = 10\text{ cm} = 0.10\text{ m}$. $f = \frac{c}{\lambda} = \frac{3.00\times10^8\text{ m/s}}{0.10\text{ m}} = 3.0\times10^9\text{ Hz}$ or 3.0 GHz.

VI. Practice Quiz

1. According to Lenz's law, the direction of an induced current in a conductor will be that which tends to produce which of the following effects?
(a) enhance the effect which produces it (b) produce a greater heating effect
(c) produce the greatest voltage (d) oppose the effect which produces it
(e) none of the above

2. A motor with an internal resistance of 10 Ω operating on a 24 V source produces a back emf of 15 V. What is the current through the motor?
(a) 3.9 A (b) 2.4 A (c) 1.5 A (d) 0.90 A (e) zero

3. A uniform 1.50 T magnetic field passes perpendicularly through the plane of a wire loop 0.300 m^2 in area. What is the magnetic flux through the loop?
(a) 5.00 Wb (b) 0.450 Wb (c) 0.250 Wb (d) 0.135 Wb (e) zero

4. A 0.200-m wire is moved perpendicular to a 0.500 T-magnetic field at a speed of 1.50 m/s. What emf is induced across the ends of the wire?
(a) 2.25 V (b) 1.00 V (c) 0.600 V (d) 0.150 V (e) zero

5. A bar magnet falls through a loop of wire with constant velocity with the north pole entering first. When the north pole enters the wire, the induced current will be (as viewed from above)
(a) clockwise. (b) counterclockwise. (c) zero. (d) to right of loop. (e) to left of loop.

6. A coil consisting of four turns of wire, each with an area of 2.0×10^{-2} m^2, is oriented perpendicularly to a magnetic field that increases uniformly from 1.0×10^{-2} T to 2.5×10^{-2} T in a time of 5.0×10^{-3} s. What is the resulting current in the coil if the resistance of the coil is 5.0 Ω?
(a) 0.030 A (b) 0.060 A (c) 0.012 A (d) 2.4×10^{-4} A (e) 4.8×10^{-2} A

7. You are designing a generator with a maximum emf 8.0 V. If the generator coil has 200 turns and a cross-sectional area of 0.030 m^2, what would be the frequency of the generator in a magnetic field of 0.030 T?
(a) 7.1 Hz (b) 7.5 Hz (c) 8.0 Hz (d) 8.5 Hz (e) 44 Hz

8. A transformer consists of a 500-turn primary coil and a 2000-turn secondary coil. If the current in the secondary is 3.0 A, what is the primary current?

(a) 0 (b) 0.75 A (c) 1.33 A (d) 12 A (e) 48 A

9. An electromagnetic wave is made up of which of the following time-varying quantities?

(a) electrons only (b) electric fields only (c) magnetic fields only

(d) both electric and magnetic fields (e) neither electric fields nor magnetic fields

10. What is the wavelength of a radio wave signal transmitted at a frequency of 7.2 MHz?

(a) 21 m (b) 42 m (c) 4.2 m (d) 29 m (e) 0.024 m

Answers to Practice Quiz:

1. d 2. d 3. b 4. d 5. b 6. e 7. a 8. d 9. d 10. b

CHAPTER 21

AC Circuits

I. Chapter Objectives

Upon completion of this chapter, you should be able to:

1. specify how voltage, current, and power vary with time in an ac circuit, and explain how the rms and peak values of these quantities are related and how they are used in calculations.
2. explain the behavior of a capacitor in ac circuits and calculate the effect on ac current (capacitive reactance).
3. explain what an inductor is and calculate the effect of inductors on ac circuits (inductive reactance).
4. calculate the currents and voltages when reactive circuit elements are present in ac circuits (series RC, RL, RLC circuits), use phase diagrams to calculate overall impedance and rms currents, and understand and use the concept of the power factor in ac circuits.
5. understand the concept of resonance in ac circuits and calculate the resonance frequency of an RLC circuit.

II. Key Terms

Upon completion of this chapter, you should be able to define and/or explain the following key terms:

peak voltage
peak current
rms (or effective) current
rms (or effective) voltage
capacitive reactance (X_C)
inductance
henry (H)
inductive reactance (X_L)
phase diagram
phasors
impedance (Z)
phase angle (ϕ)
inductive circuit
capacitive circuit
power factor
resonance frequency

The definitions and/or explanations of the most important key terms can be found in the following section: **III. Chapter Summary and Discussion**.

III. Chapter Summary and Discussion

1. Resistance in an AC Circuit

In an ac circuit, voltage and current vary sinusoidally (or voltages and currents are sine or cosine functions of time) with time. The instantaneous voltage is given by $V = V_o \sin \omega t = V_o \sin 2\pi f t$, where V_o is the **peak voltage** which is the maximum voltage value attained during a cycle of oscillation, and $\omega = 2\pi f$ is the angular frequency.

In a resistor ac circuit, the current through a resistor is equal to $I = \frac{V}{R} = \frac{V_o \sin 2\pi f t}{R} = \frac{V_o}{R} \sin 2\pi f t$ $= I_o \sin 2\pi f t$, where I_o is the **peak current** (maximum current value). The voltage across and current though the resistor are in step, or *in phase*, with each other, both reaching zero values and maxima at the same time.

The voltage and current in ac circuits are often given in terms of the **rms voltage** and **rms current**, unless otherwise specified. These are statistical averages of the **effective** *ac voltage and current*. The rms values are less than the peak values, and are related to the peak values by $V_{\text{rms}} = \frac{V_o}{\sqrt{2}} \approx 0.707\ V_o$ and $I_{\text{rms}} = \frac{I_o}{\sqrt{2}} \approx 0.707\ I_o$. The rms voltage and current are related by $V_{\text{rms}} = I_{\text{rms}} R$ and the average power dissipated by a resistor is equal to $\overline{P} = V_{\text{rms}} I_{\text{rms}} = I_{\text{rms}}^2 R$. Appliances using ac are rated in terms of the average power.

Example 21.1 A oven heating element rated at 3500 W is connected to a 220-V outlet.

(a) What are the rms and peak currents through the element?

(b) What is the resistance of the element?

Solution: Given: $\overline{P} = 3500\ \text{W}$, $V_{\text{rms}} = 220\ \text{V}$.

Find: (a) I_{rms} and I_o (b) R.

(a) From $\overline{P} = V_{\text{rms}} I_{\text{rms}}$, we have $I_{\text{rms}} = \frac{\overline{P}}{V_{\text{rms}}} = \frac{3500\ \text{W}}{220\ \text{V}} = 15.9\ \text{A}$.

The peak current is $I_o = \sqrt{2}\ I_{\text{rms}} = \sqrt{2}\ (15.9\ \text{A}) = 22.5\ \text{A}$.

(b) $R = \frac{V_{\text{rms}}}{I_{\text{rms}}} = \frac{220\ \text{V}}{15.9\ \text{V}} = 13.8\ \Omega$.

Example 21.2 A 10-Ω resistor is connected to a 120-V ac power supply.

(a) What is the peak current though the resistor?

(b) What is the average power dissipated in the resistor?

Solution: Given: $R = 10\ \Omega$, $V_{rms} = 120$ V.

Find: (a) I_o (b) $\overline{P}$.

(a) $I_{rms} = \dfrac{V_{rms}}{R} = \dfrac{120\text{ V}}{10\ \Omega} = 12$ A.

So the peak current is $I_o = \sqrt{2}\, I_{rms} = \sqrt{2}$ (12 A) = 17 A.

Note: For convenience and simplicity, the rms subscripts will be omitted from now on -- but keep in mind that rms values are given for ac quantities unless otherwise specified.

2. Capacitive Reactance

A capacitor in an ac circuit limits, or regulates the current, but does not completely prevent the flow of charge. This impeding effect is expressed in terms of **capacitive reactance** (X_C) given by $X_C = \dfrac{1}{2\pi f C} = \dfrac{1}{\omega C}$. The SI unit of capacitive reactance is ohm (Ω). Capacitive reactance is analogous to resistance and relates the voltage across and current through the capacitor by Ohm's law, $V = I X_C$.

In a purely capacitive ac circuit (capacitor is the only circuit element), the *current leads the voltage* by 90° or a quarter (1/4) cycle. This means the voltage and current are *not* in step, or in phase. When the current reaches maximum, the voltage is zero, and when the current reaches zero the voltage is maximum. In words, the voltage lags the current by 90°. A capacitor simply charges and discharges in an ac circuit and does not dissipate power or consume energy.

Example 21.3 A 10.0-μF capacitor is connected to a 120-V, 60-Hz ac source.

(a) What is the capacitive reactance?

(b) What is the rms current?

(c) What power does the capacitor dissipate?

Solution: Given: $C = 10.0\ \mu\text{F} = 10.0\times10^{-6}$ F, $V = 120$ V, $f = 60$ Hz (assume exact).

Find: (a) X_C (b) I (c) P.

(a) $X_C = \dfrac{1}{2\pi f C} = \dfrac{1}{2\pi\,(60\text{ Hz})(10.0\times10^{-6}\text{ F})} = 265\ \Omega$.

(b) From Ohm's law, $I = \dfrac{V}{X_C} = \dfrac{120\text{ V}}{265\ \Omega} = 0.453$ A.

(c) An ideal ($R = 0$) capacitor does not dissipate power, so $P = 0$.

---

3. Inductive Reactance

An inductor in an ac circuit also limits, or regulates the current, but does not completely prevent the flow of charge. This impeding effect is expressed in terms of **inductive reactance** (X_L) given by $X_L = 2\pi f L = \omega L$, where L is the **inductance** of the inductor and has an SI unit of **henry** (H). The SI unit of inductive reactance (X_L) is the ohm (Ω). Inductive reactance is analogous to resistance and relates the voltage across and current through the inductor by Ohm's law, $V = I X_L$.

In a purely inductive ac circuit (inductor is the only circuit element), the *voltage leads the current* by 90° or a quarter (1/4) cycle. This means the voltage and current are *not* in step, or in phase. When the voltage reaches maximum, the current is zero, and when the voltage reaches zero the current is maximum. In words, the current lags the current by 90°. A inductor simply stores and releases magnetic energy in an ac circuit and does not dissipate power or consume energy.

Note: The phase relationship of current and voltage for purely inductive and purely capacitive circuits are opposite. A phrase that may help you remember the relationship is *ELI* the *ICE* man. With *E* representing voltage and *I* representing current, *ELI* indicates that with an inductance (*L*) the voltage leads the current (*I*). Similarly, *ICE* tells you that with a capacitance (*C*) the current leads the voltage.

Example 21.4 A 25.0-mH inductor is connected to a 120-V, 60-Hz ac source. Find the inductive reactance and the rms current in the circuit.

Solution: Given: $L = 25.0 \text{ mH} = 25.0 \times 10^{-3}$ H, $V = 120$ V, $f = 60$ Hz.

Find: X_L and I.

The inductive reactance is $X_L = 2\pi f L = 2\pi(60 \text{ Hz})(25.0 \times 10^{-3} \text{ H}) = 9.42\ \Omega$.

From Ohm's law, the rms current is equal to $I = \dfrac{V}{X_L} = \dfrac{120 \text{ V}}{9.42\ \Omega} = 12.7$ A.

---

4. Impedance: RLC Circuits

Impedance (Z) is the generalization of current opposition to include not only resistance, but also capacitive and inductive reactances. Because of the phase difference between the voltage and currents on different circuit elements, a **phase diagram** is conveniently used. In a phase diagram, the resistance and reactance of the circuit are given vectorlike properties and their magnitudes are represented as arrows called **phasors**. On a set of *x*-

y axes, the resistance is used as a reference and is plotted on the positive x axis, since the voltage-current phase difference for a resistor is zero ($\phi = 0$). The reactances are plotted based on their phase differences. For example, the voltage lags the current by 90° in a capacitor, so the capacitive reactance is plotted on the $-y$ axis; the voltage leads the current by 90° for a inductor, so the inductive reactance is plotted on the positive y axis;

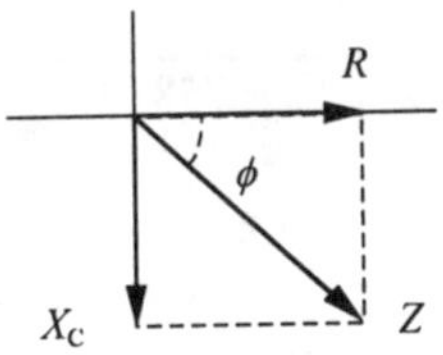

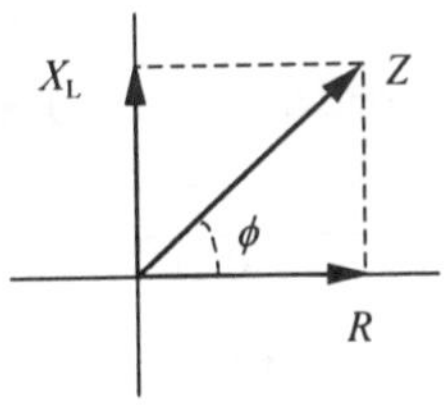

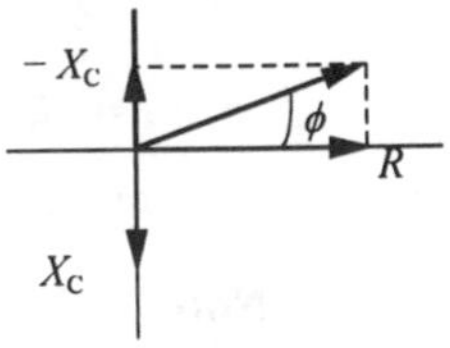

From the phase diagram, the impedance in an RLC series circuit is given by

$Z = \sqrt{R^2 + (X_L - X_C)^2} = \sqrt{R^2 + \left(2\pi f L - \dfrac{1}{2\pi f C}\right)^2}$. The angle ϕ is called the **phase angle** of the circuit and is equal to $\tan\phi = \dfrac{X_L - X_C}{R}$. If $X_L > X_C$, the angle ϕ is positive and the circuit is said to be an **inductive circuit**; if $X_C > X_L$, the angle ϕ is negative and the circuit is said to be a **capacitive circuit**.

If the circuit is an RC circuit (no inductor), then $X_L = 0$, $Z = \sqrt{R^2 + X_C^2}$ and $\tan\phi = -\dfrac{X_C}{R}$; if the circuit is an RL circuit (no capacitor), then $X_C = 0$, $Z = \sqrt{R^2 + X_L^2}$ and $\tan\phi = \dfrac{X_L}{R}$. Ohm's law $V = IZ$ is valid for all such circuits.

In an RLC series circuit, *there are no power losses associated with pure capacitor and inductor. Only the resistor dissipates power.* Therefore, the power in an ac circuit can be expressed as $P = I^2 R = I^2 Z \dfrac{R}{Z} = I^2 Z \cos\phi = I V \cos\phi$, where the term $\cos\phi$ is called the **power factor** and is defined as $\cos\phi = \dfrac{R}{Z}$ from the phase diagram. It is a measure of how close the circuit is to expending the maximum power ($\cos 0° = 1$).

Example 21.5 An RLC series circuit consists of a 250-Ω resistor, a 3.50-μF capacitor, and a 0.600-H inductor. The circuit is connected to a 120-V, 60-Hz power supply.

(a) What is the impedance?

(b) What is the phase angle? Is the circuit inductive or capacitive?

(c) What is the rms current?

(d) What is the rms voltage across each element?

(e) What is the power dissipated in the circuit?

Solution: Given: $R = 250\ \Omega$, $C = 3.50\ \mu F = 3.50 \times 10^{-6}$ F, $L = 0.600$ H, $V = 120$ V, $f = 60$ Hz.

Find: (a) Z (b) I (c) V on each element (d) P.

(a) $X_L = 2\pi f L = 2\pi(60\ \text{Hz})(0.600\ \text{H}) = 226\ \Omega$, $X_C = \dfrac{1}{2\pi f C} = \dfrac{1}{2\pi(60\ \text{Hz})(3.50\times10^{-6}\ \text{F})} = 758\ \Omega$.

So the impedance is $Z = \sqrt{R^2 + (X_L - X_C)^2} = \sqrt{(250\ \Omega)^2 + (226\ \Omega - 758\ \Omega)^2} = 588\ \Omega$.

(b) $\tan\phi = \dfrac{X_L - X_C}{R} = \dfrac{226\ \Omega - 758\ \Omega}{250\ \Omega} = -2.13$. So $\phi = -64.8°$.

Since the phase angle is negative, the circuit is a capacitive circuit.

(c) $I = \dfrac{V}{Z} = \dfrac{120\ \text{V}}{588\ \Omega} = 0.204$ A.

(d) $V_R = IR = (0.204\ \text{A})(250\ \Omega) = 51.0$ V, $V_C = IX_C = (0.204\ \text{A})(758\ \Omega) = 155$ V, and $V_L = IX_L = (0.204\ \text{A})(226\ \Omega) = 46.1$ V.

Note: $V_R + V_C + V_L = 51.0\ \text{V} + 155\ \text{V} + 46.1\ \text{V} = 252\ \text{V} > 120$ V. Is this a violation of the conservation of energy? How can the sum of the individual voltages be greater than the total voltage? (Hint: the voltages calculated here are rms, or effective voltages, and the voltages, at any instant, are *not* in phase!)

(e) Only the resistor dissipates energy. $P = I^2 R = (0.204\ \text{A})^2\ (250\ \Omega) = 10.4$ W.

Or, $P = IV\cos\phi = (0.204\ \text{A})(120\ \text{V})\cos 64.8° = 10.4$ W.

---

5. Circuit Resonance

When $X_L = X_C$ (the capacitive reactance and the inductive reactance cancel each other completely), $\phi = 0$ and $Z = R$ (completely resistive) is at minimum, so the circuit dissipates maximum power. This condition is called resonance and the frequency at which resonance occurs is called the **resonance frequency** (f_o). At resonance, $X_L = X_C$, or $2\pi f_o L = \dfrac{1}{2\pi f_o C}$, so $f_o = \dfrac{1}{2\pi\sqrt{LC}}$. Resonant circuits have a wide variety of applications such as radio and television tuning.

Example 21.6 An RLC series circuit consists of a 100-Ω resistor, a 20.0-μF capacitor, and a 0.550-H inductor. The circuit is connected to a 120-V tunable (frequency) power supply.

(a) What is the resonance frequency of the circuit?

(b) What is the impedance at resonance?

(c) What is the current in the circuit at resonance?

(d) What is the power dissipated at resonance?

(e) What are the voltages across the elements at resonance?

Solution: Given: $R = 100\ \Omega$, $C = 20.0\mu\text{F} = 20.0\times10^{-6}$ F, $L = 0.550$ H, $V = 120$ V.

Find: (a) f_o (b) Z (c) I (d) P (e) V for each element.

(a) $f_o = \dfrac{1}{2\pi\sqrt{LC}} = \dfrac{1}{2\pi\sqrt{(20.0\times10^{-6}\ \text{F})(0.550\ \text{H})}} = 48.0$ Hz.

(b) At resonance, $X_L = X_C$, so $Z = \sqrt{R^2 + (X_L - X_C)^2} = R = 100\ \Omega$ (it is at minimum).

(c) $I = \dfrac{V}{Z} = \dfrac{120\ \text{V}}{100\ \Omega} = 1.20$ A.

(d) $P = I^2 R = (1.20\ \text{A})^2\,(100\ \Omega) = 144$ W.

(e) $V_R = IR = (1.20\ \text{A})(100\ \Omega) = 120$ V.

$X_L = 2\pi f L = 2\pi(48.0\ \text{Hz})(0.550\ \text{H}) = 166\ \Omega$, so $V_L = I X_L = (1.20\ \text{A})(166\ \Omega) = 199$ V.

$X_C = \dfrac{1}{2\pi f C} = \dfrac{1}{2\pi(48.0\ \text{Hz})(20.0\times10^{-6}\ \text{F})} = 166\ \Omega$, so $V_C = IX_C = (1.20\ \text{A})(166\ \Omega) = 199$ V.

Note: The voltages across the capacitor and the inductor are out of phase (180°, see phase diagram), so they cancel out completely.

IV. Mathematical Summary

Instantaneous Voltage in an ac Circuit	$V = V_o \sin 2\pi f t$ (21.1)	Gives the instantaneous voltage in an ac circuit as a function of time
Instantaneous Current in an ac Circuit	$I = I_o \sin 2\pi f t$ (21.2) where $I_o = V_o/R$	Gives the instantaneous current in an ac circuit as a function of time
Rms (or Effective) Current	$I_{rms} = \dfrac{I_o}{\sqrt{2}} \approx 0.707\, I_o$ (21.6)	Relates the rms current and the peak current
Rms (or Effective) Voltage	$V_{rms} = \dfrac{V_o}{\sqrt{2}} \approx 0.707\, V_o$ (21.8)	Relates the rms voltage and the peak voltage
Relationship between Voltage and Current in an ac Circuit*	$V_{rms} = I_{rms} R$ (21.9)	Gives the relationship between the rms voltages and rms current
Effective Power in an ac circuit	$\overline{P} = I_{rms}^2 R$ (21.7)	Computes the effective (average) power in an ac circuit
Capacitive Reactance	$X_C = \dfrac{1}{2\pi f C} = \dfrac{1}{\omega C}$ (21.10)	Defines the capacitive reactance of a capacitor in an ac circuit
Relationship between	$V = I X_C$ (21.11)	Gives the relationship between the rms voltage and rms

Voltage and Current	(capacitor only)	current on a capacitor
Inductive Reactance	$X_L = 2\pi f L = \omega L$ (21.13)	Defines the inductive reactance of an inductor in an ac circuit
Relationship between Voltage and Current	$V = IX_L$ (21.14) (inductor only)	Gives the relationship between the instantaneous voltage and instantaneous current on an inductor
Impedance for an RLC Circuit	$Z = \sqrt{R^2 + (X_L - X_C)^2}$ (21.18) $= \sqrt{R^2 + \left(2\pi f L - \frac{1}{2\pi f C}\right)^2}$	Defines the impedance of a RLC series circuit
Phase Angle between Voltage and Current in an ac Circuit	$\tan\phi = \frac{X_L - X_C}{R}$ (21.20)	Defines the phase angle between the voltage and current in an ac circuit
Relationship between rms Current and Voltage for RC, RL, and RLC Circuits	$V = IZ$ (21.19)	Relates rms voltage and rms current in RC, RL, and RLC series ac circuits
Power Factor for an RLC Circuit	$\cos\phi = \frac{R}{Z}$ (21.22)	Defines the power factor in an ac circuit
Power in Terms of Power Factor	$P = IV\cos\phi$ (21.23) or $P = I^2 Z\cos\phi$ (21.24)	Expresses the power in an ac circuit in terms of the power factor
Resonance Frequency for an RLC Circuit	$f_o = \frac{1}{2\pi\sqrt{LC}}$ (21.25)	Computes the resonance frequency in an RC or RLC series circuit

V. Solutions of Selected Exercises and Paired Exercises

9. (a) $V_{rms} = I_{rms} R = (0.75\text{ A})(5.0\ \Omega) = 3.8\text{ V}$; $V_o = \sqrt{2}\, V_{rms} = \sqrt{2}\,(3.75\text{ V}) = 5.3\text{ V}$.

(b) $\bar{P} = I_{rms} V_{rms} = (0.75\text{ A})(3.75\text{ V}) = 2.8\text{ W}$.

13. (a) Since $I = I_o \sin 2\pi f t$, $2\pi f t = 380\,t$. So $f = \frac{380}{2\pi} = 60\text{ Hz}$.

(b) $I_{rms} = \frac{I_o}{\sqrt{2}} = \frac{2.0\text{ A}}{\sqrt{2}} = 1.4\text{ A}$.

(c) $\bar{P} = I_{rms}^2 R = (1.41\text{ A})^2 (60\ \Omega) = 1.2\times10^2\text{ W}$.

(d) $V_o = I_o R = (2.0\text{ A})(60\ \Omega) = 120\text{ V}$. So $V = (120\text{ V})\sin 380t$.

(e) $P = VI = (120\text{ V})(2.0\text{ A})\sin^2 380t$. So $P = (240\text{ W})\sin 380t$.

(f) $P = (240\text{ W})\dfrac{1-\cos 2(380t)}{2} = 120\text{ W} - (120\text{ W})\cos 2(380t)$.

The average of a sine or cosine function is zero. So $\overline{P} = 120$ W.

18. $I_{\text{rms}} = \dfrac{\overline{P}}{V_{\text{rms}}} = \dfrac{40\text{ W}}{120\text{ V}} = 0.33\text{ A}$ and $I_{\text{o}} = \sqrt{2}\, I_{\text{rms}} = \sqrt{2}\,(0.333\text{ A}) = 0.47\text{ A}$.

28. The current is zero when voltage is maximum because current lags a quarter cycle (90°) behind the voltage in an inductive ac circuit. The slope of the curve for the current is directly proportional to the slope of the curve for the change in flux over time. When a change in flux is zero, voltage is at a maximum.

30. From $X_{\text{C}} = \dfrac{1}{2\pi fC}$, $f = \dfrac{1}{2\pi C X_{\text{C}}} = \dfrac{1}{2\pi\,(25\times10^{-6}\text{ F})(25\ \Omega)} = 2.5\times10^{2}\text{ Hz}$.

36. From $X_{\text{L}} = 2\pi fL$, $L = \dfrac{X_{\text{L}}}{2\pi f} = \dfrac{90\ \Omega}{2\pi(60\text{ Hz})} = 0.24\text{ H}$.

45. $Z = \sqrt{R^2 + (X_{\text{L}} - X_{\text{C}})^2} = X_{\text{L}} - X_{\text{C}} = 2\pi fL - \dfrac{1}{2\pi fC}$,

$$I_{\text{rms}} = \frac{V_{\text{rms}}}{Z} = \frac{V_{\text{rms}}}{2\pi fL - \dfrac{1}{2\pi fC}}.$$

There is a resonance at $f = f_{\text{o}}$ (where $I_{\text{rms}} = \infty$).

46. (a) $X_{\text{C}} = \dfrac{1}{2\pi fC} = \dfrac{1}{2\pi(60\text{ Hz})(25\times10^{-6}\text{ F})} = 1.1\times10^{2}\ \Omega$.

(b) $Z = \sqrt{R^2 + (X_{\text{L}} - X_{\text{C}})^2} = \sqrt{(200\ \Omega)^2 + (0 - 106\ \Omega)^2} = 226\ \Omega$.

So $I_{\text{rms}} = \dfrac{V_{\text{rms}}}{Z} = \dfrac{120\text{ V}}{226\ \Omega} = 0.53\text{ A}$.

56. $f_{\text{o}} = \dfrac{1}{2\pi\sqrt{LC}}$,

so $C = \dfrac{1}{4\pi^2 f_{\text{o}}^2 L} = \dfrac{1}{4\pi^2\,(980\times10^{3}\text{ Hz})^2\,(0.50\times10^{-3}\text{ H})} = 5.3\times10^{-11}\text{ F}$.

66. In resonance, $\overline{P}_{\text{o}} = \dfrac{V_{\text{rms}}^2}{R} = \dfrac{(120\text{ V})^2}{50\ \Omega} = 288\text{ W}$.

Not in resonance, $X_{\text{L}} = 2\pi fL = 2\pi(60\text{ Hz})(0.15\text{ H}) = 56.5\ \Omega$,

$X_{\text{C}} = \dfrac{1}{2\pi fC} = \dfrac{1}{2\pi(60\text{ Hz})(20\times10^{-6}\text{ F})} = 133\ \Omega$.

$Z = \sqrt{R^2 + (X_L - X_C)^2} = \sqrt{(50\ \Omega)^2 + (56.5\ \Omega - 133\ \Omega)^2} = 91.4\ \Omega.$

$I_{rms} = \dfrac{V_{rms}}{Z} = \dfrac{120\ \text{V}}{91.4\ \Omega} = 1.31\ \text{A}.$

Only the resistor dissipates power. So $\bar{P} = I_{rms}^2 R = (1.31\ \text{A})^2\,(50\ \Omega) = 86.2\ \text{W}.$

Therefore, the percentage is $\dfrac{\bar{P}}{P_o} = \dfrac{86.2\ \text{W}}{288\ \text{W}} = 30\%.$

74. (a) $X_L = 2\pi f L = 2\pi(60\ \text{Hz})(0.100\ \text{H}) = 38\ \Omega.$

(b) $Z = \sqrt{R^2 + (X_L - X_C)^2} = \sqrt{(50\ \Omega)^2 + (37.7\ \Omega)^2} = 63\ \Omega.$

(c) $I_{rms} = \dfrac{V_{rms}}{Z} = \dfrac{110\ \text{V}}{62.6\ \Omega} = 1.8\ \text{A}.$

(d) Only the resistor dissipates power. So the power dissipated in the coil is 0.

77. From $X_C = \dfrac{1}{2\pi f C}$ and $X_L = 2\pi f L$, we can see that the capacitor opposes current more strongly at lower frequencies and the inductor opposes current more strongly at high frequencies. In Fig. 21.14a, the inductor in series with R_L filters out the high frequency current and so only the low frequency current reaches R_L. In Fig. 21.14b, the capacitor in series with R_L filters out the low frequency current and so only the high frequency current reaches R_L.

78. From $f_o = \dfrac{1}{2\pi\sqrt{LC}}$, $C = \dfrac{1}{4\pi^2 f_o^2 L} = \dfrac{1}{4\pi^2\,(60\ \text{Hz})(0.750\ \text{H})} = 9.4\times10^{-6}$ F for any resistor since f_o is independent of R.

VI. Practice Quiz

1. The capacitive reactance in an ac circuit will change by what factor when the frequency is tripled?

(a) 1/3 (b) 1/9 (c) 1 (d) 9 (e) 3

2. What is the rms current in an ac circuit with a peak current of 20 A?

(a) zero (b) 28 A (c) 7.1 A (d) 14 A (e) 20 A

3. What is the impedance of an ac circuit with 12.0-Ω resistance, 15.0-Ω inductive reactance, and 10.0-Ω capacitive reactance?

(a) 37.0 Ω (b) 27.7 Ω (c) 11.6 Ω (d) 21.9 Ω (e) 13.0 Ω

4. An ac series circuit has 12.0-Ω resistance, 15.0-Ω inductive reactance, and 10.0-Ω capacitive reactance. If an rms voltage of 120 V is applied, what is the power output?

(a) 435 W (b) 598 W (c) 1022 W (d) 1108 W (e) 1200 W

5. What is the phase difference between the voltages of the inductor and capacitor in a RLC series circuit.

(a) 45° (b) 90° (c) 180° (d) 360° (e) zero

6. What is the resonant frequency for a circuit containing a 10-mH inductor and a 100-μF capacitor?

(a) 80 Hz (b) 1.6×10^{2} Hz (c) 1.6×10^{5} Hz (d) 6.3×10^{-6} Hz (e) 6.3×10^{-3} Hz.

7. An ac series circuit has an impedance of 60 Ω and a resistance of 30 Ω. What is the power factor?

(a) 0.50 (b) 0.71 (c) 1.0 (d) 1.4 (e) 2.0

8. Resonance occurs in an ac series circuit when

(a) resistance equals capacitive reactance.
(b) resistance equals inductive reactance.
(c) capatitive reactance equals inductive reactance.
(d) capacitive reactance equals zero.
(e) inductive reactance equals zero.

9. A series RLC circuit consists of a 100-Ω resistor, a 10.0-μF capacitor, and a 0.250-H inductor. The circuit is connected to a 120-V, 60-Hz power supply. What is the current in the circuit?

(a) 0.31 A (b) 0.41 A (c) 0.51 A (d) 0.61 A (e) 0.71 A

10. What is the phase angle and power factor for a series RLC circuit containing a 50-Ω resistor, a 10-μF capacitor, and a 0.45-H inductor, when connected to a 60 Hz power supply?

(a) zero and 1.0 (b) 38° and 0.79 (c) −38° and 0.79 (d) 62° and 0.47 (e) −62° and 0.47

Answers to Practice Quiz:

1. a 2. d 3. e 4. c 5. c 6. b 7. a 8. c 9. d 10. e

CHAPTER 22

Geometrical Optics: Reflection and Refraction of Light

I. Chapter Objectives

Upon completion of this chapter, you should be able to:

1. define and explain the concept of wave fronts and rays.
2. explain the law of reflection, and distinguish between regular (specular) and irregular (diffuse) reflections.
3. explain refraction in terms of Snell's law and the index of refraction, and give examples of refractive phenomena.
4. describe internal reflection and give examples of fiber optic applications.
5. explain dispersion and some of its effects.

II. Key Terms

Upon completion of this chapter, you should be able to define and/or explain the following key terms:

wave front
plane wave front
ray
geometrical optics
reflection
angle of incidence
angle of reflection
law of reflection
regular (specular) reflection
irregular (diffuse) reflection
refraction
Huygen's principle
angle of refraction
Snell's law
index of refraction
critical angle
total internal reflection
fiber optics
dispersion

The definitions and/or explanations of the most important key terms can be found in the following section: **III. Chapter Summary and Discussion**.

III. Chapter Summary and Discussion

1. Wave Fronts and Rays

A **wave front** is the line (in two dimensions) or surface (in three dimensions) defined by adjacent portions of a wave that is in phase. For example, a point light source emits spherical wave fronts because the points having the same phase angle are on the surface of a sphere. For a parallel beam of light, the wave front is a **plane wave front** (in three dimensions). The distance between adjacent wave fronts is the wavelength.

A **ray** is a line drawn perpendicular to a series of wave fronts and pointing in the direction of propagation. For a spherical wave, the rays are radially outward, and for a plane wave, they are parallel to each other. The use of wave fronts and rays in describing optical phenomena such as reflection and refraction is called **geometrical optics**.

2. Reflection

The **law of reflection** states that the **angle of incidence** (the angle between the incident ray and the normal) equals the **angle of reflection** (the angle between the reflected ray and the normal), $\theta_i = \theta_r$, where θ_i is the angle of incidence and θ_r the angle of reflection. The incident ray, the normal, and the reflected ray are always in the same plane.

Note: All angles are measured *from the normal (line perpendicular) to the reflecting surface.*

Regular (specular) reflection occurs from smooth surfaces, with the reflected rays parallel to each other. **Irregular (diffuse) reflection** occurs from rough surfaces, with the reflected rays being at different angles.

Example 22.1 Two mirrors make an angle of 90° with each other. A ray is incident on mirror M_1 at an angle of 30° to the normal. Find the direction of the ray after it is reflected from mirror M_1 and M_2.

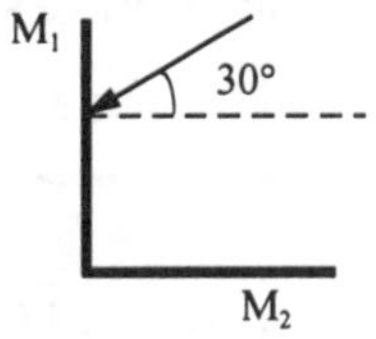

Solution: All angles should be measured from the normal to the reflecting surface. The angle of incidence of the ray at M_1 is 30°. According to the law of reflection, the angle of reflection at M_1 is also 30°. From geometry, the angle of incidence at M_2 is $90° - 30° = 60°$. Therefore, the angle of reflection at M_2 is also 60°, and the ray emerges parallel to the original incident ray.

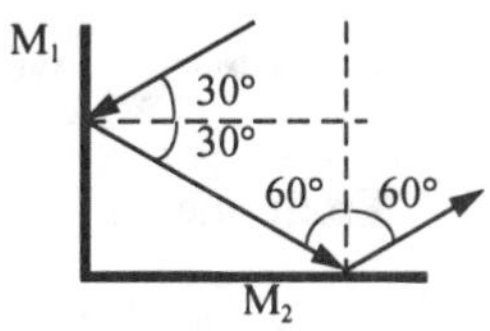

---

3. Refraction

Refraction refers to the change in direction of a wave at a boundary where it passes from one medium into another as a result of different wave speeds in different media. This phenomenon can be geometrically explained by **Huygens' principle**: every point on an advancing wave front can be considered to be a source of secondary waves, or wavelets, and the line or surface tangent to all these wavelets defines a new position of the wave front.

Snell's law relates the angle of incidence, θ_1, and angle of refraction θ_2, (the angle between the refracted ray and the normal to the boundary) to the wave speeds in the respective media: $\frac{\sin\theta_1}{\sin\theta_2} = \frac{v_1}{v_2}$. The **index of refraction** of a medium is defined as the ratio of the speed of light in vacuum to its speed in that medium, $n = \frac{c}{v}$. Snell's law can be conveniently expressed in terms of the indices of refraction, $n_1 \sin\theta_1 = n_2 \sin\theta_2$. If the second medium is more optically dense ($n_2 > n_1$), then $\theta_1 > \theta_2$, or the refracted ray is bent toward the normal; if less dense ($n_2 < n_1$), then $\theta_2 > \theta_1$ and bending is away from the normal.

When light travels from one medium to another, the frequency remains constant, but the speed and wavelength change. In terms of wavelength, the index of refraction can be rewritten as $n = \frac{\lambda}{\lambda_m}$, where λ is the wavelength in vacuum and λ_m the wavelength in the medium.

Note: Since the geometrical representation of light is used here, a diagram is very helpful (if not necessary) in problem solving. Again, all angles are measured from the normal to the boundary.

Example 22.2 A light ray travels through an air–water interface at an angle of 30° to the normal. Find the speed of light in the water and the angle of refraction.

Solution: Given: $n_1 \approx 1.00$ (air), $n_2 = 1.33$ (water), $\theta_1 = 30°$.

Find: v and θ_2.

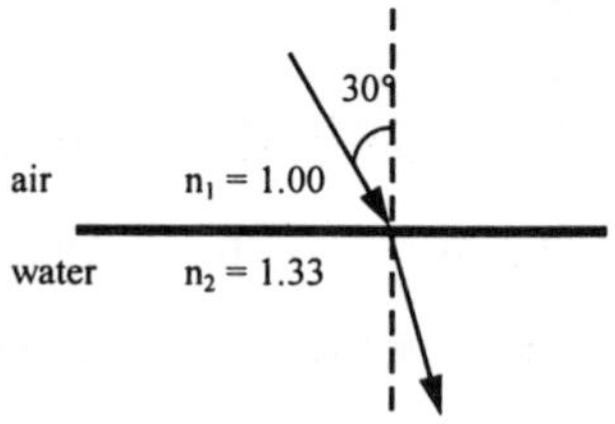

From $n = \frac{c}{v}$, we have $v_2 = \frac{c}{n_2} = \frac{3.00\times10^8\ \text{m/s}}{1.33} = 2.26\times10^8\ \text{m/s}$.

From Snell's law, $n_1 \sin\theta_1 = n_2 \sin\theta_2$,

$\sin\theta_2 = \frac{n_1 \sin\theta_1}{n_2} = \frac{(1.00)\sin 30°}{1.33} = 0.376$. So $\theta_2 = 22.1°$

Example 22.3 A beam of light traveling in air is incident on a slab of transparent material. The incident beam and the refracted beam make angles of 40° and 26° to the normal, respectively. Find the speed of light in the transparent material.

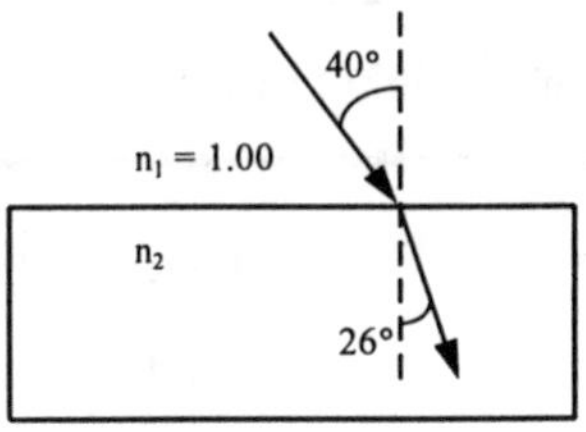

Solution: Given: $n_1 \approx 1.00$ (air), $\theta_1 = 40°$, $\theta_2 = 26°$.

Find: v_2.

We first find the index of refraction of the transparent material.

From Snell's law, $n_1 \sin \theta_1 = n_2 \sin \theta_2$,

we have $n_2 = \dfrac{n_1 \sin \theta_1}{\sin \theta_2} = \dfrac{(1.00) \sin 40°}{\sin 26°} = 1.47.$

Therefore, from $n = \dfrac{c}{v}$, the speed of light in the material is $v_2 = \dfrac{c}{n_2} = \dfrac{3.00 \times 10^8 \text{ m/s}}{1.47} = 2.04 \times 10^8$ m/s.

Example 22.4 A ray of wavelength 590 nm travels from air to water.

(a) What is the frequency of the light in air?

(b) What is the frequency of the light in water?

(c) What is the wavelength of light in water?

Solution: Given: $n_1 = 1.00$ (air), $n_2 = 1.33$ (water), $\lambda = 590 \text{ nm} = 590 \times 10^{-9}$ m.

Find: (a) f (b) f_m (c) λ_m.

(a) From $c = \lambda f$, we have $f = \dfrac{c}{\lambda} = \dfrac{3.00 \times 10^8 \text{ m/s}}{590 \times 10^{-9} \text{ m}} = 5.08 \times 10^{14}$ Hz.

(b) The frequency is a constant (same for all media). So $f_m = f = 5.08 \times 10^{14}$ Hz.

(c) From $n = \dfrac{\lambda}{\lambda_m}$, $\lambda_m = \dfrac{\lambda}{n} = \dfrac{590 \text{ nm}}{1.33} = 444$ nm.

4. Total Internal Reflection and Fiber Optics

At a certain **critical angle** (θ_c), the angle of refraction for a ray going from a medium of greater optical density to a medium of lesser optical density ($n_1 > n_2$) is 90° and the refracted ray is along the media boundary. For any angle of incidence $\theta_1 > \theta_c$, the **total internal reflection** (no refracted light) of light occurs and the surface acts like a mirror. By Snell's law, the critical angle can be calculated in terms of the indices of refraction of the two media, $\sin \theta_c = \dfrac{n_2}{n_1}$ (for $n_1 > n_2$). If the second medium is air, then $\sin \theta_c = \dfrac{1}{n}$. **Fiber optics** uses the principle of total internal reflection. Signals can travel a long distance without losing much intensity.

Note: Total internal reflection occurs only if the second medium is less dense than the first *and* if the angle of incidence exceeds the critical angle.

Example 22.5 A diver is 1.2 m beneath the surface of a still pond of water. At what angle must the diver shine a beam of light toward the surface in order for a person on a distant bank to see it?

Solution: Given: $\theta_2 = 90°$, $n_2 = 1.00$ (air), $n_1 = 1.33$ (water).

Find: θ_1.

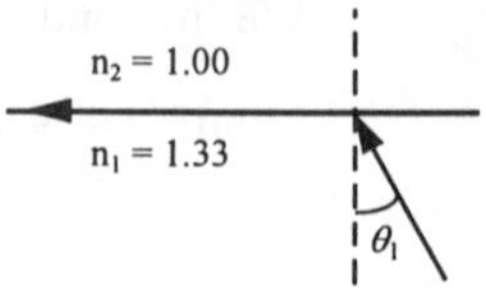

When the refracted light is along the boundary, the angle in water is equal to the critical angle. $\theta_1 = \theta_c = \sin^{-1}\dfrac{n_2}{n_1} = \sin^{-1}\dfrac{1.00}{1.33} = \sin^{-1} 0.752 = 48.8°$. Or from Snell's law,

$n_1 \sin\theta_1 = n_2 \sin\theta_2$, we have $\sin\theta_1 = \dfrac{n_2 \sin\theta_2}{n_1} = \dfrac{(1.00)\sin 90°}{1.33} = 0.752$, so $\theta_1 = 48.8°$.

For the light to reach the distant person, it has to shine at or below 48.8°, so there is no total internal reflection. Do you know why some fish-preying birds like pelicans stay very low before they try to catch fish?

Example 22.6 A 45°–45° prism is a wedge-shaped object in which the two acute angles are 45°. It is very useful for changing the direction of light rays in optical devices. If a light ray is traveling through a glass prism according to the diagram shown,what is the minimum index of refraction of the glass?

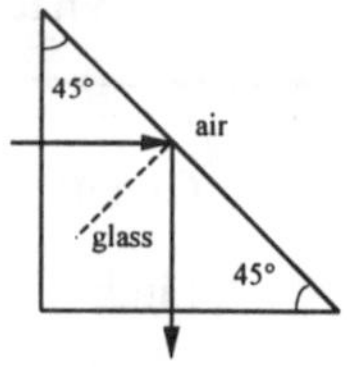

Solution: There is no refracted ray beyond the glass–air boundary, so the light must be internally reflected. For total internal reflection to occur, the index of reflection n_1 must be greater than that of air. The minimum index of reflection corresponds to an angle of incidence of 45°.

Given: $n_2 = 1.00$ (air), $\theta_c = 45°$. Find: n_1 (minimum).

$\sin\theta_c = \dfrac{n_2}{n_1}$, so the minimum n_1 is $\dfrac{n_2}{\sin\theta_c} = \dfrac{1.00}{\sin 45°} = 1.41$.

---

5. Dispersion

Dispersion is the separation of light into its component wavelengths when the light is refracted. This is caused by the fact that different wavelengths have slightly different indices of refraction. In most materials (so called normal dispersion), longer wavelengths have smaller indices of refraction. According to Snell's law, different wavelengths will have different angles of refraction and are therefore separated. A rainbow is produced by refraction, dispersion, and total internal reflection within water droplets.

Example 22.7 A beam of white light strikes a piece of glass at a 70° angle (measured from the normal). A red light of wavelength of 680 nm and a blue light of wavelength 430 nm emerge from the boundary after being dispersed. The index of refraction for the red light is 1.4505 and the index of refraction for the blue light is 1.4693.

(a) Which color of light is refracted more?

(b) What is the angle of refraction for each color?

(c) What is the angular separation between the two colors?

Solution: (a) Since the index of refraction of the blue light is greater, its angle of refraction is smaller (bent more toward the normal), and therefore it is refracted more.

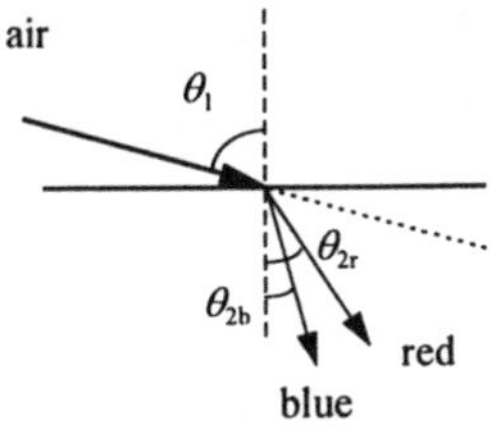

Given: $n_1 = 1.0000$ (air), $n_{2r} = 1.4505$, $n_{2b} = 1.4693$, $\theta_1 = 70°$.

Find: (b) θ_{2r} and θ_{2b} (c) $\Delta\theta$.

(b) From Snell's law, $n_1 \sin\theta_1 = n_2 \sin\theta_2$,

for red: $\sin\theta_{2r} = \dfrac{n_1 \sin\theta_1}{n_{2r}} = \dfrac{(1.0000)\sin 70°}{1.4505} = 0.64784$, so $\theta_{2r} = 40.379°$;

for blue: $\sin\theta_{2b} = \dfrac{(1.0000)\sin 70°}{1.4693} = 0.63955$, so $\theta_{2b} = 39.758°$.

(c) $\Delta\theta = \theta_{2r} - \theta_{2b} = 40.379° - 39.758° = 0.621°$.

IV. Mathematical Summary

Law of Reflection	$\theta_i = \theta_r$ (22.1)	Relates the angles of incidence and reflection
Index of Refraction	$n = \dfrac{c}{v} = \dfrac{\lambda}{\lambda_m}$ (22.3–4)	Defines the index of refraction of a material
Snell's Law	$\dfrac{\sin\theta_1}{\sin\theta_2} = \dfrac{v_1}{v_2}$ (22.2) or $n_1 \sin\theta_1 = n_2 \sin\theta_2$ (22.5)	Relates the angles of incidence and refraction, and the speeds of light in the media (or indices of refraction)
Critical Angle for Boundary between Two Materials	$\sin\theta_c = \dfrac{n_2}{n_1}$ (22.6) where $n_1 > n_2$	Computes the critical angle between two materials for total internal reflection
Critical Angle for Material–Air Boundary	$\sin\theta_c = \dfrac{1}{n}$ (22.7)	Computes the critical angle between material–air boundary for total internal reflection

V. Solutions of Selected Exercises and Paired Exercises

9. $\tan\theta_i = \frac{2.5\text{ m}}{3.0\text{ m}} = 0.833$,

so $\theta_i = \theta_r = \tan^{-1} 0.833 = 40°$.

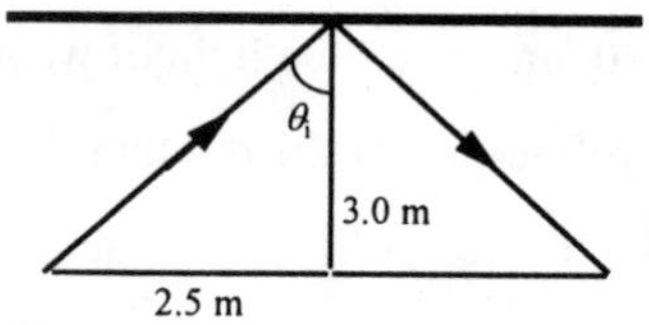

10. Since the object distance is equal to the image distance, the camera should be focused to $1.5\text{ m} + 1.5\text{ m} = 3.0\text{ m}$.

14. Use similar triangles. $\frac{d + 0.45\text{ m} + 0.45\text{ m}}{8.50\text{ m}} = \frac{0.45\text{ m}}{0.30\text{ m}}$, so $d = 11.9\text{ m}$.

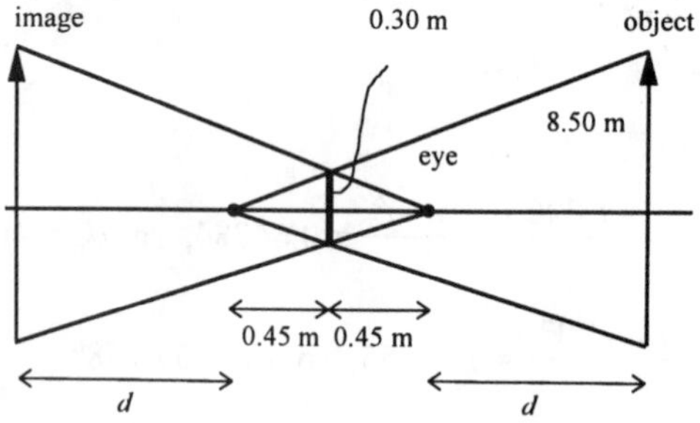

23. Initially light has to travel along a straight line to reach our eyes and the top of the container blocks the light from the bottom of the coin. When water is added, the light coming out of the water is bent into the air with a larger angle of refraction and so the light reaches our eyes.

27. From $n_1 \sin\theta_1 = n_2 \sin\theta_2$, $\sin\theta_1 = \frac{n_2 \sin\theta_2}{n_1} = \frac{(1.46)\sin 30°}{1} = 0.73$. So $\theta_1 = 47°$.

Therefore, the angle of reflection is also 47° according to the law of reflection.

34. $\frac{v_B}{v_A} = \frac{c/n_B}{c/n_A} = \frac{n_A}{n_B} = \frac{4/3}{5/4} = \frac{16}{15}$.

38. The first refraction is at the air–glass interface and the second refraction is at the glass–water interface.

From $n_1 \sin\theta_1 = n_2 \sin\theta_2$, we have $\sin\theta_2 = \frac{n_1 \sin\theta_1}{n_2} = \frac{(1)\sin 40°}{1.50} = 0.429$. So $\theta_2 = 25.4°$.

The critical angle at the glass–water interface is $\theta_c = \sin^{-1}\frac{n_3}{n_2} = \sin^{-1}\frac{1.33}{1.50} = \sin^{-1} 0.887 = 62.5°$.

Therefore, the angle of incidence at the glass–water interface is smaller than the critical angle and thus there is no total internal reflection. Yes, the fish is illuminated.

48. The altitude is measured from the horizon and so the angle from the normal is 90° – altitude.

From $n_1 \sin \theta_1 = n_2 \sin \theta_2$, we have $\sin \theta_1 = \dfrac{n_2 \sin \theta_2}{n_1} = \dfrac{(1.33) \sin (90° - 50°)}{1} = 0.855.$

So $\theta_1 = 58.7°$. Therefore, the actual altitude is $90° - 59° = 31°$.

52. From $n_1 \sin \theta_1 = n_2 \sin \theta_2$, we have $\sin \theta_1 = \dfrac{n_2 \sin \theta_2}{n_1} = \dfrac{(1.33) \sin 45°}{1} = 0.940.$ So $\theta_1 = 70.1°$.

Therefore, $d = (1.75 \text{ m}) \tan 70.1° = 4.84 \text{ m}.$

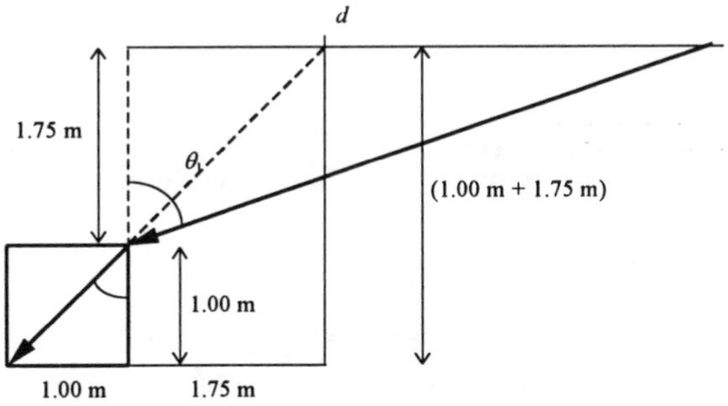

53. $\theta_2 = \tan^{-1} \dfrac{0.50 \text{ m}}{0.75 \text{ m}} = \tan^{-1} 0.667 = 33.7°.$

$n_1 \sin \theta_1 = n_2 \sin \theta_2,$

so $\sin \theta_1 = \dfrac{n_2 \sin \theta_2}{n_1} = \dfrac{(1.33) \sin 33.7°}{1} = 0.738.$

Therefore, $\theta_1 = 47.5°.$

Thus, $d = (1.8 \text{ m}) \tan \theta_1 = (1.8 \text{ m}) \tan 47.5° = 2.0 \text{ m}.$

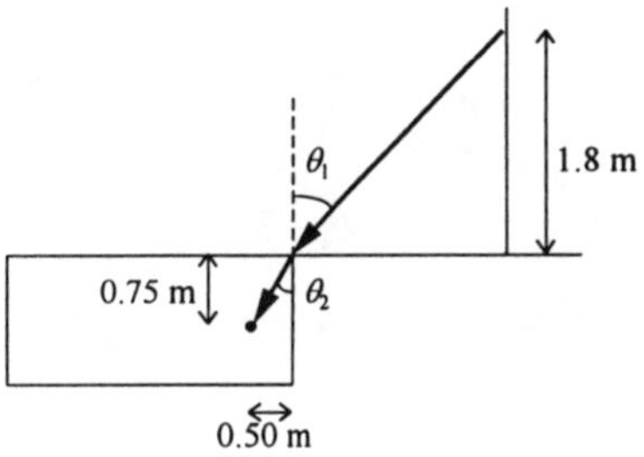

55. To see the whole newspaper print, the minimum angle of refraction is 45°.

The maximum $n_1 \sin\theta_1$ is 1 sin 90° = 1.

$n_2 \sin \theta_2 = 1.66 \sin 45° = 1.17 > 1.$

No, it is impossible.

56. (a) For refraction at the prism–prism interface, $(1.60) \sin 45° = (1.40) \sin \theta_2,$

so $\theta_2 = \sin^{-1} 0.808 = 53.9°.$

For the prism–air interface,

$\theta_3 = 180° - [135° + (90° - \theta_2)] = \theta_2 - 45° = 53.9° - 45° = 8.9°.$

So $(1.40) \sin 8.9° = (1) \sin \theta,$

therefore, $\theta = \sin^{-1} 0.217 = 12.5°.$

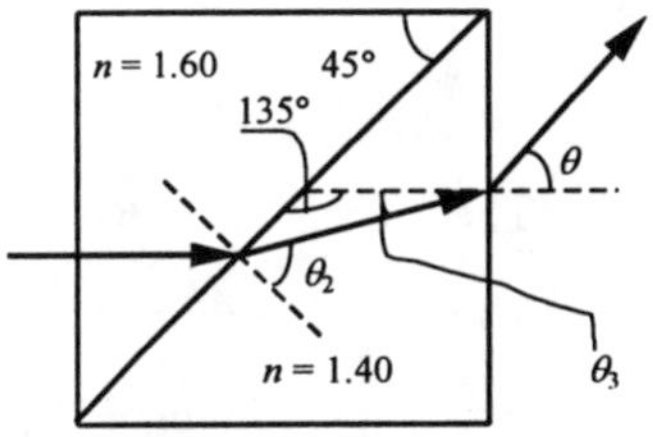

(b) $\theta_c = \sin^{-1}\dfrac{n_2}{n_1} = \sin^{-1}\dfrac{1.40}{1.60} = 61.0°$.

This is the angle required at the interface.

For the air–prism interface,

$\theta_2 = 180° - [135° + (90° - \theta_c)] = \theta_c - 45° = 61.0° - 45° = 16.0°$.

So (1) $\sin\theta_1 = (1.60)\sin 16°$,

therefore, $\theta_1 = \sin^{-1} 0.441 = 26.2°$

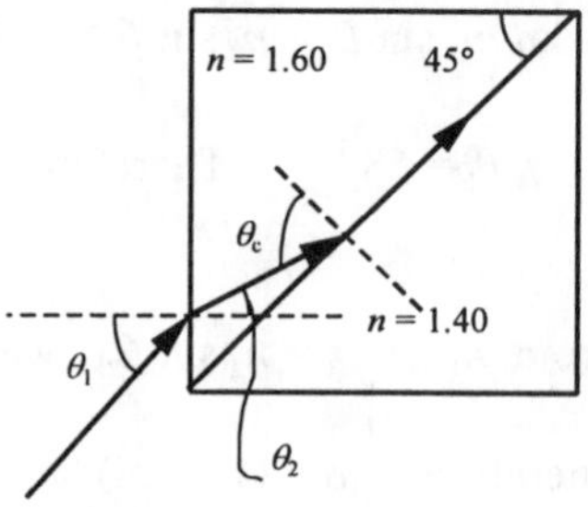

64. From $n_1 \sin\theta_1 = n_R \sin\theta_R = n_B \sin\theta_B$, we have $\sin\theta_R = \dfrac{n_1 \sin\theta_1}{n_R} = \dfrac{(1)\sin 45°}{1.445} = 0.4893$.

So $\theta_R = 29.298°$, $\sin\theta_B = \dfrac{(1)\sin 45°}{1.470} = 0.4810$. Therefore, $\theta_B = 28.752°$.

Hence $\Delta\theta = 29.298° - 28.752° = 0.546° = 0.00953$ rad.

(b) Since $\lambda_m = \dfrac{\lambda}{n}$, $\dfrac{\lambda_{mR}}{\lambda_{mB}} = \dfrac{\lambda_R/n_R}{\lambda_B/n_B} = \dfrac{(680\text{ nm})/1.445}{(400\text{ nm})/1.470} = 1.73$.

66. $n_1 \sin\theta_1 = n_2 \sin\theta_2$,

so $\sin\theta_2 = \dfrac{n_1 \sin\theta_1}{n_2} = \dfrac{(1)\sin 50°}{1.33} = 0.576$.

Therefore, $\theta_2 = 35.2°$.

Thus, $x = (15\text{ cm})\tan\theta_2 = (15\text{ cm})\tan 35.2° = \boxed{11\text{ cm}}$.

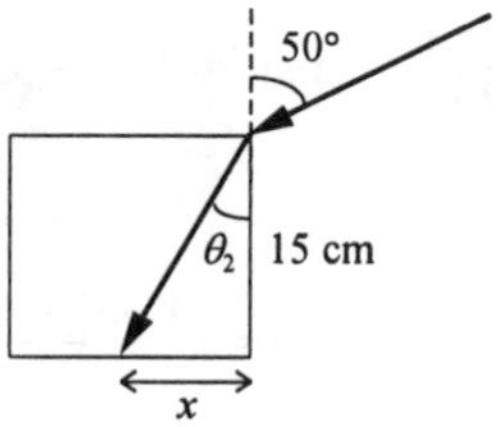

71. (a) From $n_1 \sin\theta_1 = n_2 \sin\theta_2$, $\sin\theta_2 = \dfrac{n_1 \sin\theta_1}{n_2} = \dfrac{(1)\sin 40°}{1.52} = 0.423$. So $\theta_2 = 25°$.

(b) $v = \dfrac{c}{n} = \dfrac{3.00\times10^8\text{ m/s}}{1.52} = 1.97\times10^8$ m/s.

(c) $\lambda_m = \dfrac{\lambda}{n} = \dfrac{550\text{ nm}}{1.52} = 362$ nm.

74. (a) From $d' = \dfrac{d}{n}$, we have $d = d'n = (4.5\text{ m})(1.33) = 6.0$ m.

Since $n_1 \sin\theta_1 = n_2 \sin\theta_2$, $\sin\theta_2 = \dfrac{n_1 \sin\theta_1}{n_2} = \dfrac{(1)\sin 12°}{1.33} = 0.156$.

So $\theta_2 = 8.99°$.

Therefore, $x = (6.0\text{ m})\tan\theta_2 = (6.0\text{ m})\tan 8.99° = 0.95$ m.

Thus the coordinates are (0.95 m, −6.0 m).

(b) As determined in (a), $\theta_2 = 9.0°$.

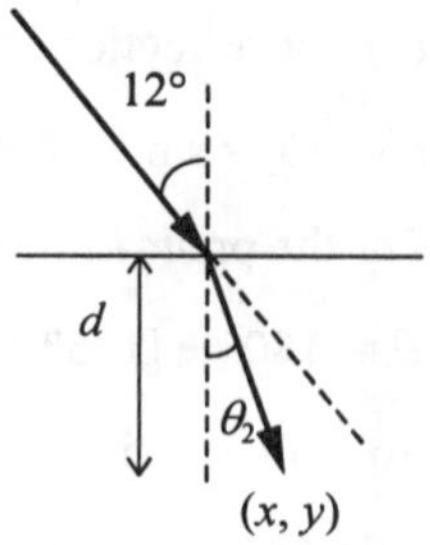

VI. Practice Quiz

1. If a material has an index of refraction of 1.461, what is the speed of light through this liquid?

   (a) 4.38×10^8 m/s (b) 2.05×10^8 m/s (c) 2.341×10^8 m/s (d) 4.461×10^8 m/s (e) 1.461×10^8 m/s

2. A light ray in air is incident on an air to glass interface boundary at an angle of 45° and is refracted in the glass at an angle of 30° with the normal. What is the index of refraction of the glass?

   (a) 2.13 (b) 1.74 (c) 1.23 (d) 1.31 (e) 1.41

3. An optical fiber is made of clear plastic with index of refraction of $n = 1.50$. For what angles of incidence will light remain within the plastic "guide"?

   (a) > 23.4° (b) > 32.9° (c) > 38.3° (d) > 40.3° (e) > 41.8°

4. A certain kind of glass has an index of refraction of 1.65 for blue light and an index of refraction of 1.61 for red light. If a beam of white light (containing all colors) is incident at an angle of 30°, what is the angle between the refracted red and blue light?

   (a) 0.22° (b) 0.35° (c) 0.45° (d) 1.90° (e) 1.81°

5. A ray of white light, incident upon a glass prism, is dispersed into its various color components. Which one of the following colors experiences the greatest refraction?

   (a) orange (b) yellow (c) red (d) blue (e) green

6. Which one of the following describes what will generally happen to a light ray incident on an air–to–glass boundary?

   (a) total reflection (b) total refraction (c) partial reflection, partial refraction

   (d) partial reflection, total refraction (e) total reflection, partial refraction

7. Light enters air from water. The angle of refraction will be

   (a) greater than or equal to the angle of incidence. (b) less than or equal to the angle of incidence.

   (c) equal to the angle of incidence. (d) greater than the angle of incidence.

   (e) less than the angle of incidence.

8. A monochromatic light source emits a wavelength of 490 nm in air. When passing through a liquid, the wavelength reduces to 429 nm. What is the index of refraction of the liquid?

   (a) 1.14 (b) 1.26 (c) 1.33 (d) 1.49 (e) 1.52

9. A fiber optic cable ($n = 1.50$) is submerged in water. What is the critical angle for light to stay inside the cable?

(a) 27.6° (b) 41.8° (c) 45.0° (d) 62.5° (e) 83.1°

10. An oil film ($n = 1.47$) floats on a water surface. If a ray of light is incident on the air–oil boundary at an angle of 37° to the normal, what is the angle of refraction at the oil–water boundary?

(a) 18.1° (b) 24.2° (c) 26.9° (d) 32.4° (e) 37.0°

Answers to Practice Quiz:

1. b 2. e 3. e 4. c 5. d 6. c 7. d 8. a 9. d 10. c

CHAPTER 23

Mirrors and Lenses

I. Chapter Objectives

Upon completion of this chapter, you should be able to:

1. describe the characteristics of plane mirrors and explain apparent right-left reversals.
2. distinguish between converging and diverging spherical mirrors, describe images and their characteristics, and determine these image characteristics using ray diagrams and the spherical mirror equation.
3. distinguish between converging and diverging lenses, describe images and their characteristics, and find image characteristics using ray diagrams and the thin-lens equation.
4. describe some common lens aberrations and explain how they can be reduced or corrected.

*5. describe and apply the lens maker's equation, and explain how its application differs from that of the thin-lens equation.

II. Key Terms

Upon completion of this chapter, you should be able to define and/or explain the following key terms:

plane mirror
virtual image
real image
lateral magnification
spherical mirror
concave (converging) mirror
convex (diverging) mirror
center of curvature
radius of curvature
focal point
focal length
parallel ray (mirror)
chief (radial) ray (mirror)
focal ray (mirror)
spherical mirror equation
magnification factor (spherical mirror)
spherical aberration (mirror)
lens
converging (biconvex) lens
diverging (biconcave) lens
parallel ray (lens)
chief ray (lens)
focal ray (lens)
thin-lens equation
magnification factor (lens)
spherical aberration
chromatic aberration
astigmatism
***lens maker's equation**
***diopters**

The definitions and/or explanations of the most important key terms can be found in the following section: **III. Chapter Summary and Discussion.**

III. Chapter Summary and Discussion

1. Plane Mirrors

The images of objects formed by optical systems (mirrors and/or lenses) can be either real or virtual. A **real image** is one formed by light rays that converge at and pass through the image location, and can be seen or formed on a screen. A **virtual image** is one for which light rays *appear* to emanate from the image, but do not actually do so. Virtual images cannot be seen or formed on screens.

The images of objects can also be *upright, inverted, magnified, unmagnified,* or *reduced in size.* All these characteristics can be determined from the **lateral magnification,** $M = \frac{\text{image height}}{\text{object height}} = \frac{h_i}{h_o}$. For real objects (these are the only ones with which we will deal), if M is negative, the image is real, inverted; if M is positive, the image is virtual and upright; if $|M| > 1$, the image is magnified; if $|M| = 1$, the image is unmagnified; and if $|M| < 1$, the image is reduced. These image characteristics are very important in geometrical optics.

A **plane mirror** is a mirror with a flat surface. A mirror forms an image based on the law of reflection. The characteristics of the images formed by a plane mirror are virtual, upright, and unmagnified ($M = +1$). Also, the image formed by a plane mirror appears to be at a distance behind the mirror that is equal to the distance of the object in front of the mirror and has right-left or front-back reversal.

Example 23.1 An arrow is placed in front of a plane mirror as shown. Sketch the image of the arrow formed by the plane mirror.

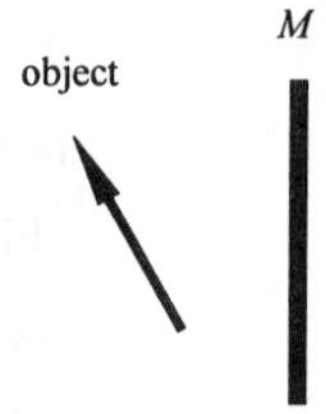

Solution: The image is virtual, upright, and unmagnified. The object distance (distance from object to mirror) is equal to the image distance (distance from image to mirror) for a plane mirror. By sketching the images of each individual point on the object, we obtain the image.

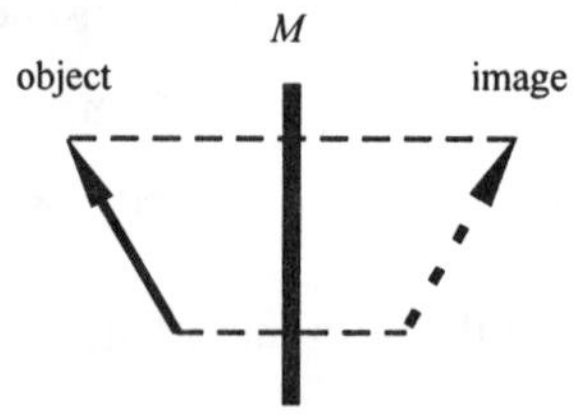

---

2. Spherical Mirrors

A **spherical mirror** is a section of a sphere. Either the outside (convex) surface or the inside (concave) surface of the spherical section may be the reflecting surface. A **concave mirror** is called a **converging mirror** and a **convex mirror** is called a **diverging mirror**. These terms refer to the reflections of rays parallel to a mirror's *optical axis*, which is a line through the center of the spherical mirror that intersects the mirror at the *vertex* of the spherical section.

The **center of curvature** (C) of a spherical mirror is the point on the optic axis that corresponds to the center of the sphere of which the mirror forms a section. The **radius of curvature** (R) is the distance from the vertex to the center of curvature. The **focal point** (F) of a spherical mirror is the point at which parallel rays converge or appear to diverge. The **focal length** (f) of a spherical mirror is the distance from the focal point to the vertex of the spherical section. It is equal to one-half of the radius of curvature, $f = R/2$.

The images formed by spherical mirrors can be studied from geometry (**ray diagrams**). The ray diagram is as important as the free-body diagram in the application of Newton's laws. Three important rays are used to determine the images: a **parallel ray** is a ray incident along a path parallel to the optic axis and reflected through the focal point F (or appear to go though); a **chief (radial) ray** is a ray incident through the center of curvature C (or appear to go though) and reflected back along its incident path though C; a **focal ray** is a ray which passes through (or appear to go through) the focal point and is reflected parallel to the optic axis.

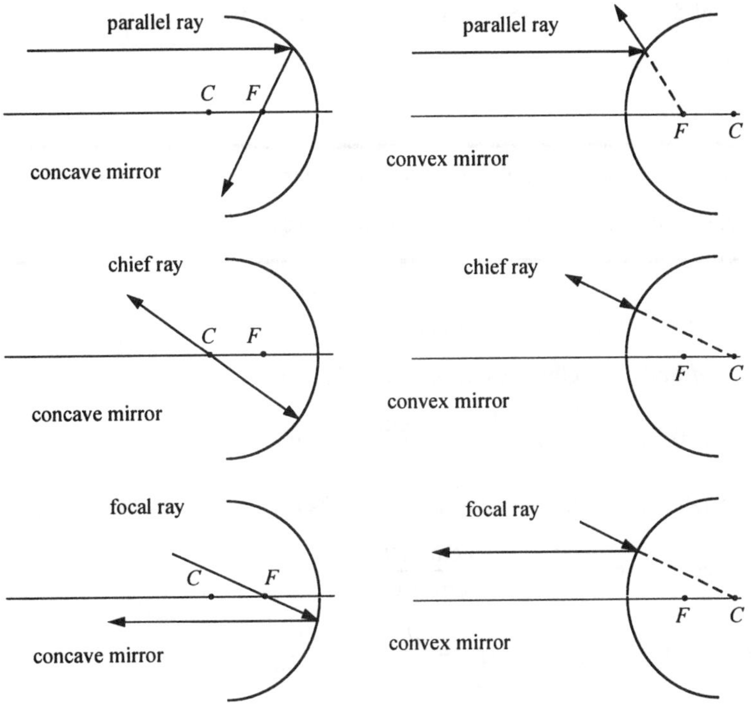

Example 23.2 An object is placed 30-cm in front of a concave mirror of radius 20 cm. Determine the characteristics of the image with a ray diagram.

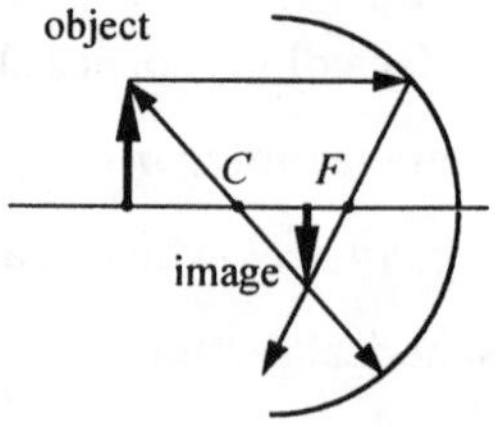

Solution: The focal length of the mirror is $f = \frac{R}{2} = \frac{20 \text{ cm}}{2} = 10 \text{ cm}$.

We use the parallel ray (parallel in, through focal point out) and chief ray (through center of curvature in, through center of curvature out). These two rays originate from the tip of the object and cross after being reflected by the mirror. The point where they cross is the image point, so the image is real (because the two rays converged), inverted (the image is upside down), and reduced (the image is smaller than the object), as determined from the ray diagram.

Try to draw the focal ray from the tip of the object here. Will it cross the image point after being reflected?

The images formed by spherical mirrors can also be studied from equations. The **spherical mirror equation** is given by $\frac{1}{d_o} + \frac{1}{d_i} = \frac{1}{f}$, or $d_i = \frac{d_o f}{d_o - f}$, where d_o is the object distance (from the object to the vertex), d_i the image distance (from the image to the vertex), and f the focal length. The lateral magnification can be written in terms of the object and image distances as $M = \frac{h_i}{h_o} = -\frac{d_i}{d_o}$.

Note: d_i and f should be used as algebraic quantities, that is, they have signs (positive or negative). For a real object (positive d_o), if a mirror is concave, f is positive; if a mirror is convex, f is negative; if the image is real (formed on the same side of the mirror as the object), d_i is positive; if the image is virtual (formed behind the mirror), d_i is negative.

Example 23.3 Find the location and describe the characteristics of the image formed by a concave mirror of radius 20 cm if the object distance is (a) 30 cm, (b) 20 cm, (c) 15 cm, and (d) 5.0 cm.

Solution: Given: $R = 20$ cm, (a) $d_o = 30$ cm, (b) $d_o = 20$ cm, (c) $d_o = 15$ cm, (d) $d_o = 5.0$ cm.

Find: d_i and characteristics in (a), (b), (c), and (d).

The focal length of the mirror is $f = \frac{R}{2} = \frac{20 \text{ cm}}{2} = 10 \text{ cm}$.

(a) The object distance is greater than two focal lengths (or the radius of curvature), $d_o > 2f = R$.

The image distance is $d_i = \frac{d_o f}{d_o - f} = \frac{(30 \text{ cm})(10 \text{ cm})}{30 \text{ cm} - 10 \text{ cm}} = 15 \text{ cm}$.

The lateral magnification is equal to $M = -\frac{d_i}{d_o} = -\frac{15 \text{ cm}}{30 \text{ cm}} = -\frac{1}{2}$.

So the image is *real* (d_i positive), *inverted* (M is negative), and *reduced* ($|M| < 1$) for $d_o > 2f = R$.

(b) The object distance is equal to two focal lengths (or the radius of curvature), $d_o = 2f = R$.

$$d_i = \frac{(20\text{ cm})(10\text{ cm})}{20\text{ cm} - 10\text{ cm}} = 20\text{ cm} \quad \text{and} \quad M = -\frac{20\text{ cm}}{20\text{ cm}} = -1.$$

So the image is *real* (d_i positive), *inverted* (M is negative), and *unmagnified* ($|M| = 1$) for $d_o = 2f = R$.

(c) The object distance is greater than one focal length and smaller than two focal lengths, $f < d_o < 2f = R$.

$$d_i = \frac{(15\text{ cm})(10\text{ cm})}{15\text{ cm} - 10\text{ cm}} = 30\text{ cm} \quad \text{and} \quad M = -\frac{30\text{ cm}}{15\text{ cm}} = -2.$$

So the image is *real* (d_i positive), *inverted* (M is negative), and *magnified* ($|M| > 2$) for $f < d_o < 2f = R$.

(d) The object distance is smaller than one focal length, $d_o < f = R/2$.

$$d_i = \frac{(5.0\text{ cm})(10\text{ cm})}{5.0\text{ cm} - 10\text{ cm}} = -10\text{ cm} \quad \text{and} \quad M = -\frac{-10\text{ cm}}{5.0\text{ cm}} = -2.$$

So the image is *virtual* (d_i negative), *upright* (M is positive), and *magnified* ($|M| > 1$) for $d_o < f = R/2$.

Hence a concave mirror can form a variety of images depending on the object distance.

Example 23.4 When a person stands 40 cm in front of a cosmetic mirror (concave mirror), the erect image is twice the size of the object. What is the focal length of this mirror?

Solution: Since the image is erect and twice the size, the lateral magnification is equal to +2.0 ($M = 2.0$).

Given: $d_o = 40$ cm, $M = +2.0$. Find: f.

From the spherical mirror equation, $\frac{1}{d_o} + \frac{1}{d_i} = \frac{1}{f}$, we first need to know the image distance d_i before we can find the focal length f. Since $M = -\frac{d_i}{d_o}$, $d_i = -Md_o = -(+2.0)(40\text{ cm}) = -80$ cm.

Therefore, $\frac{1}{f} = \frac{1}{40\text{ cm}} + \frac{1}{-80\text{ cm}} = \frac{1}{80\text{ cm}}$, or $f = 80$ cm.

Example 23.5 Show that the image formed by a convex mirror is always (no matter where is the object is) virtual, upright, and reduced.

Solution: For a convex mirror, the focal length $f = -|f|$ is negative.

From the spherical mirror equation, $d_i = \frac{d_o f}{d_o - f} = \frac{d_o(-|f|)}{d_o - (-|f|)} = -\frac{d_o|f|}{d_o + |f|}$.

Since d_o is always positive, d_i is always negative, so the image is virtual.

The lateral magnification is equal to $M = -\frac{d_i}{d_o} = -\frac{-\frac{d_o|f|}{d_o+|f|}}{d_o} = \frac{|f|}{d_o + |f|}$, is positive.

Since $d_o + |f|$ is always larger than $|f|$, M is smaller than 1, so the image is upright and reduced.

---

3. Lenses

A lens forms an image based on the law of refraction (Snell's law). A spherical **biconvex lens is a converging lens** and a spherical **biconcave lens** is a **diverging lens**. For these lenses, the focal length is *not* equal to one-half the radius of curvature of the surface. Only thin lenses are studied here to ignore the lateral displacement of the ray after refracted from the two lens surfaces.

The images formed by thin lenses can also be studied from geometry (**ray diagrams**). The general rules for drawing ray diagrams for lenses are similar to those for spherical mirrors with some modifications. Three important rays are used to determine the images: a **parallel ray** is a ray incident along a path parallel to the optic axis and refracted through the focal point F (or appears to go though); a **chief (radial) ray** is a ray incident through the center of the lens (or appears to go though) and refracted undeviated; a **focal ray** is a ray which passes through (or appears to go through) the focal point and is refracted parallel to the optic axis.

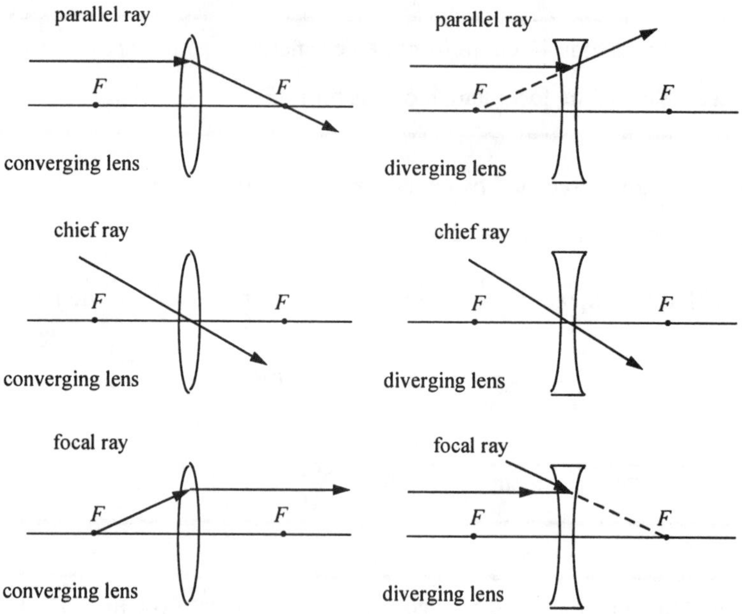

Example 23.6 An object is placed 5.0 cm in front of a thin lens of focal length 15 cm. Determine the characteristics of the image with a ray diagram.

Solution: We use the chief ray (through center of lens in, undeviated out) and the focal ray (through focal point in, parallel out). These two rays originate from the tip of the object and *appear* to cross after being refracted by the lens. The point where they appear to cross is the image point, so the image is virtual (because the two rays do not cross), upright (the image is right side

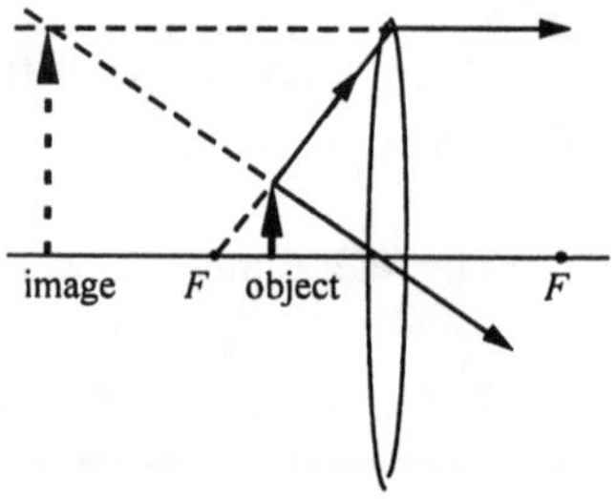

up), and magnified (the image is larger than the object), as determined from the ray diagram.

Try to draw the parallel ray from the tip of the object here. Will it cross the image point after being refracted?

The images formed by thin lenses can also be studied from equations. The **thin lens equation** is identical in form to the spherical mirror equation, $\frac{1}{d_o} + \frac{1}{d_i} = \frac{1}{f}$, or $d_i = \frac{d_o f}{d_o - f}$, where d_o is the object distance (from the object to the center of the lens), d_i the image distance (from the image to the center of the lens), and f the focal length. The lateral magnification is also defined the same way as for spherical mirrors, $M = \frac{h_i}{h_o} = -\frac{d_i}{d_o}$.

Note: d_i and f should be used as algebraic quantities, that is, they have signs (positive or negative). For real objects (positive d_o), if a lens is convex, f is positive; if a lens is concave, f is negative; if the image is real (formed on the opposite side of the lens from the object), d_i is positive; if the image is virtual (formed on the same side as the object), d_i is negative.

Example 23.7 Find the location and describe the characteristics of the image formed by a convex lens of focal length 10 cm if the object distance is (a) 30 cm, (b) 20 cm, (c) 15 cm, and (d) 5.0 cm.

Solution: Since the equation and lateral magnification for thin lenses are identical to the equation and the lateral magnification for spherical mirrors, where $2f$ is used instead of R, the numerical answers to this Example are identical to those for Example 23.3. The only difference is on which side of the lens the images are formed. In mirrors, real images are formed on the same side of the mirror as the objects and in thin lenses, real images are formed on the opposite side of lens from the objects.

Example 23.8 When an object is 40 cm in front of a converging thin lens the inverted image is half the size of the object. What is the focal length of this lens?

Solution: Since the image is inverted and half the size, the lateral magnification is $-\frac{1}{2}$, $(M = -0.50)$.

Given: $d_o = 40$ cm, $M = -0.50$. Find: f.

From the this lens equation, $\frac{1}{d_o} + \frac{1}{d_i} = \frac{1}{f}$, we first need to know the image distance d_i before we can find the focal length f. Since $M = -\frac{d_i}{d_o}$, $d_i = -M\,d_o = -(-0.50)(40 \text{ cm}) = 20$ cm.

Therefore, $\frac{1}{f} = \frac{1}{40 \text{ cm}} + \frac{1}{20 \text{ cm}} = \frac{3}{40 \text{ cm}}$, or $f = \frac{40 \text{ cm}}{3} = 13$ cm.

Example 23.9 Determine the location and describe the characteristics of the image of an object placed at 30 cm in front of a diverging lens with a focal length of 10 cm.

Solution: The focal length of a diverging lens is negative.

Given: $f = -10$ cm, $d_o = 30$ cm.

Find: d_i and image characteristics.

From the thin lens equation, $d_i = \dfrac{d_o f}{d_o - f} = \dfrac{(40 \text{ cm})(-10 \text{ cm})}{30 \text{ cm} - (-10 \text{ cm})} = -10$ cm.

Since the image distance is negative, the image is virtual.

The lateral magnification is $M = -\dfrac{d_i}{d_o} = -\dfrac{-10 \text{ cm}}{30 \text{ cm}} = 0.33$, so the image is upright and reduced.

Many optical systems such as telescopes and compound microscopes use more than one lens. When two or more lenses are used in combination, the overall image produced may be determined by considering the lenses individually in sequence. That is, the image formed by the first lens is the object for the second lens, and so on. If the first lens produces an image in front of the second lens, the image is treated as a real object for the second lens. If, however, the lenses are close enough together that the image from the first lens is not formed in front of the second lens, then a modification must be made in the sign convention. In this case, the image from the first lens is treated as a *virtual* object for the second lens, and the object distance for it is taken to be *negative* in the lens equation. The total magnification (M_t) of a compound lens system is the product of the magnification factors (absolute value) of all the component lenses. For a two lens system, $M_t = |M_1| \times |M_2|$.

4. Lens Aberrations

Rays passing through a lens may not exactly follow the particular rays in a ray diagram because of aberrations. **Spherical aberration** occurs when parallel rays passing through different regions of a lens do not come together on a common image plane. (The thin lens equation is an approximation for rays very close to the optical axis.) **Chromatic aberration** results from the fact that different wavelengths (different colors) have different indices of refraction, and so different angles of refraction, resulting in different image planes for different colors. **Astigmatism** results when a cone (circular cross-section) of light from an off-axis source falls on a lens surface and forms an elliptically illuminated area, giving rise to two images. All these aberrations can be minimized, if not eliminated, by compounding lens systems. For example, some of the common 50-mm camera lenses are actually made up of seven lenses to reduce aberrations.

*5. Lens Maker's Equation

The **lens maker's equation** is a general equation for determining the focal length of a spherical lens with sides having different (or equal) radii of curvature. In general, the focal length of a lens in air ($n_{air} = 1$) can be calculated from $\frac{1}{f} = (n-1)\left(\frac{1}{R_1} - \frac{1}{R_2}\right)$, where n is the index of refraction of the lens material, and R_1 and R_2 are the radii of the first (front) and second (back) surfaces. The sign convention for the radii are as follows: if the center of curvature C is on the side of the lens from which light emerges or on the back side of the lens, R is positive; if C is on the side of the lens from which light incidents or on the front side of the lens, R is negative. If the lens of a material with an index of refraction n is in a fluid with an index of refraction n_m, the lens maker's equation is modified to $\frac{1}{f} = \left(\frac{n}{n_m} - 1\right)\left(\frac{1}{R_1} - \frac{1}{R_2}\right)$.

The power of a lens is defined as $P = \frac{1}{f}$, where f is the focal length in meters. The unit of lens power is expressed in **diopters** (D).

Example 23.10 A double convex lens made of glass ($n = 1.5$) has a radius of curvature of 40 cm on the front side and 30 cm on the back side. Find the power and the focal length of the lens?

Solution: The center of curvature of the front surface is on the back side of the lens, so its radius is positive. The center of curvature of the back surface is on the front side of the lens, so its radius is negative, according to the sign convention.

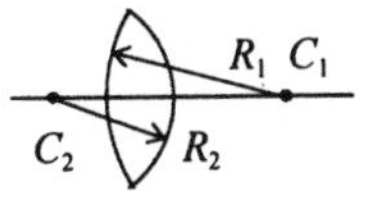

Given: $n = 1.5$, $R_1 = 40$ cm, $R_2 = -30$ cm.

Find: f.

$$P = \frac{1}{f} = (n-1)\left(\frac{1}{R_1} - \frac{1}{R_2}\right) = (1.5-1)\left(\frac{1}{0.40\text{ m}} - \frac{1}{-0.30\text{ m}}\right) = 2.9\text{ m}^{-1} = 2.9\text{ D}.$$

So $f = \frac{1}{P} = \frac{1}{2.9\text{ D}} = 0.34\text{ m} = 34\text{ cm}.$

IV. Mathematical Summary

Focal Length for Spherical Mirror	$f = \frac{R}{2}$ (23.2)	Defines the focal length of a spherical mirror in terms of its radius of curvature
Spherical Mirror Equation	$\frac{1}{d_o} + \frac{1}{d_i} = \frac{1}{f}$ (22.3) or $d_i = \frac{d_o f}{d_o - f}$ (22.4)	Relates object distance, image distance, and focal length of a spherical mirror
Thin Lens Equation (where $f \neq R/2$)	$\frac{1}{d_o} + \frac{1}{d_i} = \frac{1}{f}$ or $d_i = \frac{d_o f}{d_o - f}$ (22.6)	Relates object distance, image distance, and focal length of a thin-lens
Magnification Factor (mirror and lens)	$M = -\frac{d_i}{d_o}$ (23.5 and 7)	Defines the lateral magnification as a ratio of the size of the image to the size of the object
Total Magnification with a Two-Lens System	$M_t = \lvert M_1 \rvert \times \lvert M_2 \rvert$ (23.8)	Relates the total magnification of a two lens system to the magnifications of the individual lenses
Sign Convention for Spherical Mirrors and Lenses	Positive *f* for concave mirror and convex lens (converging). negative *f* for convex mirror and concave lens (diverging). d_o always positive. Positive d_i for real image; negative d_i for virtual image. Positive *M* for upright image; negative *M* for inverted image.	Defines the sign convention for focal length *f*, object distance d_o, image distance d_i, and the lateral magnification *M*
*Lens Maker's Equation	$\frac{1}{f} = (n-1)\left(\frac{1}{R_1} - \frac{1}{R_2}\right)$ (23.9)	Calculates the focal length of a thin-lens
*Lens Power in Diopters (*f* in meters)	$P = \frac{1}{f}$ (23.10)	Defines the power of a lens in terms of its focal length
*Sign Convention for Lens Maker's Equation	$+R$ for *C* on back side of lens $-R$ for *C* on front side of lens $R = \infty$ for a plane (flat) surface $+f$ for converging lens $-f$ for diverging lens	Defines the sign convention for the radii of curvature of a lens

V. Solutions of Selected Exercises and Paired Exercises

13. The first image by the north mirror is 3.0 m behind the north mirror.

This image is 3.0 m + 3.0 m + 5.0 m = 11 m from the south mirror. So the second image by the south mirror is 11 m behind the south mirror.

The first image by the south mirror is 5.0 m behind the south mirror.

This image is 5.0 m + 5.0 m + 3.0 m = 13 m from the north mirror. So the second image by the north mirror is 13 m behind the north mirror.

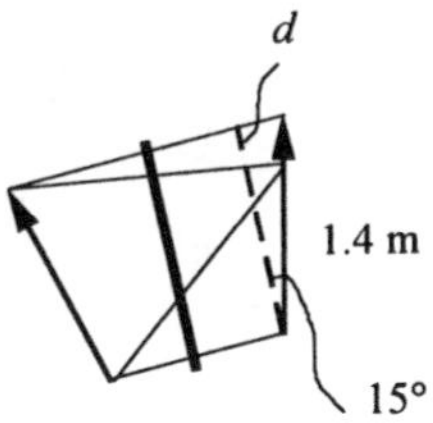

14. The projection of the student's height in the direction parallel to the mirror is

$d = (1.4\text{ m})\cos 15° = 1.35\text{ m}.$

From Example 23.2, the height of the mirror is, therefore, $\dfrac{1.35\text{ m}}{2} = 0.68\text{ m}.$

28. (a) $f = \dfrac{R}{2} = \dfrac{30\text{ cm}}{2} = 15\text{ cm},$

$d_o = 40\text{ cm}.\quad d_i = \dfrac{d_o f}{d_o - f} = \dfrac{(40\text{ cm})(15\text{ cm})}{40\text{ cm} - 15\text{ cm}} = 24\text{ cm}.$

$M = -\dfrac{d_i}{d_o} = -\dfrac{24\text{ cm}}{40\text{ cm}} = -0.60.$ So $h_i = Mh_o = -0.60\ (3.0\text{ cm}) = -1.8\text{ cm} = 1.8\text{ cm}$, real and inverted.

(b) $d_i = \dfrac{(30\text{ cm})(15\text{ cm})}{30\text{ cm} - 15\text{ cm}} = 30\text{ cm}.\quad M = -\dfrac{30\text{ cm}}{30\text{ cm}} = -1.0.$

So $h_i = -1.0\ (3.0\text{ cm}) = -3.0\text{ cm} = 3.0\text{ cm}$, real and inverted.

(c) $d_i = \dfrac{(15\text{ cm})(15\text{ cm})}{15\text{ cm} - 15\text{ cm}} = \infty.$ The characteristics of the image are not defined.

(d) $d_i = \dfrac{(5.0\text{ cm})(15\text{ cm})}{5.0\text{ cm} - 15\text{ cm}} = -7.5\text{ cm}.\quad M = -\dfrac{-7.5\text{ cm}}{5.0\text{ cm}} = +1.5.$

So $h_i = +1.5\ (3.0\text{ cm}) = 4.5\text{ cm}$, virtual and upright.

34. $f = \dfrac{R}{2} = \dfrac{40\text{ mm}}{2} = 20\text{ mm},\quad d_o = 15\text{ mm}.\quad d_i = \dfrac{d_o f}{d_o - f} = \dfrac{(15\text{ mm})(20\text{ mm})}{15\text{ mm} - 20\text{ mm}} = -60\text{ mm}.$

$M = -\dfrac{d_i}{d_o} = -\dfrac{-60\text{ mm}}{15\text{ mm}} = +4.0.$ So $h_i = Mh_o = +4.0\ (2.0\text{ mm}) = 8.0\text{ mm}.$

36. (a) $f = \dfrac{R}{2} = \dfrac{30\text{ cm}}{2} = 15\text{ cm},\quad d_o = 20\text{ cm}.\quad d_i = \dfrac{d_o f}{d_o - f} = \dfrac{(20\text{ cm})(15\text{ cm})}{20\text{ cm} - 15\text{ cm}} = 60\text{ cm}.$

$M = -\dfrac{d_i}{d_o} = -\dfrac{60\text{ cm}}{20\text{ cm}} = -3.0$, real and inverted.

(b)

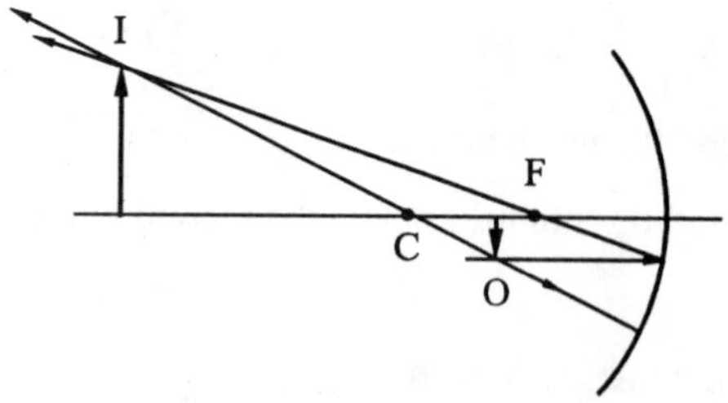

39. (a) $d_i = \dfrac{d_o f}{d_o - f} = \dfrac{f}{1 - f/d_o}$. $|M| = \dfrac{d_i}{d_o} = \dfrac{f}{d_o - f}$.

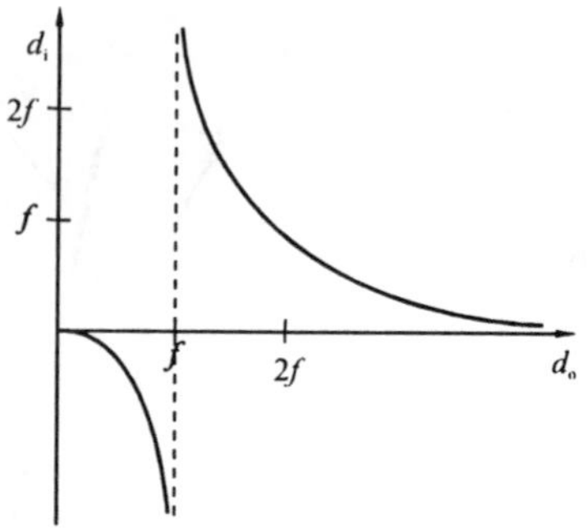

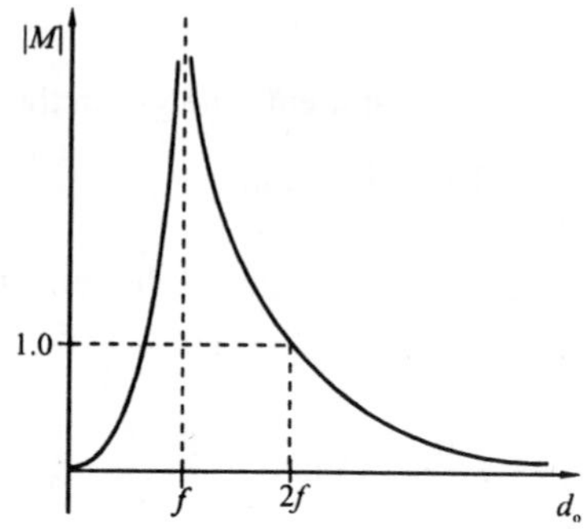

(b) $d_i = \dfrac{d_o(-f)}{d_o + f} = \dfrac{-f}{1 + f/d_o}$. $|M| = \dfrac{d_i}{d_o} = \dfrac{-f}{d_o + f}$.

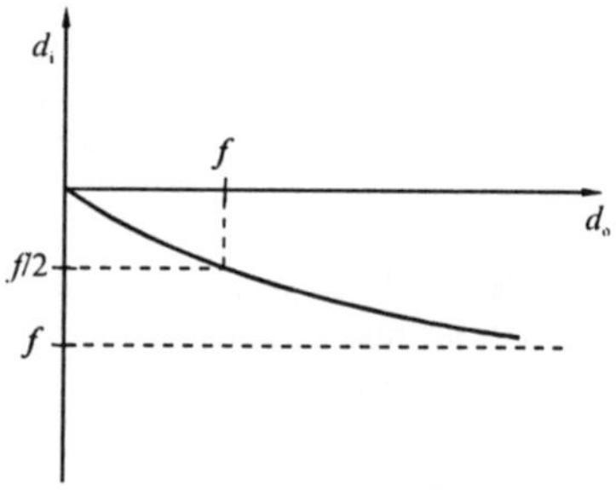

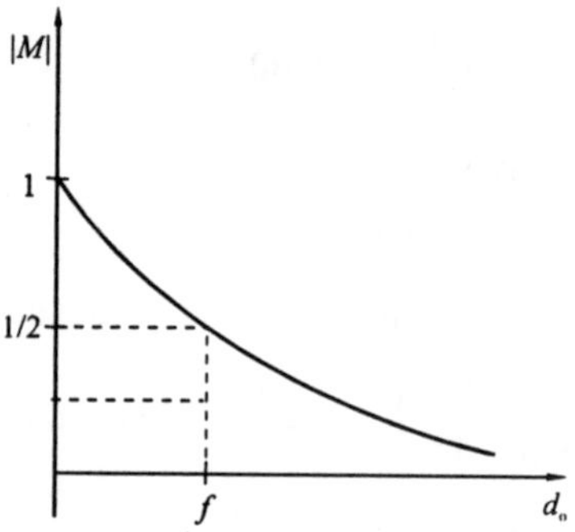

43. The two shaded triangles are similar to each other.

$$\frac{d_i - R}{R - d_o} = -\frac{h_i}{h_o},$$

where the negative sign is due to the inverted image.

Also $M = -\dfrac{d_i}{d_o} = \dfrac{h_i}{h_o}$. So $\dfrac{d_i - R}{R - d_o} = \dfrac{d_i}{d_o}$,

or $d_o d_i - d_o R = d_i R - d_o d_i$, i.e., $d_i R + d_o R = 2 d_o d_i$.

Dividing by $d_o\, d_i\, R$ on both sides gives $\dfrac{1}{d_o} + \dfrac{1}{d_i} = \dfrac{2}{R} = \dfrac{1}{R/2} = \dfrac{1}{f}$, according to the mirror equation.

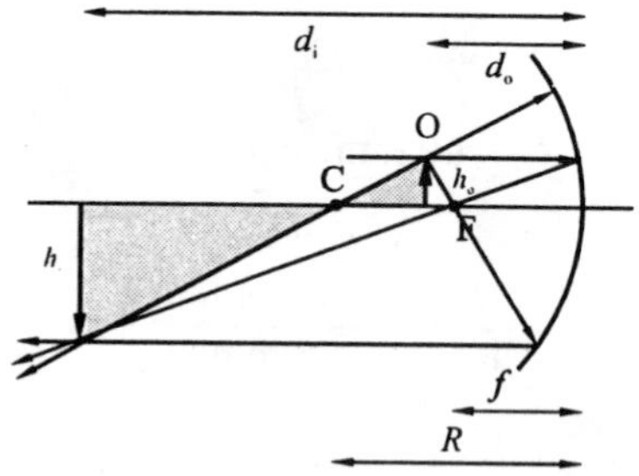

53. (a) $f = -18.0$ cm, $d_o = 10$ cm.

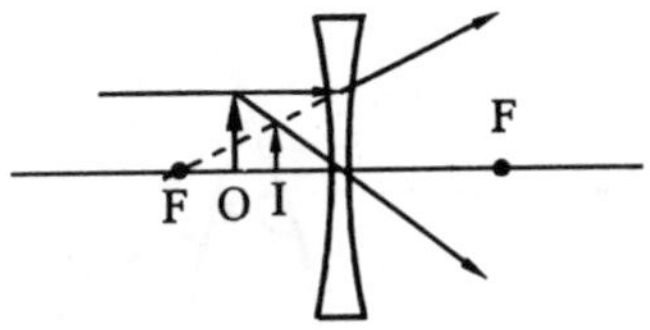

$$d_i = \frac{d_o f}{d_o - f} = \frac{(10 \text{ cm})(-18.0 \text{ cm})}{10 \text{ cm} - (-18.0 \text{ cm})} = -6.4 \text{ cm}.$$

$$M = -\frac{d_i}{d_o} = -\frac{-6.43 \text{ cm}}{10 \text{ cm}} = 0.64, \text{ virtual and upright.}$$

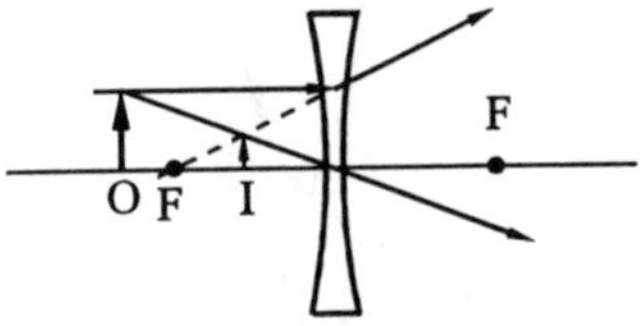

$$\text{(b) } d_i = \frac{(25 \text{ cm})(-18.0 \text{ cm})}{25 \text{ cm} - (-18.0 \text{ cm})} = -10.5 \text{ cm}.$$

$$M = -\frac{-10.5 \text{ cm}}{25 \text{ cm}} = 0.42, \text{ virtual and upright.}$$

58. (a) $\frac{1}{f} = \frac{1}{d_o} + \frac{1}{d_i} = \frac{1}{6.0 \text{ cm}} + \frac{1}{400 \text{ cm}} = 0.169 \text{ m}^{-1}$, so $f = 5.9$ cm.

(b) $M = -\frac{d_i}{d_o} = -\frac{400 \text{ cm}}{6.0 \text{ cm}} = -66.7.$

So $h_i = Mh_o = -66.7 \text{ (1.0 cm)} = -66.7 \text{ cm} = 67$ cm, inverted.

68. $f = \frac{R}{2} = \frac{90 \text{ cm}}{2} = 45$ cm, $d_i = -30$ cm (virtual image).

$$d_o = \frac{d_i f}{d_i - f} = \frac{(-30 \text{ cm})(45 \text{ cm})}{-30 \text{ cm} - 45 \text{ cm}} = 18 \text{ cm}.$$

$M = -\frac{d_i}{d_o} = -\frac{-30 \text{ cm}}{18 \text{ cm}} = +1.67.$ So $h_o = \frac{h_i}{M} = \frac{45 \text{ cm}}{1.67} = 27$ cm.

Therefore, the object is 27 cm tall and at 18 cm.

76. From $P = \frac{1}{f} = (n-1)\left(\frac{1}{R_1} - \frac{1}{R_2}\right)$, we have $\left(\frac{1}{R_1} - \frac{1}{R_2}\right) = \frac{P}{n-1} = \frac{1.5 \text{ D}}{1.6 - 1} = 2.5$ D.

So $\frac{1}{R_2} = \frac{1}{R_1} - 2.5 \text{ D} = \frac{1}{0.20 \text{ m}} - 2.5 \text{ D} = 2.5$ D.

Therefore, $R_2 = \frac{1}{2.5 \text{ D}} = 0.40 \text{ m} = 40$ cm.

80. (a) It must be a convex mirror since a concave mirror can only form magnified virtual images.

(b) $M = +0.50$ (virtual), $d_o = 7.0$ cm. $d_i = -Md_o = -0.50 \text{ (7.0 cm)} = -3.5$ cm.

$\frac{1}{f} = \frac{1}{d_o} + \frac{1}{d_i} = \frac{1}{7.0 \text{ cm}} + \frac{1}{-3.5 \text{ cm}} = -\frac{1}{7.0 \text{ cm}}$, so $f = -7.0$ cm.

Therefore, $R = 2|f| = 14$ cm.

89. (a) Since the index of refraction of the lens is greater than that of air, the angle of refraction is smaller than the angle of incidence at the air–lens interface and greater than the angle of incidence at the lens–air interface. So both refractions bend the incident light toward the axis.

(a)

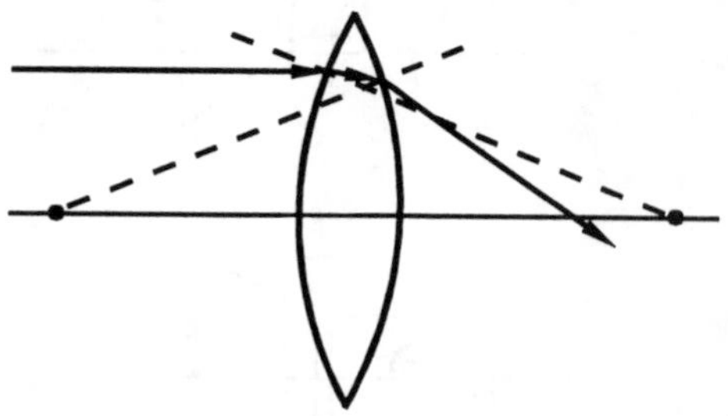

(b)

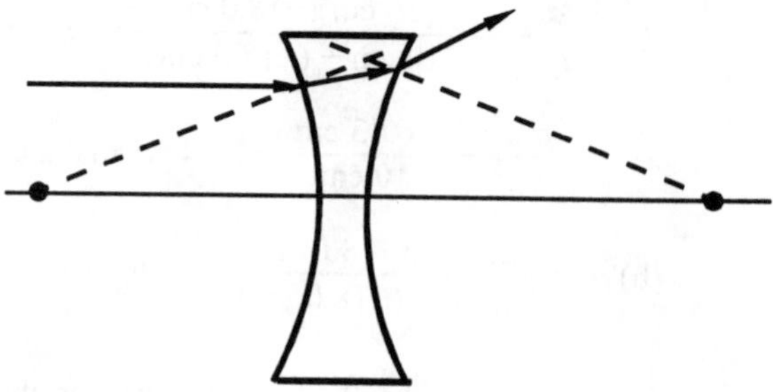

92. $d_o = 30$ cm, $d_i = 20$ cm (real because on a screen).

$$\frac{1}{f} = \frac{1}{d_o} + \frac{1}{d_i} = \frac{1}{30 \text{ cm}} + \frac{1}{20 \text{ cm}} = \frac{5}{60 \text{ cm}}, \quad \text{so} \quad f = \frac{60 \text{ cm}}{5} = 12 \text{ cm}.$$

Therefore, $R = 2f = 24$ cm.

VI. Practice Quiz

1. If you stand 5.0 ft in front of a plane mirror, how far away would you see yourself in the mirror?
(a) 2.5 ft (b) 5.0 ft (c) 7.5 ft (d) 10 ft (e) 20 ft

2. A concave mirror with a radius of curvature of 20 cm creates a real image 30 cm from the mirror. What is the object distance?
(a) 20 cm (b) 15 cm (c) 7.5 cm (d) 5.0 cm (e) 2.5 cm

3. A person's face is 30 cm in front of a concave shaving mirror. If the image is an upright image 1.5 times as large as the object, what is the mirror's focal length?
(a) 12 cm (b) 20 cm (c) 50 cm (d) 70 cm (e) 90 cm

4. Which of the following best describes the image of a plane mirror?
(a) virtual, inverted, and magnified (b) real, inverted, and diminished
(c) virtual, upright, and not magnified (d) virtual, upright, and magnified
(e) real, upright, and unmagnified

5. An object is placed at a distance of 40 cm from a thin lens. If a virtual image forms at a distance of 50 cm from the lens, on the same side as the object, what is the focal length of the lens?
(a) 200 cm (b) 90 cm (c) 75 cm (d) 45 cm (e) 22 cm

6. Which one of the following best describes the image of a concave mirror when the object distance from the mirror is less than the focal length?

(a) virtual, upright, and magnified (b) real, inverted, and diminished

(c) virtual, upright, and diminished (d) real, inverted, and magnified

(e) virtual, inverted, and magnified

7. An object is placed at a distance of 30 cm from a thin convex lens. The lens has a focal length of 10 cm. What are the values, respectively, of the image distance and lateral magnification?

(a) 65 cm, 2.0 (b) 15 cm, 2.0 (c) 25 cm, 1.0 (d) 60 cm, –0.50 (e) 15 cm, –0.50

8. The ray diagram below is an example of a

(a) camera

(b) projector

(c) magnifying glass

(d) microscope

(e) telescope

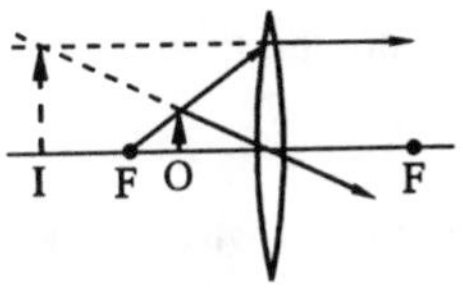

9. What causes chromatic aberration?

(a) ray is too far from the optic axis.

(b) different color has the same index of refraction.

(c) a circular cone of beam becomes an elliptical beam.

(d) too many lenses are used in the optical system.

(e) different wavelengths have different wave speeds.

10. A biconvex lens is formed by using a piece of plastic ($n = 1.7$). The radius of the front surface is 20 cm and the radius of the back surface is 30 cm. What is the focal length of the lens?

(a) 17 cm, (b) 86 cm (c) 50 cm (d) −17 cm (e) −86 cm

Answers to Practice Quiz:

1. d 2. b 3. e 4. e 5. a 6. a 7. e 8. c 9. e 10. a

CHAPTER 24

Physical Optics: The Wave Nature of Light

I. Chapter Objectives

Upon completion of this chapter, you should be able to:

1. explain how Young's experiment demonstrates the wave nature of light, and compute the wavelength of light from experimental results.
2. describe how thin films produce colorful displays and give some examples of practical applications of thin-film interference.
3. define diffraction and give examples of diffractive effects.
4. explain light polarization and give examples of polarization, both in the environment and in commercial applications.
5. define scattering and explain why the sky is blue and sunsets are red.

II. Key Terms

Upon completion of this chapter, you should be able to define and/or explain the following key terms:

physical (wave) optics
Young's double-slit experiment
thin-film interference
optical flats
Newton's rings
diffraction
diffraction grating
Bragg's law
polarization
polarizing (Brewster's) angle
birefringence
dichroism
transmission axis (polarization direction)
optical activity
LCD (liquid crystal display)
scattering
Rayleigh scattering

The definitions and/or explanations of the most important key terms can be found in the following section: **III. Chapter Summary and Discussion**.

III. Chapter Summary and Discussion

1. Young's Double-Slit Experiment

Physical (wave) optics treats light as a wave in the study of some light phenomena, such as interference, diffraction, and polarization. These effects can not be successfully explained with geometric optics, in which light is treated as several rays that follow straight line paths.

Young's double-slit experiment not only demonstrates the wave nature of light, but also allows the measurement of its wavelength. In this experiment, a light source is incident on two closely spaced narrow slits. The light waves emanating from these two slits can be considered as two in-phase spherical sources and they interfere when they arrive at a screen. On the screen, the interference pattern consists of equally spaced bright fringes separated by equally spaced dark fringes. As for sound waves (Chapter 14), the condition for interference is determined by the *path difference* (Δ) between the two waves, or the difference in distance traveled by the two waves. If $\Delta = n\lambda$ for $n = 0, 1, 2, 3, \ldots$, the interference is constructive (bright fringe); if $\Delta = \frac{m\lambda}{2}$ for $m = 1, 3, 5, \ldots$, the interference is destructive (dark fringe). For two slits separated by a distance d, the path difference is $\Delta = d \sin\theta$, so the condition for constructive interference in Young's double-slit experiment is $d \sin\theta = n\lambda$ for $n = 0, 1, 2, 3, \ldots$, where n is called the order number. The zeroth-order fringe ($n = 0$) corresponds to the central maximum, the first-order fringe ($n = 1$) is the first bright fringe on either side of the central maximum (there are two first-order fringes), and so on.

For a small angle θ, we have $\sin\theta \approx \tan\theta \approx \frac{y}{L}$, where y is the distance from the central maximum on the screen and L is the distance from the slits to the screen. The distance of the nth bright fringe (y_n) from the central maximum on either side is $y_n \approx \frac{nL\lambda}{d}$, the wavelength of light is $\lambda \approx \frac{y_n d}{nL}$ for $n = 1, 2, 3, \ldots$, and the separation between adjacent bright fringes is $y_{n+1} - y_n = \frac{L\lambda}{d}$. (The dark fringes are separated by this distance also.)

Example 24.1 Light of wavelength 575 nm falls on a double-slit and the third-order bright fringe is seen at an angle of 6.5°. Find the separation between the double slits.

Solution: Given: $\lambda = 575 \text{ nm} = 575\times10^{-9} \text{ m}$, $n = 3$, $\theta_3 = 6.5°$. Find: d.

From $d\sin\theta = n\lambda$, we have $d = \frac{n\lambda}{\sin\theta} = \frac{(3)\,\lambda}{\sin\theta_3} = \frac{(3)(575\times10^{-9}\text{ m})}{\sin 6.5°} = 1.5\times10^{-5}\text{ m} = 15\ \mu\text{m}$.

Example 24.2 In a Young's double-slit experiment, if the separation between the two slits is 0.050 mm and the distance from the slits to a screen is 2.5 m, find the spacing between the first-order and second-order bright fringes for light with wavelength of 600 nm.

Solution: Given: $d = 0.050 \text{ mm} = 0.050\times10^{-3} \text{ m}$, $L = 2.5 \text{ m}$, $\lambda = 600 \text{ nm} = 600\times10^{-9} \text{ m}$.

Find: $y_2 - y_1$.

From $y_n = \dfrac{nL\lambda}{d}$, $y_2 - y_1 = \dfrac{(2)L\lambda}{d} - \dfrac{(1)L\lambda}{d} = \dfrac{L\lambda}{d} = \dfrac{(2.5 \text{ m})(600\times10^{-9} \text{ m})}{0.050\times10^{-3} \text{ m}} = 3.0\times10^{-2} \text{ m} = 3.0 \text{ cm}.$

2. Thin-Film Interference

The waves reflected from the two faces of a thin-film can interfere, and this **thin-film interference** depends on the reflective phases of the two reflected waves. Light reflected off a material whose index of refraction is greater than that of the one it is in ($n_2 > n_1$) undergoes a *180° phase change*. If $n_2 < n_1$, there is *no phase change* on reflection. For a particular film, light reflected from the film surfaces may interfere constructively and destructively, depending on the film thickness. A practical application of thin-film interference is the nonreflecting coatings for lenses (destructive interference for reflection of certain wavelengths). In a nonreflecting coating, the index of refraction of the film is usually between those of air and glass, so there are 180° phase changes in each reflection. The minimum thickness of the film is given by $t = \dfrac{\lambda}{4\,n}$, where n is the index of refraction of the film.

When a perfect spherical lens is placed on an **optical flat**, a very flat (usually as smooth as $\lambda/20$) piece of glass, circular fringes are observed. These fringes are called **Newton's rings**. If a lens has irregularities, the fringes are distorted. This is a simple, yet effective, method to check the quality of lenses in optical industries.

Example 24.3 A transparent material with an index of refraction 1.30 is used to coat a piece of glass with an index of refraction 1.50. What is the minimum thickness of the film in order to minimize the reflected light with a wavelength of 500 nm if the light is incident perpendicularly?

Solution: The index of refraction of the film is between those of air and glass, so both reflections from the two surfaces of the film have 180° phase shifts. We can use Equation (24.8).

Given: $\lambda = 500 \text{ nm}$, $n \text{ (film)} = 1.30$. Find: t (minimum).

The minimum film thickness is $t = \dfrac{\lambda}{4\,n} = \dfrac{500 \text{ nm}}{4(1.30)} = 96.2 \text{ nm}.$

3. Diffraction

Diffraction is the deviation or bending of light around objects, edges or corners. Generally, the smaller the size of the opening or object compared to the wavelength of light, the greater the diffraction. The diffraction pattern from a single slit of width w consists of a broader central maximum and some narrower side maxima (the width of the central maximum is twice that of the side maxima). Between two maxima, there is a region of destructive interference (dark fringes). In diffraction, the dark fringes rather than the bright fringes are analyzed .

The condition for the dark fringes is given by $w \sin \theta = m\lambda$ for $m = 1, 2, 3, \ldots$, where θ is the angle for a particular minimum designated by $m = 1, 2, 3, \ldots$ on either side of the central bright fringe (there is no $n = 0$. Why?) For a small angle approximation, the position of the dark fringes from the center of the central bright fringe on a screen can be calculated from $y_{\mathrm{m}} = \dfrac{mL\lambda}{w}$, where L is the distance from the single slit to the screen. The width of the mth side maximum is the distance between the mth dark and the (m+1)th dark, $y_{\mathrm{m+1}} - y_{\mathrm{m}}$. It is evident from this equation that

- for a given slit width (w), the greater the wavelength (λ), the wider the diffraction pattern;
- for a given wavelength (λ), the narrower the slit width (w), the wider the diffraction pattern;
- the width of the central maximum is twice the width of the side maxima.

Example 24.4 Light of wavelength 580 nm is incident on a slit of width 0.300 mm. An observing screen is placed 2.00 m from the slit. Find the positions of the first-order dark fringes and the width of the central maximum.

Solution: The central maximum is the region between the first-order dark fringes on either side of this maximum, so its width is simply $2y_1$.

Given: $\lambda = 580 \text{ nm} = 580 \times 10^{-9} \text{ m}$, $w = 0.300 \text{ mm} = 0.30 \times 10^{-3} \text{ m}$, $L = 2.00 \text{ m}$,

Find: y_1 and $2y_1$.

From $y_{\mathrm{m}} = \dfrac{mL\lambda}{w}$, we have $y_1 = \dfrac{(1)L\lambda}{w} = \dfrac{(1)(2.00 \text{ m})(580 \times 10^{-9} \text{ m})}{0.300 \times 10^{-3} \text{ m}} = 3.87 \times 10^{-3} \text{ m} = 3.87 \text{ mm}$.

So the width of the central maximum is $2y_1 = 2(3.87 \text{ mm}) = 7.74 \text{ mm}$.

Example 24.5 Light of wavelength 687 nm is incident on a single slit 0.75 mm wide. At what distance from the slit should a screen be placed if the second dark fringe in the diffraction pattern is to be 1.7 mm from the center of the screen?

Solution: Given: $\lambda = 687\text{ nm} = 687 \times 10^{-9}\text{ m}$, $w = 0.75\text{ mm} = 0.75 \times 10^{-3}\text{ m}$, $m = 2$, $y_2 = 1.7\text{ mm} = 0.85 \times 10^{-3}\text{ m}$.

Find: L.

From $y_m = \dfrac{mL\lambda}{w}$, we have $L = \dfrac{y_m w}{m\lambda} = \dfrac{(1.7 \times 10^{-3}\text{ m})(0.75 \times 10^{-3}\text{ m})}{(2)(687 \times 10^{-9}\text{ m})} = 0.93\text{ m}$.

A **diffraction grating** consists of a larger number of closely spaced narrow slits. It combines multiple slit interference and single slit diffraction, and is very useful for dispersing different wavelengths or colors. (In general, it gives a larger dispersion than a prism.) The sharp (narrow, or well defined) bright fringes for a diffraction grating are given by $d \sin\theta = n\lambda$ for $n = 0, 1, 2, 3, \ldots$, where d is the spacing (or grating's spacing) between adjacent grating slits, which can be obtained from the number of lines per unit length of the grating, $d = \dfrac{1}{N}$. The number of spectral orders produced by a grating depends on the wavelength and on the grating's spacing d. Since $\sin\theta$ cannot be greater than 1, $\sin\theta = \dfrac{n\lambda}{d} \leq 1$, so the order number is therefore limited as follows: $n \leq \dfrac{d}{\lambda}$.

Note: In general, the small angle approximation cannot be used in a diffraction grating because of the grating's larger dispersion power. The bright fringes are usually separated by bigger distances here and, therefore, the angles do not satisfy the small angle approximation.

Example 24.6 Monochromatic light is incident on a grating that is 75 mm wide and ruled with 50,000 lines. The second-order maximum is seen at 32.5°. What is the wavelength of the incident light?

Solution: Given: $n = 2$, $\theta_2 = 32.5°$, $N = \dfrac{50{,}000\text{ lines}}{75\text{ mm}} = 667\text{ lines/mm}$.

Find: λ.

The grating spacing is $d = \dfrac{1}{N} = \dfrac{1}{667\text{ lines/mm}} = 1.50 \times 10^{-3}\text{ mm} = 1.50 \times 10^{-6}\text{ m}$.

From $d \sin\theta = n\lambda$, we have $\lambda = \dfrac{d \sin\theta}{n} = \dfrac{(1.50 \times 10^{-6}\text{ m}) \sin 32.5°}{2} = 4.03 \times 10^{-7}\text{ m} = 403\text{ nm}$.

Example 24.7 Monochromatic light of wavelength 500 nm is incident normally on a diffraction grating. If the third-order maximum of the diffraction pattern is observed at 32.0°,

(a) what is the grating's spacing?

(b) how many total number of visible maxima can be seen?

Solution: Given: $\lambda = 500 \text{ nm} = 500\times10^{-9}\text{ m}$, $n = 3$, $\theta_3 = 32.0°$.

Find: (a) d (b) total number of visible maxima.

(a) From $d\sin\theta = n\lambda$, we have $d = \dfrac{n\lambda}{\sin\theta} = \dfrac{(3)(500\times10^{-9}\text{ m})}{\sin 32.0°} = 2.83\times10^{-6}\text{ m}$.

(b) The maximum value of the angle θ is 90° (sin 90° = 1), so the maximum order number is

$n_{\text{max}} = \dfrac{d}{\lambda} = \dfrac{2.83\times10^{-6}\text{ m}}{500\times10^{-9}\text{ m}} = 5.67$. Since n is an integer, $n_{\text{max}} = 5$.

Therefore, 11 maxima are seen including the center one ($n = 0$, two for $n = 1$, two for $n = 2$, two for $n = 3$, two for $n = 4$, and two for $n = 5$).

The regular atomic spacing in a crystalline solid acts as a diffraction grating for light of much shorter wavelength than visible light such as X-rays. By measuring the diffraction angle, which is equal to the incidence angle also, of an X-ray beam with known wavelength, the distance between the crystal's internal planes (d) can be determined from **Bragg's law**, $2d\sin\theta = n\lambda$ for $n = 1, 2, 3, \ldots$.

4. Polarization

Polarization is the preferential orientation of the electromagnetic field vectors that make up a light wave, and is evidence that light is a transverse wave. Light with some partial preferential orientation of the field vectors is said to be *partially polarized*. If the field vectors oscillate in *one* plane (or *one* direction), the light is then *plane (linearly) polarized*. Light can be polarized by reflection, double refraction (**birefringence**), selective absorption (**dichroism**), and scattering.

Light that is partially reflected and partially refracted is partially polarized. The maximum polarization (the reflected ray is linearly polarized) occurs when the reflected and the refracted rays are 90° apart. The angle of incidence for this maximum polarization is called the **polarizing** or **Brewster's angle**. The polarizing angle for a material of index of refraction n in air is given by $\tan\theta_p = n$ and $\theta_p = \tan^{-1} n$.

In some materials, the anisotropy of the speed of light with direction (speed of light is different in different directions) gives rise to different indices of refraction in different directions. Such materials are called *birefringent* or *double refracting*. When a beam of unpolarized light is incident on a birefringent crystal, it is doubly refracted and separated in two components, or rays. These two rays are linearly polarized in mutually perpendicular directions.

Some crystals, such as tourmaline and herapathite, exhibit the interesting property of absorbing one of the polarized components more than the other. This property is called **dichroism**. If the dichroic material is sufficiently thick, the more strongly absorbed component may be completely absorbed, resulting in a linearly polarized beam. Polaroid films (used in sunglasses) are synthetic polymer materials, which allow light of one polarization direction to pass. This direction is called the **transmission axis**, or **polarization direction**. If two films (the first one is called the polarizer and the second one is called the analyzer) are placed with their transmission axes parallel to each other, light can pass through both films. If the two transmission axes are perpendicular to each other, little or no light can pass. In general, the intensity of the transmitted light is given by $I = I_o \cos^2 \theta$, where θ is the angle between the transmission axes of the polarizer and analyzer, I_o the original intensity, and I the intensity after the analyzer. This expression is known as *Malus's law.*

Scattering is the process of particles (like air molecules or dust particles) absorbing light and reradiating polarized light. The scattering of sunlight by air molecules causes the sky to look blue and sunsets to look red because the shorter wavelength (blue) is scattered more efficiently than long wavelength (red). This is called **Rayleigh scattering** and the scattering intensity is found to be proportional to $\frac{1}{\lambda^4}$.

Some transparent materials have the ability to rotate the polarization direction of the linearly polarized light. This is called the **optical activity** and is due to the molecular structure of the material. Some liquid crystals are optically active, and this property forms the basis of the common **liquid crystal display (LCD)**.

Example 24.8 How far above the horizon is the Moon when its image reflected in calm water is completely polarized?

Solution: The polarizing angle is the angle of incidence (measured from the normal) when the reflected ray is linearly polarized. The angle above the horizon is the angle measured from the surface and is equal to $(90° - \theta_p)$.

Given: $n = 1.33$ (water). Find: $90° - \theta_p$.

From $\tan \theta_p = n$, we have $\theta_p = \tan^{-1} n = \tan^{-1} 1.33 = 53.1°$.

So the angle above the horizon is $90° - 53.1° = 36.9°$.

Example 24.9 Unpolarized light is passed through a polarizer-analyzer combination. The transmission axes of the polarizer and analyzer are at 30° to each other. What percentage of the light gets through the filters?

Solution: When unpolarized light is incident on the polarizer, only one of the two components can pass; that is, only 50% of the intensity gets through the polarizer. So before the light reaches the analyzer, the intensity is already reduced to half. From Malus's law, $I = I_o \cos^2 \theta$, $\frac{I}{I_o} = \cos^2 \theta = \cos^2 30° = 0.750$ or 75.0%.

Therefore, only $\frac{75.0\%}{2} = 37.5\%$ of the original intensity passes through the polarizer-analyzer.

IV. Mathematical Summary

Path Difference for Constructive Interference	$\Delta = n\lambda$ for $n = 0, 1, 2, 3, \ldots$ (24.1)	Defines the path difference condition for constructive interference
Path Difference for Destructive Interference	$\Delta = \frac{m\lambda}{2}$ for $m = 1, 3, 5, \ldots$ (24.2)	Defines the path difference condition for destructive interference
Bright Fringe Condition (double-slit interference)	$d\sin\theta = n\lambda$ for $n = 0, 1, 2, \ldots$ (24.3)	Gives the positions of the bright fringes in double-slit interference experiment
Wavelength Measurement (double-slit interference)	$\lambda = \frac{y_n d}{n L}$ for $n = 1, 2, 3, \ldots$ (24.7)	Determine the wavelength in double-slit interference experiment
Width of Bright Fringes (double-slit interference)	$y_{n+1} - y_n = \frac{L\lambda}{d}$ (Example 24.1)	Computes the separation between adjacent bright fringes in double-slit interference experiment
Nonreflecting Film Thickness (minimum)	$t = \frac{\lambda}{4n}$ (24.8)	Gives the minimum thickness of a thin film coating to minimize reflection
Dark Fringe Condition (single-slit diffraction)	$w\sin\theta = m\lambda$ for $m = 1, 2, 3, \ldots$ (24.9)	Gives the positions of the dark fringes in single slit diffraction experiment
Lateral Displacement of Dark Fringes (single-slit diffraction)	$y_m = \frac{mL\lambda}{w}$ for $m = 1, 2, 3, \ldots$ (24.10)	Gives the lateral position of the mth dark fringe in single-slit diffraction experiment
Width of Bright Fringes (single-slit diffraction)	$\Delta = y_{m+1} - y_m = \frac{L\lambda}{w}$ (24.12)	Calculates the width of the bright fringe or the separation between two adjacent dark fringes

Interference Maxima for a Diffraction Grating	$d\sin\theta = n\lambda$ for $n = 0, 1, 2, \ldots$ (24.13) where $d = \frac{1}{N}$ and N is the number of lines per unit length	Determines the positions of the bright fringes for a diffraction grating
Limit of Order Number	$n \le \frac{d}{\lambda}$ (24.14)	Gives the highest order number for a diffraction grating
Bragg's Law	$2d\sin\theta = n\lambda$ for $n = 1, 2, 3, \ldots$ (24.15)	Relates the wavelength of X-ray, diffraction angle, and the distance between crystalline planes of a crystal
Brewster's (polarizing) Angle	$\tan\theta_p = n$ and $\theta_p = \tan^{-1} n$ (24.16)	Calculates the polarizing or Brewster's angle in a medium to air boundary

V. Solutions of Selected Exercises and Paired Exercises

6. Phase difference $= \Delta\theta = \frac{2\pi\,\Delta}{\lambda}$, so $\Delta = \frac{\Delta\theta\,\lambda}{2\pi}$.

(a) $\Delta = \frac{\pi/3\ \lambda}{2\pi} = \lambda/6$. (b) $\Delta = \frac{\pi/2\ \lambda}{2\pi} = \lambda/4$. (c) $\frac{\pi\,\lambda}{2\pi} = \lambda/2$.

12. (a) $\Delta y = \frac{L\lambda}{d} = \frac{(1.50\text{ m})(640\times10^{-9}\text{ m})}{1.0\times10^{-3}\text{ m}} = 9.6\times10^{-4}\text{ m}$.

(b) $\Delta y = \frac{(1.50\text{ m})(640\times10^{-9}\text{ m})}{1.5\times10^{-3}\text{ m}} = 6.4\times10^{-4}\text{ m}$.

14. The distance is equal to $y_3 - y_o = 3\Delta y = \frac{3L\lambda}{d} = \frac{3(1.5\text{ m})(680\times10^{-9}\text{ m})}{0.25\times10^{-3}\text{ m}} = 1.2\times10^{-2}\text{ m}$.

23. Coat a piece of glass with a material whose index of refraction is between those of air and glass. The coating would be $\lambda/2$ thick, to make the path difference between the reflected rays one full wavelength. Both rays undergo 180° phase shift, and are in phase for a path difference of one wavelength, thus interfere constructively, i.e., the incident light is reflected, not transmitted, and you have a reflecting surface.

26. From $t = \frac{\lambda}{4\,n}$, we have $\Delta t = \frac{1}{4\,n}\Delta\lambda = \frac{1}{4(1.38)}(300\text{ nm}) = 54\text{ nm} = 5.4\times10^{-8}\text{ m}$.

30. (a) The two rays for interference are the reflections from the bottom surface of the top plate and the top surface from the bottom plate. The reflection from the top surface of the bottom plate has 180° phase shifts, so the condition for constructive interference for reflection is $\Delta = 2t = \dfrac{\lambda}{2}$,

so $t = \dfrac{\lambda}{4} = \dfrac{680 \text{ nm}}{4} = 170 \text{ nm}$.

(b) Constructive for transmission is the same as destructive for reflection.

So $\Delta = 2t = \lambda$, and $t = \dfrac{\lambda}{2} = \dfrac{680 \text{ nm}}{2} = 340 \text{ nm}$.

40. $d = \dfrac{1}{10{,}000 \text{ lines/cm}} = 1.0\times10^{-4} \text{ cm} = 1.0\times10^{-6} \text{ m}$.

Since $d \sin \theta = m\,\lambda$, $m_{\text{max}} = \dfrac{d \sin 90°}{\lambda} = \dfrac{d}{\lambda} = \dfrac{1.0\times10^{-6} \text{ m}}{560\times10^{-9} \text{ m}} = 1.8$.

So there are three orders of maxima for $m = 0$ or ±1.

44. $d = \dfrac{1}{8000 \text{ lines/cm}} = 1.25\times10^{-4} \text{ cm} = 1.25\times10^{-6} \text{ m}$. Since $d \sin \theta = m\lambda$, $\theta = \sin^{-1} \dfrac{m\,\lambda}{d}$.

For blue, $\theta_{1\text{b}} = \sin^{-1} \dfrac{400\times10^{-9} \text{ m}}{1.25\times10^{-6} \text{ m}} = 18.7°$.

For red, $\theta_{1\text{r}} = \sin^{-1} \dfrac{700\times10^{-9} \text{ m}}{1.25\times10^{-6} \text{ m}} = 34.1°$. So $\Delta\theta = \theta_{1\text{r}} - \theta_{1\text{b}} = 15.4°$.

56. We see the rainbow by the scattering of light from the water droplets. The light is partially polarized in the horizontal direction, so the axis of the analyzer should be in the horizontal direction. We can never block out the polarized light completely because it is only partially polarized.

57. When the axes are perpendicular it darkens and when the axes are parallel it lightens.

(a) Twice. (b) Four times. (c) None. (d) Six times.

64. Snell's law must be written as $\dfrac{\sin \theta_1}{\sin \theta_2} = \dfrac{\sin \theta_1}{\cos \theta_1} = \tan \theta_1 = \dfrac{n_2}{n_1} = n$.

So $\theta_{\text{p}} = \tan^{-1} \dfrac{n_2}{n_1}$.

72. (a) $\lambda = \dfrac{y_{\text{n}}\, d}{n\, L} = \dfrac{(0.60\times10^{-2} \text{ m})(0.25\times10^{-3} \text{ m})}{(3)(1.0 \text{ m})} = 5.0\times10^{-7} \text{ m}$.

$f = \dfrac{c}{\lambda} = \dfrac{3.00\times10^{8} \text{ m/s}}{5.0\times10^{-7} \text{ m}} = 6.0\times10^{14} \text{ Hz}$.

(b) $\Delta y = \dfrac{\lambda\, L}{d} = \dfrac{(5.0\times10^{-7} \text{ m})(1.0 \text{ m})}{0.25\times10^{-3} \text{ m}} = 2.0\times10^{-3} \text{ m}$.

81. $d = \dfrac{1}{9000 \text{ lines/cm}} = 1.11\times10^{-4}\text{ cm} = 1.11\times10^{-6}\text{ m}.$

Since $d\sin\theta = m\lambda$, $m_{\text{max}} = \dfrac{d\sin 90°}{\lambda} = \dfrac{d}{\lambda}.$

For red, $m_{\text{max}} = \dfrac{1.11\times10^{-6}\text{ m}}{700\times10^{-9}\text{ m}} = 1.6.$ So $m_{\text{max}} = 1$ for red.

For violet, $m_{\text{max}} = \dfrac{1.11\times10^{-6}\text{ m}}{400\times10^{-9}\text{ m}} = 2.8.$ So $m_{\text{max}} = 2$ for violet.

VI. Practice Quiz

1. If a wave from one slit of a Young's double-slit experiment arrives at a point on the screen one-half wavelength behind the wave from the other slit, what is observed at that point?
(a) bright fringe (b) dark fringe (c) gray fringe (d) multi-colored fringe (e) none of the above

2. A monochromatic light is incident on a Young's double-slit setup that has a slit separation of 3.00×10^{-5} m. The resultant bright fringe separation is 2.15×10^{-2} m on a screen 1.20 m from the double slit. What is the separation between the third order bright fringe and the zeroth order fringe?
(a) 8.60×10^{-2} m (b) 7.35×10^{-2} m (c) 6.45×10^{-2} m (d) 4.30×10^{-2} m (e) 2.15×10^{-2} m

3. What is the minimum thickness of a nonreflecting coating ($n = 1.30$) on a glass lens ($n = 1.50$) for wavelength 500 nm?
(a) 250 nm (b) 192 nm (c) 167 nm (d) 96.2 nm (e) 83.3 nm

4. What will happen to the width of the central maximum if the width of the slit increases in a single slit experiment?
(a) decrease (b) increase (c) remain unchanged (d) does not depend on the separation
(e) cannot be determined because not enough information is given

5. Light of wavelength 610 nm is incident on a slit 0.20-mm wide and the diffraction pattern is produced on a screen that is 1.5 m from the slit. What is the width of the central maximum?
(a) 0.34 cm (b) 0.68 cm (c) 0.92 cm (d) 1.22 cm (e) 1.35 cm

6. A beam of unpolarized light in air strikes a flat piece of glass at an angle of incidence of 57.3°. If the reflected beam is completely polarized, what is the index of refraction of the glass?
(a) 0.54 (b) 0.84 (c) 1.52 (d) 1.54 (e) 1.56

7. What is the process to obtain polarized light in a dichroic material like Polaroid film?

(a) reflection (b) refraction (c) double refraction (d) selective absorption (e) scattering

8. When the transmission axes (polarization directions) of two Polaroid sheets are perpendicular to each other, what is the percentage of the incident light which will pass the two sheets?

(a) 0% (b) 25% (c) 50% (d) 75% (e) 100%

9. White light is spread out into spectral hues by a diffraction grating. If the grating has 2000 lines per centimeter, at what angle will red light ($\lambda = 640$ nm) appear in the first-order?

(a) 0° (b) 3.57° (c) 7.35° (d) 11.2° (e) 13.4°

10. A helium–neon laser ($\lambda = 632.8$ nm) is used to calibrate a diffraction grating. If the first-order maximum occurs at 20.5°, what is the line spacing (grating's spacing)?

(a) 1.81×10^{-6} m (b) 2.20×10^{-6} m (c) 3.62×10^{-6} m (d) 5.42×10^{-6} m (e) 7.23×10^{-6} m

Answers to Practice Quiz:

1. b 2. c 3. d 4. a 5. c 6. e 7. d 8. a 9. c 10. a

CHAPTER 25

Optical Instruments

I. Chapter Objectives

Upon completion of this chapter, you should be able to:

1. describe the optical workings of the eye, and explain some common vision defects and how they are corrected.
2. distinguish between lateral and angular magnifications, and describe simple and compound microscopes, and their magnifications.
3. distinguish between refractive and reflective telescopes, and describe the advantages of each.
4. describe the relationship of diffraction and resolution, and state and explain Rayleigh's criterion.
5. relate color vision and light, and give examples of visual images and illusions.

II. Key Terms

Upon completion of this chapter, you should be able to define and/or explain the following key terms:

crystalline lens
retina
rods
cones
accommodation
far point
near point
nearsightness
farsightness
astigmatism
magnifying glass (simple microscope)
angular magnification
compound microscope
objective
eyepiece (ocular)
refracting telescope
astronomical telescope
terrestrial telescope
reflecting telescope
resolution
Rayleigh criterion
resolving power
additive primary colors
additive method of color production
complementary colors
subtractive method of color production
subtractive primary pigments

The definitions and/or explanations of the most important key terms can be found in the following section: **III. Chapter Summary and Discussion**.

III. Chapter Summary and Discussion

1. The Human Eye

The **crystalline lens** in the eye is a converging lens composed of microscopic glassy fibers. Though muscle action, the shape of the lens is adjusted (this adjustment is called **accommodation**) and sharp images are formed on the **retina**, a light-sensitive surface. The photo-sensitive **rod** and **cone** cells of the retina are responsible for twilight (black and white) vision and color vision, respectively. The extremes of the range over which distinct vision (sharp focus) is possible are known as the far point and the near point. The **far point** is the greatest distance at which the eye can see objects clearly, and is for the normal eye taken to be infinity. The **near point** is the position closest to the eye at which objects can be seen clearly, and depends on the extent the lens can be deformed (thickened) by accommodation, which varies with age even to the normal eye.

There are three common vision defects. **Nearsightness** is the condition of being able to see nearby objects clearly, but not distant objects. This is caused by the focal length of the crystalline lens being too short (the lens is too converging, image is focused in front of the retina). A person with nearsightness will have the far point not at infinity, but at a nearer point. This vision defect can be corrected by using a diverging lens. **Farsightness** is the condition of being able to see distant objects clearly, but not nearby objects. This is caused by the focal length of the crystalline lens being too long (the lens is too diverging, image is focused behind the retina). A person with farsightness will have the near point not at the normal position, but at some point farther from the eye. This vision defect can be corrected by using a converging lens. **Astigmatism** is caused by a refractive surface, most usually the cornea or crystalline lens, being out of spherical shape. As a result, the eye has different focal lengths in different planes. This condition may be corrected with lenses that have greater curvature in the plane in which the cornea or crystalline lens has deficient curvature.

Example 25.1 A person cannot see clearly objects more than 70.0 cm away. What power of lens should be prescribed if the glass is to be worn 1.00 cm in front of the eye? Is the lens converging or diverging?

Solution: The normal eye has the far point at infinity. To correct for nearsightness, the image of an object at infinity must be formed at 70.0 cm in front of the eye, so the object distance is infinity and the image distance is −(70.0 cm − 1.00 cm) = −69.0 cm. The image distance is negative because the image is on the same side as the object (virtual imege).

Given: $d_o = \infty$, $d_i = -0.690$ m. Find: $P = \dfrac{1}{f}$.

$$P = \frac{1}{f} = \frac{1}{d_o} + \frac{1}{d_i} = \frac{1}{\infty} + \frac{1}{-0.690\text{ m}} = 0 - 1.45\text{ m}^{-1} = -1.45\text{ D}.$$

Since the power of the lens is negative, it is a diverging lens.

---

2. Microscopes

A **magnifying glass (simple microscope)** is a single converging lens which allows one to view an object clearly when it is brought closer than the near point. In such a position, an object subtends a greater angle and therefore appears larger, or magnified. The magnification of an object viewed through a magnifying glass is expressed in terms of **angular magnification** (m), which is defined as the ratio of the *angular* size of the object viewed through the magnifying glass (θ) to the angular size of the object viewed without the magnifying glass (θ_o), or $m = \dfrac{\theta}{\theta_o}$. If the image is at the near point (normally 25 cm from the eye), the magnification of the magnifying glass is $m = 1 + \dfrac{25\text{ cm}}{f}$, where f is the focal length of the converging lens. If the image is formed at infinity (eye in relaxed position), the magnification is then $m = \dfrac{25\text{ cm}}{f}$. Note that the shorter the focal length, the greater the magnification. A magnifying glass provides limited magnification because for very short focal lengths, the image becomes distorted.

A **compound microscope** consists of a pair of converging lenses, each of which contributes to the magnification. A compound microscope provides greater magnification than can be attained with a single lens. A converging lens having a relative short focal length ($f_o < 1$ cm) is known as the **objective**. It forms a real, inverted, and magnified image of an object positioned *slightly beyond* its focal point. The other lens, called the **eyepiece** (or **ocular**), has longer focal length (f_e is a few centimeters) and is positioned so that the image formed by the objective falls just *inside* its focal point. This lens forms a magnified virtual image that is viewed by the observer. The total angular magnification of the combination is $m_t = \dfrac{(25\text{ cm})\,L}{f_o f_e}$, where L is the separation between the two lenses, and f_o, f_e, and L are in centimeters. When finding the magnification of a compound microscope, make sure to convert f_o, f_e, and L into the same units, usually centimeters because the near point is usually so expressed.

Example 25.2 A person uses a converging lens of focal length 5.0 cm as a magnifying glass.

(a) What is the maximum possible angular magnification?

(b) What is the magnification if the person's eye is relaxed?

Solution: Given: $f = 5.0$ cm. Find: (a) m_{max} (b) $m_{relaxed}$.

(a) The maximum magnification is attained when the image is at the near point.

$$m = 1 + \frac{25\text{ cm}}{f} = 1 + \frac{25\text{ cm}}{5.0\text{ cm}} = 6.0\times.$$

(b) When the eyes are relaxed, the image is at infinity, so $m = \dfrac{25\text{ cm}}{f} = 5.0\times.$

Example 25.3 A compound microscope has an objective with a focal length of 5.0 mm and an eyepiece of focal length 4.0 cm. If the two lenses are separated by 20 cm, what is the total angular magnification?

Solution: Given: $f_o = 5.0 \text{ mm} = 0.50 \text{ cm}$, $f_e = 4.0 \text{ cm}$, $L = 20 \text{ cm}$.

Find: m_t.

$$m_t = \frac{(25 \text{ cm})\, L}{f_o f_e} = \frac{(25 \text{ cm})(20 \text{ cm})}{(0.50 \text{ cm})(4.0 \text{ cm})} = 250\times.$$

3. Telescopes

A **refracting telescope** uses a converging lens to collect and converge light from a distant object while a **reflecting telescope** uses a mirror to collect and converge light. The principle of a refracting telescope is similar to that of a compound microscope. The major components are objective and eyepiece lenses. The objective is a large converging lens with a long focal length. The movable eyepiece has a relatively short focal length. The image formed by the objective acts like the object for the eyepiece and a magnified image is seen. If the eyepiece of a refracting telescope is also a converging lens, the final image is inverted and the setup is known as an **astronomical telescope.** If the final image is upright, the setup is known as a **terrestrial telescope.** There are several ways to achieve an upright image. One of these is to use a diverging lens as the eyepiece (Galilean telescope). The separation of the objective and the eyepiece is equal to the sum of the focal lengths of the two lenses, $L = f_o + f_e$. The angular magnification of a refracting telescope for final image at infinity is given by $m = -\dfrac{f_o}{f_e}$

Example 25.4 A student constructs a astronomical telescope with a magnification 10. If the telescope has a converging lens of focal length 50 cm,

(a) what should be the focal length of the eyepiece?

(b) what is the resulting length of the telescope?

Solution: An astronomical telescope has an inverted final image, so the magnification is negative.

Given: $m = -10$, $f_o = 50 \text{ cm}$.

Find: (a) f_e (b) L.

(a) From $m = -\dfrac{f_o}{f_e}$, we have $f_e = -\dfrac{f_o}{m} = -\dfrac{50 \text{ cm}}{-10} = 5.0 \text{ cm}$.

(b) The length of a refracting telescope is equal to the sum of the focal lengths of the two lenses.

$L = f_o + f_e = 50 \text{ cm} + 5.0 \text{ cm} = 55 \text{ cm}$.

4. Diffraction and Resolution

The diffraction of light places a limitation on our ability to distinguish objects that are close together when using microscopes or telescopes. In general, images of two sources can be resolved if the center of the central maximum of one falls at or beyond the first minimum (dark fringes) of the other. This generally accepted limiting condition for the **resolution** of two diffracted images is known as the **Rayleigh criterion**. The limiting, or minimum, angle of resolution (θ_{min}) for a slit of width d is given by $\theta_{\text{min}} = \frac{\lambda}{d}$ (where θ_{min} is a pure number and therefore must be expressed in radians). Thus, the images of two sources will be *distinctly* resolved if the angular separation of the sources is greater than λ/d. For *circular* apertures, the minimum angle of resolution is $\theta_{\text{min}} = \frac{1.22\,\lambda}{D}$, where D is the diameter of the aperture.

For a microscope, it is convenient to specify the **resolving power**, or the actual separation (s) between two point sources. Since the objects are usually near the focal point of the objective, the minimum distance between two points whose images can be just resolved is $s = f\theta_{\text{min}} = \frac{1.22\,\lambda f}{D}$.

Note: The term resolving power used here is actually a distance, not power as used in our everyday lives.

Example 25.5 A binary star system in the constellation Orion has an angular separation between the two stars of 10^{-5} rad. If the wavelength of the light from the system is $\lambda = 500$ nm, what is the smallest aperture (diameter) telescope that can just resolve the two stars? (Ignore atmospheric blurring.)

Solution: Given: $\theta_{\text{min}} = 10^{-5}$ rad, $\lambda = 500 \text{ nm} = 500\times10^{-9}$ m.

Find: D.

From $\theta_{\text{min}} = \frac{1.22\,\lambda}{D}$, we have $D = \frac{1.22\,\lambda}{\theta_{\text{min}}} = \frac{1.22(500\times10^{-9}\text{ m})}{10^{-5}\text{ rad}} = 6.1\times10^{-2}\text{ m} = 6.1\text{ cm}$.

Due to atmospheric blurring, the diameter of the telescope is much larger than the value predicted here.

Example 25.6 A compound microscope is designed to resolve objects which are 0.010 mm apart. If the focal length of the objective is 4.0 cm and the wavelength of light used is 550 nm, what is the diameter of the aperture of the objective?

Solution: The 0.010 mm distance is the resolving power of the microscope.

Given: $s = 0.010\text{ mm} = 0.010\times10^{-3}\text{ m}$, $f = 4.0\text{ cm} = 0.040\text{ m}$, $\lambda = 550\text{ nm} = 550\times10^{-9}\text{ m}$.

Find: D.

From $s = \dfrac{1.22\,\lambda f}{D}$, we have $D = \dfrac{1.22\,\lambda f}{s} = \dfrac{1.22(550\times10^{-9}\text{ m})(0.040\text{ m})}{0.010\times10^{-3}\text{ m}} = 2.7\times10^{-3}\text{ m} = 2.7\text{ mm}$.

In reality, the aperture of the microscope must be larger than the predicted value to take other laboratory conditions into account, such as air movement, temperature fluctuations, etc.

5. Color and Optical Illusions

The **additive primary colors** (additive primaries) are red, blue, and green. The mixing of light of the additive primaries is called the **additive method of color production**. When light of the three additive primaries is mixed in proper portion, the mixture appears white to the eye. Pairs of color combinations that appear white to the eye are called **complementary colors**. The complementary color of red is cyan, of blue is yellow, and of green is magenta, etc.

The **subtractive primary pigments** (subtractive primaries) are cyan, yellow, and magenta. A mixture of absorbing pigments results in the subtraction of colors, and one sees the color of light that is not absorbed or subtracted. This is called **subtractive method of color production.** When the three subtractive primaries are mixed in the proper portion, the mixture appears black (all colors are absorbed) to the eye.

IV. Mathematical Summary

Angular Magnification	$m = \dfrac{\theta}{\theta_o}$ (25.1)	Defines the angular magnification
Magnification of a Magnifying Glass with Image at Near Point (25 cm)	$m = 1 + \dfrac{25\text{ cm}}{f}$ (25.3)	Computes the angular magnification of a magnifying glass when the image is at near point (25 cm)
Magnification of a Magnifying Glass with Image at Infinity	$m = \dfrac{25\text{ cm}}{f}$ (25.4)	Computes the angular magnification of a magnifying glass when the image is at infinity
Magnification of a Compound Microscope (with L and f's in cm)	$M_t = M_o m_e = \dfrac{(25\text{ cm})L}{f_o f_e}$ (25.5)	Computes the magnification of a compound microscope consisting of an objective and an eyepiece

Magnification of a Refracting Telescope	$m = -\frac{f_o}{f_e}$ (25.6)	Calculates the magnification of a refracting telescope consisting of an objective and an eyepiece
Resolution for a Slit	$\theta_{min} = \frac{\lambda}{d}$ (25.7)	Defines the minimum angle of resolution for a slit width d
Resolution for a Circular Aperture	$\theta_{min} = \frac{1.22\,\lambda}{D}$ (25.8)	Defines the minimum angle of resolution for a circular aperture of diameter D
Resolving Power	$s = f\,\theta_{min} = \frac{1.22\,\lambda f}{D}$ (25.9)	Defines the resolving power of a circular lens of focal length f

V. Solutions of Selected Exercises and Paired Exercises

6. $d_o = \infty$, $d_i = -300$ cm (image is on the object side).

$$\frac{1}{f} = \frac{1}{d_o} + \frac{1}{d_i} = \frac{1}{\infty} + \frac{1}{-300\text{ cm}} = \frac{1}{-300\text{ cm}},$$ so $f = -300$ cm, diverging.

10. $d_o = 25$ cm, $d_i = -80$ cm (image on object side).

$$\frac{1}{f} = \frac{1}{d_o} + \frac{1}{d_i} = \frac{1}{25\text{ cm}} + \frac{1}{-80\text{ cm}} = 0.0275\text{ cm}^{-1},$$ so $f = 36$ cm, converging.

18. Top: $d_o = \infty$, $d_i = -500\text{ cm} = -5.0\text{ m}$ (image on object side).

$$P = \frac{1}{f} = \frac{1}{d_o} + \frac{1}{d_i} = \frac{1}{\infty} + \frac{1}{-5.0\text{ m}} = -0.20\text{ D}.$$

Bottom: $d_o = 25\text{ cm} = 0.25\text{ m}$, $d_i = -70\text{ cm} = -0.70\text{ m}$ (image on object side).

$$P = \frac{1}{0.25\text{ m}} + \frac{1}{-0.70\text{ m}} = +2.6\text{ D}.$$

21. (a) $d_o = \infty$, $d_i = -4.0$ m (image on object side).

$$P = \frac{1}{f} = \frac{1}{d_o} + \frac{1}{d_i} = \frac{1}{\infty} + \frac{1}{-4.0\text{ m}} = -0.25\text{ D}.$$

(b) Without the contact lens, an object at 4.0 m will form an image on the retina ($d_i = d_r$).

With the crystalline lens having a focal length f_e, we have $\frac{1}{f_e} = \frac{1}{4.0\text{ m}} + \frac{1}{d_r}$.

With the lens pair (f_c is the contact lens focal length): $\frac{1}{f} = \frac{1}{f_e} + \frac{1}{f_c} = \frac{1}{\infty} + \frac{1}{d_r}$.

Eliminating d_r, $P = \frac{1}{f_c} = \frac{1}{d_r} - \frac{1}{4.0\text{ m}} - \frac{1}{d_r} = -\frac{1}{4.0\text{ m}} = -0.25\text{ D}.$

28. (a) $m = 1 + \frac{25\text{ cm}}{f} = 1 + \frac{25\text{ cm}}{15\text{ cm}} = 2.7\times$. (b) $m = \frac{25\text{ cm}}{f} = \frac{25\text{ cm}}{15\text{ cm}} = 1.7\times$.

34. From $M_t = \frac{(25\text{ cm})L}{f_o f_e}$, we have $P_o = \frac{1}{f_o} = \frac{M_t f_e}{(0.25\text{ m})L} = \frac{(360)(0.0080\text{ cm})}{(0.25\text{ cm})(0.15\text{ cm})} = +77\text{ D}$.

39. From thin lens equation: $d_o = \frac{d_i f}{d_i - f}$, $\frac{d_i}{d_o} = \frac{d_i - f}{f} = \frac{-(D-d)-f}{f}$.

By small angle approximation: $m = \frac{\theta_i}{\theta_o} = \frac{y_i/D}{y_o/25\text{ cm}} = \frac{y_i}{y_o}\frac{25\text{ cm}}{D}$.

By similar triangles: $\frac{y_i}{y_o} = \frac{-d_i}{d_o}$, the – sign is introduced because d_i is negative (virtual image).

So $m = \frac{(D-d)+f}{f}\frac{25\text{ cm}}{D} = \frac{25}{f}(1 - \frac{d}{D}) + \frac{25}{D}$.

48. Since $L = f_o + f_e + 4f_i$, $f_i = \frac{L - f_o - f_e}{4} = \frac{75\text{ cm} - 45\text{ cm} - 15\text{ cm}}{4} = 3.8\text{ cm}$.

56. The central maximum of one pattern falls on the first minimum of the other. The angular position of the first minimum is determined by $w \sin\theta \approx w\,\theta = m\,\lambda = (1)\,\lambda$.

So $\theta_{\min} = \frac{\lambda}{w} = \frac{680\times10^{-9}\text{ m}}{0.35\times10^{-3}\text{ m}} = 1.9\times10^{-3}\text{ rad}$.

60. $\theta_{\min} = \frac{1.22\,\lambda}{D} = \frac{1.22\,(550\times10^{-9}\text{ m})}{8.0\times10^{-3}\text{ m}} = 8.4\times10^{-5}\text{ rad}$.

64. (a) $\theta_{\min} = \frac{1.22\,\lambda}{D} = \frac{1.22\,(570\times10^{-9}\text{ m})}{0.0250\text{ m}} = 2.78\times10^{-5}\text{ rad}$.

(b) $s = f\,\theta_{\min} = (30.0\text{ mm})(2.78\times10^{-5}\text{ rad}) = 8.34\times10^{-4}\text{ mm}$.

70. Since white is obtained by adding colors, it cannot be obtained by the subtractive method. That method subtracts colors, and the one we see is the one that is not absorbed.

Black objects do not absorb all wavelengths of light. We see the objects because we perceive the extremely faint light as black. (Think of twilight vision.)

76. The far point is at the image location with glasses. $d_o = \infty$, $P = \frac{1}{f} = -0.15\text{ D}$.

$\frac{1}{d_o} + \frac{1}{d_i} = \frac{1}{f} = P$, so $d_i = f = \frac{1}{P} = \frac{1}{-0.15\text{ D}} = -6.7\text{ m}$. Therefore, the far point is 6.7 m.

82. The near point is at the image location with glasses.

$$d_o = 25\text{ cm} = 0.25\text{ m},\quad f = \frac{1}{P} = \frac{1}{2.8\text{ D}} = 0.357\text{ m}.$$

$$d_i = \frac{d_o f}{d_o - f} = \frac{(0.25\text{ m})(0.357\text{ m})}{0.25\text{ m} - 0.357\text{ m}} = -0.834\text{ m}.$$ So the near point is 83 cm.

VI. Practice Quiz

1. The closest distance at which the normal eye can see objects clearly is

(a) the near point. (b) the far point. (c) nearsightness. (d) farsightness. (e) astigmatism.

2. Which one of the following is not a primary color of light?

(a) red (b) yellow (c) blue (d) green (e) both (c) and (d)

3. A nearsighted person wears glasses whose lenses have power of −0.15 D. What is the person's far point if the glasses are very close to the eyes?

(a) 1.5 m (b) 3.3 m (c) 6.0 m (d) 6.7 m (e) infinity

4. A magnifying glass has a focal length of 6.0 cm. What is the angular magnification if the image is viewed by a relaxed eye?

(a) 2.2× (b) 3.2× (c) 4.2× (d) 5.2× (e) 6.2×

5. A compound microscope has a 18-cm barrel and an eyepiece with a focal length of 8.0 mm. What is the focal length of the objective to give a total magnification of 240×?

(a) 0.094 cm (b) 0.13 cm (c) 1.5 cm (d) 1.9 cm (e) 2.3 cm

6. A person is designing a 10× telescope. If the telescope is limited to a length of 20 cm, what is the approximate focal length of the objective?

(a) 15 cm (b) 16 cm (c) 17 cm (d) 18 cm (e) 19 cm

7. The 2.4-m reflecting Hubble Space Telescope has been placed into Earth orbit by the space shuttle. What angular resolution could this telescope achieve by the Rayleigh criterion if the wavelength is 500 nm?

(a) 7.5×10^{-7} rad (b) 6.8×10^{-7} rad (c) 5.7×10^{-7} rad (d) 3.6×10^{-7} rad (e) 2.5×10^{-7} rad

8. An individual's eye pupil changes from a diameter of 3.5 mm to 1.5 mm as the illumination is increased. By what factor does the minimum angle of resolution change?

(a) 0.43 (b) 0.65 (c) 1.0 (d) 2.0 (e) 2.3

9. If a farsighted person has a near point of 120 cm, what power of glasses should be prescribed?

(a) +0.83 D (b) −0.83 D (c) +3.2 D (d) −3.0 D (e) +4.0 D

10. A refracting telescope has an angular magnification *m*. If the objective focal length is doubled and the eyepiece focal length is halved, what is the new magnification?

(a) 4*m* (b) 2*m* (c) *m* (d) *m*/2 (e) *m*/4

Answers to Practice Quiz:

1. a 2. b 3. d 4. c 5. e 6. d 7. e 8. e 9. c 10. a

CHAPTER 26

Relativity

I. Chapter Objectives

Upon completion of this chapter, you should be able to:

1. summarize the concepts of classical relativity and relative velocities, define inertial and noninertial reference frames, and explain the reasoning behind the ether hypothesis.
2. explain the general concept and operation of the Michelson-Morley experiment and the effect on the ether concept.
3. explain how the two postulates of relativity imply the relativity of simultaneity, and how the relativity of simultaneity leads to length contraction.
4. understand the concepts of time dilation and length contraction, and calculate the relationship between time and intervals and lengths observed in different inertial frames.
5. understand the need for more general expressions for kinetic energy, mass, momentum, and total energy when objects move near the speed of light; and use the relativistically correct expressions to calculate energy and momentum in particle interactions.
6. explain the principle of equivalence and specify some of the predictions of general relativity.

*7. understand the necessity for a relativistic velocity addition formula and apply it to simple relative velocity calculations.

II. Key Terms

Upon completion of this chapter, you should be able to define and/or explain the following key terms:

relativity
inertial reference frame
principle of classical (Newtonian) relativity
ether
Michelson-Morley experiment
interferometer
principle of relativity
constancy of the speed of light
special theory of relativity
length contraction
twin (clock) paradox
relativistic kinetic energy
rest mass
relativistic mass
relativistic momentum
rest energy
mass-energy equivalence
general theory of relativity
principle of equivalence

event

time dilation

proper time

proper length

black hole

Schwarzschild radius

event horizon

The definitions and/or explanations of the most important key terms can be found in the following section: **III. Chapter Summary and Discussion**.

III. Chapter Summary and Discussion

1. Classical Relativity

Relativity is a branch of physics that deals with fast moving (close to the speed of light) particles and observations from two different reference frames. Classical (nonrelativistic) physics is simply the approximations of relativity at low speeds.

An **inertial reference frame** is a nonaccelerating frame in which Newton's first law holds. In an inertial frame, an *isolated* object (one on which there is no net force) is stationary or moves with constant velocity. The **principle of classical (Newtonian) relativity** states: the laws of mechanics are the same in all inertial reference frames. The speed of light was predicted by Maxwell equations to be 3.00×10^8 m/s in vacuum. But, relative to what reference frame does light have this speed? Classically the speed of light measured from different frames of reference would be expected to differ -- possibly to be even greater than 3.00×10^8 m/s if you were approaching the light wave -- by simple vector addition of velocities, so it was reasoned that the speed of light *must* be referenced to a particular frame. This gives rise to a unique reference frame associated with the proposed medium of transport of electromagnetic waves called the **ether**; an absolute reference frame. As a result, it would seem that Maxwell's equations (which govern light and its speed) did *not* satisfy the Newtonian principle as did the laws of mechanics.

2. The Michelson-Morley Experiment

The **Michelson-Morley experiment** was an attempt to detect the ether by using the interference of light in an extremely sensitive **interferometer**. In the interferometer, a beam of light is divided into two beams, the first traveling in a direction perpendicular to the velocity of the Earth and the second traveling in a direction parallel to the velocity of the Earth. If the ether existed, interference fringe *shifts* should be observed, but the experiment had *null* results, that is, *no fringe shift at all* was detected. This experiment basically ruled out the existence of the ether, although attempts were made to explain the null result of the experiment in terms of other theories.

3. The Postulates of Special Relativity and the Relativity of Simultaneity

The null result, or the failure to detect the ether, was resolved by the **special theory of relativity** formulated by Albert Einstein. His theory had two basic postulates:

Postulate I (**principle of relativity**): all the laws of physics (not just mechanics) are the same in all inertial reference frames.

Postulate II (**constancy of the speed of light**): the speed of light c in vacuum is the same as measured in any inertial reference frame.

The first postulate implies that all inertial reference frames are equivalent, with physical laws being the same in all of them. The second postulate implies that two observers in different inertial reference frames would measure the speed of light to be c independent of the speed of the source and/or the observer.

An **event** is a "happening" in the language of relativity. We need to specify the location *and* time to tell an event. Events that are simultaneous in a particular inertial reference frame may *not* be simultaneous as measured in a different inertial frame. Simultaneity is thus a relative, not an absolute, concept. The nonintuitive results of length contraction and time dilation (discussed next) are just two of the many relativistic results that follow directly from the two postulates and the relativity of simultaneity. They are nonintuitive because we are used to dealing with speeds much slower than light, where lengths and time intervals are absolute.

4. The Relativity of Length and Time: Length Contraction and Time Dilation

Two results of the postulates of the special theory of relativity are time dilation and length contraction. **Time dilation** refers to a phenomenon in which a fast moving clock is measured to run more slowly than clock at rest in the observer's own frame of reference. Since the time interval is relative and there are two time intervals, the time measured in a frame at rest with respect to the clock is called the **proper time** (t_o). The dilated time (t) and the proper time are related by the equation $t = \dfrac{t_o}{\sqrt{1-(v/c)^2}}$, where v is the relative speed between the two frames and c is the speed of light. Many relativistic equations can be written more simply if we represent the expression $\dfrac{1}{\sqrt{1-(v/c)^2}}$ as a factor, gamma (γ): $\gamma = \dfrac{1}{\sqrt{1-(v/c)^2}}$. Gamma ($\gamma$) is always greater than or equal to 1. Therefore, the time dilation equation can be written as $t = \gamma t_o$.

Note: t and t_o are time intervals (Δt and Δt_o), not instants of time. The Δ's are dropped simply for simplicity.

Length contraction refers to a phenomenon in which the length of a fast moving object relative to an observer in an inertial reference frame is less (in the dimension of relative motion) than if the object were at rest in the observer's frame. The length measured in a frame at rest with respect to the object is called the **proper length** (L_o). The contracted length (L) and the proper length are related by $L = \frac{L_o}{\gamma} = L_o\sqrt{1-(v/c)^2}$.

Note: It is very important to identify the proper time and the proper length in problem solving. They are the time and length measured in a frame at rest with respect to the clock and object.

Time dilation gives rise to another popular relativistic topic -- the so called **twin (clock) paradox**. A twin on a space journey relative to the Earth would age more slowly than the other Earth-based twin. The twin paradox has been experimentally verified with atomic clocks on Earth.

Example 26.1 One 20-year old twin brother takes a space trip with a speed of $0.80c$ for 30 years according to a clock on the spaceship. Upon returning to the Earth, what is his own age and the age of the Earth-based twin?

Solution: The time 30 years is measured by the traveling twin with his own clock, so it is the proper time.

Given: $t_o = 30$ y, $v = 0.80c$.

Find: (a) Age of traveling twin (b) age of Earth-based twin.

(a) The age of the traveling twin is simply $20\text{ y} + 30\text{ y} = 50\text{ y}$.

(b) According to the Earth-based twin, 30 years on the spaceship is

$$t = \gamma t_o = \frac{t_o}{\sqrt{1-(v/c)^2}} = \frac{(30\text{ y})}{\sqrt{1-(0.80)^2}} = 50\text{ y on the Earth.}$$

So the age of the Earth-based twin is $20\text{ y} + 50\text{ y} = 70\text{ y}$.

Example 26.2 A meterstick is moving toward you with a speed of $0.85c$. By what percentage will its length change compared to the meterstick at rest?

Solution: A meterstick measures 1.0 m long at rest, so this is the proper length.

Given: $L_o = 1.0$ m, $v = 0.85c$.

Find: L.

From length contraction, $L = \frac{L_o}{\gamma} = L_o\sqrt{1-(v/c)^2} = (1.0\text{ m})\sqrt{1-(0.85)^2} = 0.53\text{ m}$.

So the meterstick appears to be $\frac{0.53\text{ m}}{1.0\text{ m}} = 0.53 = 53\%$ as compared to the meterstick at rest.

Example 26.3 The closest star to our solar system is Alpha Centauri, which is 4.30 light years away. A spaceship with a constant velocity of $0.800c$ relative to the Earth travels toward the star.

(a) What distance does the space ship travel according to a passenger on the ship?

(b) How much time would elapse on a clock on board the spaceship?

(c) How much time would elapse on a clock on the Earth?

Solution: The 4.30 light-year distance is measured by observers on the Earth, so it is the proper length. The proper time is the time measured with a clock on the spaceship.

Given: $L_o = 4.30$ ly, $v = 0.800c$.

Find: (a) L (b) t_o (c) t.

(a) From length contraction, $L = \dfrac{L_o}{\gamma} = L_o\sqrt{1-(v/c)^2} = (4.30\text{ ly})\sqrt{1-(0.800)^2} = 2.58$ ly.

(b) The proper time (the time measured on the spaceship) is $t_o = \dfrac{L}{v} = \dfrac{2.58\text{ ly}}{0.800c} = \dfrac{2.58c\text{ y}}{0.800c} = 3.23$ y.

(c) The dilated time (the time measured by a Earth-based clock) is then

$$t = \gamma t_o = \frac{t_o}{\sqrt{1-(v/c)^2}} = \frac{3.23\text{ y}}{\sqrt{1-(0.800)^2}} = 5.38\text{ y}.$$

Or, $\Delta t = \dfrac{L_o}{v} = \dfrac{4.30\text{ ly}}{0.800c} = \dfrac{4.30c\text{ y}}{0.800c} = 5.38$ y. Two different ways, the same result.

5. Relativistic Kinetic Energy, Mass, Momentum, and Total Energy

Other ramifications of the special theory of relativity are relativistic mass, and changes to the classical expressions for momentum, and energy. **Relativistic mass** (m) of a particle increases with speed, $m = \gamma m_o = \dfrac{m_o}{\sqrt{1-(v/c)^2}}$, where m_o is the **rest** (or proper) **mass**. As a result, the **relativistic momentum** is given by

$p = mv = \gamma m_o v = \dfrac{m_o v}{\sqrt{1-(v/c)^2}}$. The **relativistic kinetic energy** is given by $K = mc^2 - m_o c^2 = (\gamma - 1)m_o c^2$, or $K = E - E_o$, where $E_o = m_o c^2$ is the **rest energy** and $E = mc^2 = \gamma m_o c^2 = \gamma E_o$ is the **total relativistic energy**. The expression $E_o = m_o c^2$ is known as the famous **mass-energy equivalence** because it points out that mass is also a form of energy. In particular, a particle has rest energy E_o associated with its rest mass m_o. For speeds below 10% of the speed of light, the nonrelativistic formulas ($K = \frac{1}{2}m v^2$, etc.) are acceptable and will only cause less than 1% error.

Example 26.4 An electron is traveling with a speed of $0.95c$. What is its relativistic mass?

Solution: Given: $m_o = 9.11\times10^{-31}$ kg, $v = 0.95c$.

Find: m.

$$m = \frac{m_o}{\sqrt{1-(v/c)^2}} = \frac{9.11\times10^{-31}\text{ kg}}{\sqrt{1-(0.95)^2}} = 2.9\times10^{-30}\text{ kg}.$$

Example 26.5 The kinetic energy of a proton is 80% of its total energy.

(a) What is the speed of the proton?

(b) What is the magnitude of the momentum of the proton?

Solution: Given: $K = 0.80E$, $m_o = 1.67\times10^{-27}$ kg.

Find: (a) v (b) p.

(a) Since $E = K + E_o = 0.80E + E_o$, $E_o = 0.20E$. From $E = \gamma E_o$, $\gamma = \dfrac{E}{E_o} = \dfrac{1}{0.20} = 5.0$.

Since $\dfrac{1}{\gamma} = \sqrt{1-(v/c)^2} = 0.20$, $v = c\sqrt{1-(0.20)^2} = 0.98c$.

(b) $p = mv = \gamma m_o v = (5.0)(1.67\times10^{-27}\text{ kg})(0.98)(3.00\times10^8\text{ m/s}) = 2.5\times10^{-18}\text{ kg·m/s}$.

6. The General Theory of Relativity

The special theory of relativity applies only to inertial reference frames, *not* noninertial (accelerating) systems. The **general theory of relativity** considers accelerating frames and is essentially a gravitational theory. Its **principle of equivalence** states that an inertial reference frame in a uniform gravitational field is equivalent to a reference frame in the absence of a gravitational field that has a constant acceleration with respect to that inertial frame. It basically means that no experiment performed in a closed system can distinguish between the effects of a gravitational field and the effects of an acceleration.

Some of the predictions of the general theory of relativity are gravitational light bending, gravitational lensing, black holes, and the gravitational red shift. One way a **black hole** can form is from the gravitational collapse of a massive star after it is done producing light. Such an object has a density so great and a gravitational field so intense that nothing (including light) can escape it. The critical radius around a black hole from which light will not be able to escape is given by the **Schwarzschild radius,** $R = \dfrac{2GM}{c^2}$, where G is the universal gravitational constant and M the mass within the radius. The boundary of a sphere of radius R defines what is called the **event horizon**. Any event occurring within this horizon is invisible to an observer outside, since light cannot escape.

Example 26.6 What radius must our Sun have in order for light to not be able to escape from it?

Solution: Given: $M = 2.0\times10^{30}$ kg, $G = 6.67\times10^{-11}$ N·m²/kg². Find: R.

$$R = \frac{2GM}{c^2} = \frac{2(6.67\times10^{-11}\ \text{N}\cdot\text{m}^2/\text{kg}^2)(2.0\times10^{30}\ \text{kg})}{(3.00\times10^8\ \text{m/s})^2} = 3.0\times10^3\ \text{m} = 3.0\ \text{km}.$$

*7. Relativistic Velocity Addition

When dealing with relative velocities for fast moving particles, the relativistic velocity addition must be used, $u = \dfrac{v + u'}{1 + \dfrac{v\,u'}{c^2}}$. The meanings of the symbols in the equation are:

v = velocity of object 1 with respect to an inertial observer (e.g., on Earth)

u' = velocity of object 2 with respect to object 1

u = velocity of object 2 with respect to an inertial observer (e.g., on Earth).

Example 26.7 A spaceship moves away from the Earth with a speed of $0.80c$. The spaceship then fires a missile with a speed of $0.50c$ relative to the ship. What is the velocity of the missile measured by observers on the Earth if

(a) the missile is fired away from the Earth?

(b) the missile is fired toward the Earth?

Solution: Take away from the Earth as the positive direction for velocity.

Given: $v = 0.80c$, (a) $u' = +0.50c$, (b) $u' = -0.50c$.

Find: u (a) and (b).

$$\text{(a) } u = \frac{v + u'}{1 + \dfrac{v\,u'}{c^2}} = \frac{0.80c + 0.50c}{1 + (0.80)(0.50)} = 0.93c \text{ away from the Earth.}$$

According to the classical theory, $u = v + u' = 0.80c + 0.50c = 1.3c$!

$$\text{(b) } u = \frac{0.80c + (-0.50c)}{1 + (0.80)(-0.50)} = 0.50c \text{ away from the Earth.}$$

According to the classical theory, $u = 0.80c + (-0.50c) = 0.30c$!

IV. Mathematical Summary

Time Dilation	$t = \gamma t_o = \dfrac{t_o}{\sqrt{1-(v/c)^2}}$ (26.8)	Computes the dilated time in terms of proper time
Length Contraction	$L = \dfrac{L_o}{\gamma} = L_o\sqrt{1-(v/c)^2}$ (26.10)	Computes the contracted length in terms of proper length
Relativistic Kinetic Energy	$K = \gamma m_o c^2 - m_o c^2 = E_o(\gamma - 1)$ (26.11)	Defines the relativistic kinetic energy
Relativistic Mass	$m = \gamma m_o = \dfrac{m_o}{\sqrt{1-(v/c)^2}}$ (26.12)	Defines the relativistic mass
Relativistic Momentum	$\mathbf{p} = m\mathbf{v} = \gamma m_o \mathbf{v}$ (26.13)	Defines the relativistic momentum
Rest Energy	$E_o = m_o c^2$ (26.15)	Defines the rest energy
Relativistic Total Energy	$E = K + E_o$ $E = K + m_o c^2 = m c^2$ (26.16)	Computes the total relativistic energy
Total Energy and Rest Energy	$E = \gamma E_o$ (26.17)	Relates total relativistic energy and rest energy
Schwarzschild Radius	$R = \dfrac{2GM}{c^2}$ (26.19)	Defines the radius of event horizon
*Relativistic Velocity Addition	$u = \dfrac{v + u'}{1 + \dfrac{v u'}{c^2}}$ (26.20)	Computes relativistic velocity addition

V. Solutions of Selected Exercises and Paired Exercises

8. Take eastward as positive.

(a) $v_{BA} = v_B - v_A = -65$ km/h $- 90$ km/h $= -155$ km/h $= 155$ km/h westward.

(b) $v_{BA} = -65$ km/h $- 90$ km/h $= -25$ km/h $= 25$ km/h westward.

10. (a) The relative velocity to ground is 200 km/h + (−35 km/h) = 165 km/h.

(b) The relative velocity to ground is 200 km + 25 km/h = 225 km/h.

12. (a) Perpendicular to the shore, $t = \dfrac{y}{v_\perp} = \dfrac{50 \text{ m}}{0.45 \text{ m/s}} = \boxed{1.1\times10^2 \text{ s}}$.

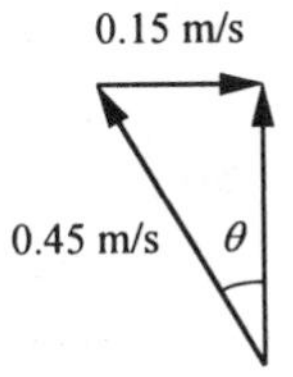

(b) Along the shore, $x = v_{||}\, t = (0.15 \text{ m/s})(111 \text{ s}) = \boxed{17 \text{ m}}$.

(c) $\theta = \sin^{-1}\left(\dfrac{0.15}{0.45}\right) = \boxed{19° \text{ up stream}}$.

24. We are finding the proper time in another reference frame.

From $t = \dfrac{t_o}{\sqrt{1 - v^2/c^2}}$, we have $t_o = t\sqrt{1 - v^2/c^2} = (5.0 \text{ min})\sqrt{1 - 0.90^2} = 2.2 \text{ min}.$

28. (a) The time observed on Earth is $t = \dfrac{4.30 \text{ ly}}{0.60c} = 7.17 \text{ y}.$

From $t = \dfrac{t_o}{\sqrt{1 - v^2/c^2}}$, we have $t_o = t\sqrt{1 - v^2/c^2} = (7.17 \text{ y})\sqrt{1 - 0.60^2} = 5.7 \text{ y}.$

(b) As found in (a), $t = 7.2$ y.

32. The diameter is still 8.25 m because it is perpendicular to the velocity.

$L = L_o\sqrt{1 - v^2/c^2} = (35.0 \text{ m})\sqrt{1 - 2.44^2/3.00^2} = 20.4 \text{ m}.$

So the dimensions are length 20.4 m and diameter 8.25 m.

40. (a) From the figure, it can be seen directly that

$x = x' + v\,t;$

$y = y';$

$z = z'.$

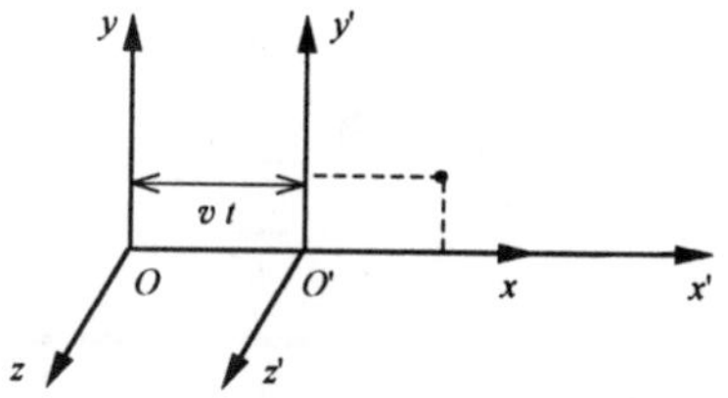

(b) $x = x'\sqrt{1 - v^2/c^2} + v\,t,$

so $x' = \dfrac{x - v\,t}{\sqrt{1 - v^2/c^2}} = \gamma(x - v\,t);$

$y' = y; \quad z' = z.$

44. $E_o = m_o c^2 = (9.11\times10^{-31} \text{ kg})(3.00\times10^8 \text{ m/s})^2 = 8.20\times10^{-14} \text{ J} \approx 0.511 \text{ MeV}.$

$E = \gamma E_o = \dfrac{1}{\sqrt{1 - v^2/c^2}}E_o = \dfrac{1}{\sqrt{1 - 0.500^2}}(0.511 \text{ MeV}) = 0.590 \text{ MeV}.$

52. $E_o = m_o c^2 = (9.11\times10^{-31} \text{ kg})(3.00\times10^8 \text{ m/s})^2 = 8.20\times10^{-14} \text{ J} \approx 0.511 \text{ MeV}.$

From $E = mc^2 = \dfrac{m_o c^2}{\sqrt{1 - v^2/c^2}}$,

we have $v = c\sqrt{1 - m_o^2/m^2} = c\sqrt{1 - (0.511)^2/(2.8)^2} = 0.983c.$

So $p = \dfrac{m_o v}{\sqrt{1 - v^2/c^2}} = \dfrac{(9.11\times10^{-31} \text{ kg})(0.983)(3.00\times10^8 \text{ m/s})}{\sqrt{1 - 0.983^2}} = 1.5\times10^{-21} \text{ kg·m/s}.$

54. (a)

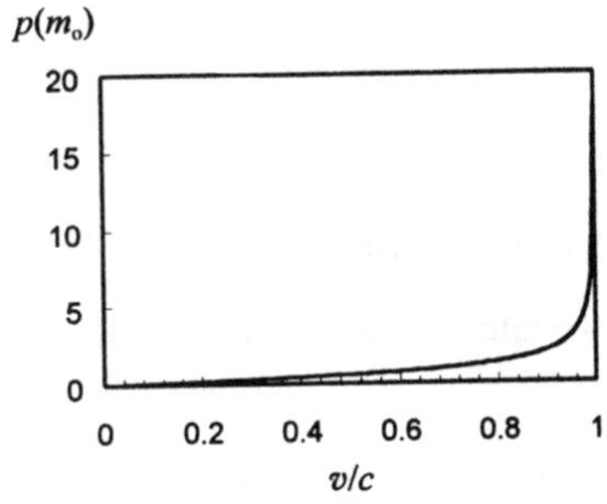

(b)

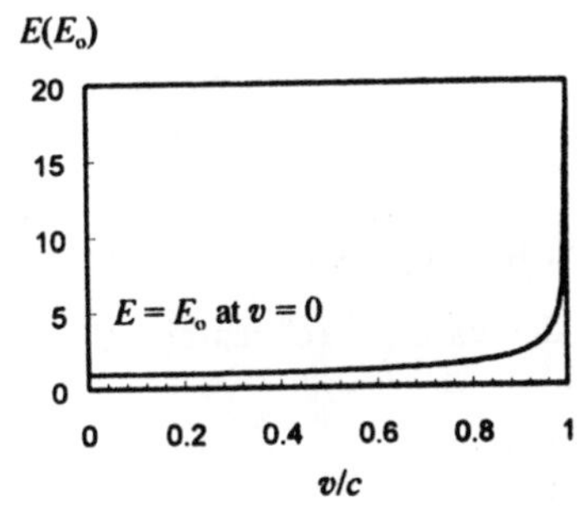

(c)

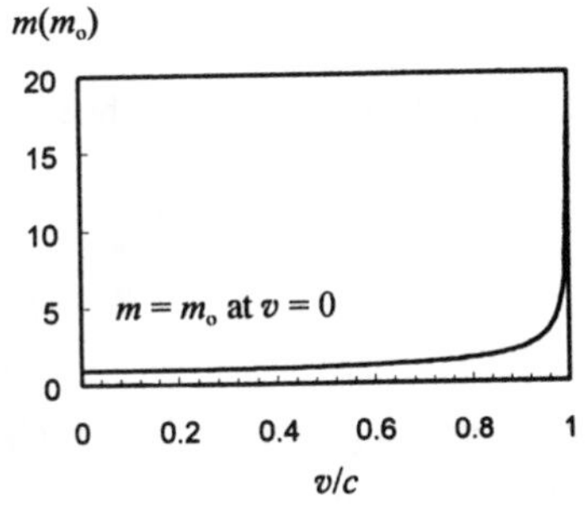

64. (a) The minimum energy is the rest energy, which for a proton is

$E_o = m_o c^2 = (1.67\times10^{-27}\text{ kg})(3.00\times10^8\text{ m/s})^2 = 1.502\times10^{-10}\text{ J} = 939\text{ MeV} > 600\text{ MeV}.$

(b) $E = K + E_o = 600\text{ MeV} + 939\text{ MeV} = 1539\text{ MeV}.$

From $E = \dfrac{E_o}{\sqrt{1 - v^2/c^2}}$, we have $v = c\sqrt{1 - E_o^2/E^2} = c\sqrt{1 - (939)^2/(1539)^2} = 0.792c.$

(c) $p = \dfrac{m_o v}{\sqrt{1 - v^2/c^2}} = \dfrac{(1.67\times10^{-27}\text{ kg})(0.792)(3.00\times10^8\text{ m/s})}{\sqrt{1 - 0.792^2}} = 6.50\times10^{-19}\text{ kg·m/s}.$

74. $u = \dfrac{v + u'}{1 + v u'/c^2} = \dfrac{0.40c + (-0.15c)}{1 + (0.40)(-0.15)} = 0.27c$, in the same direction as spacecraft

82. $E = m_o c^2 = (1.0\text{ kg})(3.00\times10^8\text{ m/s})^2 = 9.0\times10^{16}\text{ J} = (9.0\times10^{16}\text{ W·s})(1\text{ kW}/1000\text{ W})(3600\text{ s/h})$

$= 2.5\times10^{10}\text{ kWh}.$

88. (a) Magnetic force provides centripetal force.

Since $F_m = qvB = m\dfrac{v^2}{R}$, $B = \dfrac{vm}{qR} = \dfrac{(0.90)(3.00\times10^8\text{ m/s})(1.67\times10^{-27}\text{ kg})}{(1.6\times10^{-19}\text{ C})(1.5\text{ m})} = 1.9\text{ T}.$

(b) $m = \dfrac{m_o}{\sqrt{1 - v^2/c^2}}$, $B = = \dfrac{(0.90)(3.00\times10^8\text{ m/s})(1.67\times10^{-27}\text{ kg})}{(1.6\times10^{-19}\text{ C})(1.5\text{ m})\sqrt{1 - 0.90^2}} = 4.3\text{ T}.$

(c) the classical expression does not account for the relativistic mass of the proton.

VI. Practice Quiz

1. As the speed of a particle approaches the speed of light, the mass of the particle

   (a) increases. (b) decreases. (c) remains the same. (d) approaches zero. (e) none of the above.

2. What is the momentum in kg·m/s of a proton when it is moving with a speed of $0.60c$?

   (a) 1.2×10^{-19} (b) 1.5×10^{-19} (c) 3.0×10^{-19} (d) 3.8×10^{-19} (e) 5.0×10^{-19}

3. A fast spaceship is traveling with a speed of $0.80c$. How fast would light travel from the headlights of the ship relative to a stationary observer?

   (a) $0.20c$ (b) $0.80c$ (c) c (d) $0.90c$ (e) $1.8c$

4. A spaceship takes a nonstop journey to a planet and returns in 10 h according to a clock on the spaceship. If the speed of the spaceship is $0.80c$, how much time has elapsed on the Earth?

   (a) 6.0 h (b) 6.3 h (c) 10 h (d) 16 h (e) 17 h

5. The length of a spaceship is 10 m when it is at rest. If the spaceship travels past you with a speed of $0.70c$, what length does it appear to you?

   (a) 5.5 m (b) 7.1 m (c) 10 m (d) 14 m (e) 18 m

6. What is the total energy of an electron moving with a speed of $0.95c$?

   (a) 2.6×10^{-13} J (b) 8.2×10^{-14} J (c) 1.1×10^{-13} J (d) 1.2×10^{-14} J (e) 3.7×10^{-14} J

7. What was the result of the Michelson-Morley experiment?

   (a) verified the existence of the ether (b) detected the interference fringe shifts

   (c) proved light is a wave (d) detected no fringe shift

   (e) verified the gravitational light bending

8. In a nuclear power plant, 10^{17} J of energy are available from mass conversion. How much mass was lost?

   (a) 0.1 kg (b) 1 kg (c) 10 kg (d) 90 kg (e) 100 kg

9. What is the speed of a proton if its total mass is twice its rest mass?

   (a) $0.50c$ (b) $0.58c$ (c) $0.71c$ (d) $0.75c$ (e) $0.86c$

10. A space traveler moves away from the Earth with a speed of $0.30c$ when a missile is fired with a speed of $0.80c$ toward the Earth relative to the ship. How fast does the missile appear to travel toward the Earth?

(a) $0.40c$ (b) $0.50c$ (c) $0.66c$ (d) $0.89c$ (e) c

Answers to Practice Quiz:

1. a 2. d 3. c 4. e 5. b 6. a 7. d 8. b 9. e 10. c

CHAPTER 27

Quantum Physics

I. Chapter Objectives

Upon completion of this chapter, you should be able to:

1. define blackbody radiation and use Wien's law to calculate the wavelength of maximum radiation, and understand how Planck's hypothesis paved the way for quantum ideas.
2. describe the photoelectric effect, explain how it can be understood by assuming that light energy is carried by particles, and summarize the properties of photons.
3. understand how the photon model for light explains the scattering of light from various materials (the Compton effect) and calculate scattered wavelength in the Compton effect.
4. understand how the Bohr model of the hydrogen atom explains atomic emission and absorption spectra, calculate the energies and wavelengths of emitted photons for various transitions in hydrogen, and understand how the generalized concept of atomic energy levels can be used to explain other atomic phenomena.
5. understand some of the practical applications of the quantum hypothesis, in particular, the laser.

II. Key Terms

Upon completion of this chapter, you should be able to define and/or explain the following key terms:

thermal radiation	**absorption spectrum**
blackbody	**Balmer series**
Wien's displacement law	**principal quantum number**
Planck's constant	**ground state**
quantum	**excited state**
photon	**binding energy**
photoelectric effect	**lifetime**
stopping potential	**fluorescence**
work function	**laser**
threshold frequency	**metastable state**
Compton effect	**stimulated emission**

dual nature of light

Bohr theory of the hydrogen atom

emission spectrum

population inversion

holography

The definitions and/or explanations of the most important key terms can be found in the following section: **III. Chapter Summary and Discussion.**

III. Chapter Summary and Discussion

1. Quantization: Planck's Hypothesis

One of the problems scientists were having at the end of twentieth century was how to explain **thermal radiation**, the continuous spectra of radiation emitted by hot objects. A **blackbody** is an ideal system that absorbs and emits all radiation that falls on it. The wavelength of maximum radiation (spectral component), λ_{max} is inversely proportional to its absolute temperature, *T*, and obeys Wien's displacement law, $\lambda_{max} T = 2.90\times10^{-3}$ m·K. Classically, the intensity of the blackbody radiation at a particular wavelength is predicted to be proportional to $\frac{1}{\lambda^4}$ (Chapter 11), which leads to what is sometimes called the *ultraviolet catastrophe* -- "ultraviolet" because disagreement between theory and experiment occurs for short wavelengths beyond the violet end of the visible spectrum, and "catastrophe" because it predicts the emitted intensity at these short wavelengths will be infinitely large.

Note: Wien's displacement law relates the wavelength of the *most intense* radiation and the absolute temperature (Kelvin), not maximum radiation and temperature. λ_{max} does not mean it is the maximum wavelength emitted.

Max Planck successfully explained the spectrum of blackbody radiation by proposing a radical hypothesis. According to **Planck's hypothesis**, the energy of a thermal oscillator is *quantized.* That is, an oscillator can have only *discrete*, or particular, amounts of energy, rather than a continuous distribution of energies. The smallest quantum amount of energy is given by $E = hf$, where $h = 6.63\times10^{-34}$ J·s is called the **Planck's constant**. This is called a **quantum** of energy. The thermal oscillator of a blackbody can have only multiples of this quantum of energy, $E_n = n(hf)$.

Example 27.1 What is the most intense color of light emitted by a giant star of surface temperature 4400 K? What is the color of the star?

Solution: Given: $T = 4400$ K.

Find: λ_{max}.

From Wien's displacement law, $\lambda_{max} T = 2.90\times10^{-3}$ m·K,

we have $\lambda_{max} = \frac{2.90\times10^{-3}\text{ m·K}}{T} = \frac{2.90\times10^{-3}\text{ m·K}}{4400\text{ K}} = 659$ nm.

This is in the red end of the visible spectrum.

2. Quanta of Light: Photons and the Photoelectric Effect

A quantum or packet of light energy is referred to as a **photon**, and each photon has an energy $E = hf$, where f is the frequency of light. This idea suggests that light may sometimes behave as discrete quanta, or "particles" of energy, rather than as continuous waves.

Einstein used the photon theory to explain what is called the **photoelectric effect**, another area in which classical physical description was inadequate. Some materials are *photosensitive*, that is, when light strikes the surface of the material, electrons are emitted and a current may be established. However, if the frequency of light is below a certain cutoff value, no *photoelectrons* are emitted no matter how strong the light is and how long light is incident on the material! If the frequency is higher than the certain cutoff value, photoelectrons are emitted instantaneously, no matter how low the light intensity is. The intensity of light has only to do with the number of photoelectrons, not whether they are emitted. Classical physics cannot explain these observations.

To study the maximum kinetic energy (K_{max}) of the photoelectrons, a **stopping potential** (V_o) is applied to a beam of electrons. When there is no photocurrent, the maximum kinetic energy is related to the stopping potential by $K_{max} = eV_o$, where eV_o is the work needed to stop the most energetic electrons. The minimum energy needed to free the electrons from the material is called the **work function** (ϕ_o). According to energy conservation, $hf = K_{max} + \phi_o$, that is, the absorbed photon goes into the work of freeing the electron, and the rest is carried off by the emitted electron as kinetic energy. The **threshold frequency**, the lowest frequency to emit photoelectrons, corresponds to photoelectrons having zero kinetic energy. $hf_o = 0 + \phi_o$, so $f_o = \frac{\phi_o}{h}$.

Note: We are often given the wavelength of light (in nm) rather than the frequency, and energy. The energy of the photon in units of eV can be calculated quickly from $E = \frac{1.24\times10^3\text{ eV·nm}}{\lambda}$.

Example 27.2 What is the photon energy of red light having wavelength of 640 nm?

Solution: Given: $\lambda = 640 \text{ nm} = 640 \times 10^{-9}$ m. Find: E.

The frequency of light is calculated from $f = \dfrac{c}{\lambda} = \dfrac{3.00 \times 10^8 \text{ m/s}}{640 \times 10^{-9} \text{ m}} = 4.69 \times 10^{14}$ Hz.

So the energy of a photon is $E = hf = (6.63 \times 10^{-34} \text{ J·s})(4.69 \times 10^{14} \text{ Hz}) = 3.11 \times 10^{-19} \text{ J} = 1.94$ eV.

Or, $E = \dfrac{1.24 \times 10^3 \text{ eV·nm}}{\lambda} = \dfrac{1.24 \times 10^3 \text{ eV·nm}}{640 \text{ nm}} = 1.94$ eV.

Example 27.3 A metal has a work function of 4.5 eV. Find the maximum kinetic energy of the photoelectrons if the wavelength of light is (a) 300 nm (b) 250 nm.

Solution: Given: $\phi_o = 4.5$ eV. (a) $\lambda = 300$ nm (b) $\lambda = 250$ nm.

Find: K_{max} in (a) and (b).

(a) From energy conservation, $E = hf = K_{max} + \phi_o$,

so $K_{max} = E - \phi_o = \dfrac{1.24 \times 10^3 \text{ eV·nm}}{\lambda} - \phi_o = \dfrac{1.24 \times 10^3 \text{ eV·nm}}{300 \text{ nm}} - 4.5 \text{ eV} = -0.37$ eV.

Since kinetic energy cannot be negative, the result indicates there is no photoelectron ejected.

(b) $K_{max} = \dfrac{1.24 \times 10^3 \text{ eV·nm}}{250 \text{ nm}} - 4.5 \text{ eV} = 0.46$ eV.

Example 27.4 When light of wavelength 350 nm is incident on a metal surface, the stopping potential of the photoelectrons is 0.500 V.

(a) What is the work function of the metal?

(b) What is the threshold frequency?

(c) What is the maximum kinetic energy of the electrons?

Solution: Given: $\lambda = 350$ nm, $V_o = 0.500$ V or $eV_o = 0.500$ eV

Find: (a) ϕ_o (b) f_o (c) K_{max}.

(a) According to energy conservation, $E = hf = K_{max} + \phi_o = eV_o + \phi_o$,

so $\phi_o = E - eV_o = \dfrac{1.24 \times 10^3 \text{ eV·nm}}{\lambda} - eV_o = \dfrac{1.24 \times 10^3 \text{ eV·nm}}{350 \text{ nm}} - 0.500 \text{ eV} = 3.04$ eV.

(b) $f_o = \dfrac{\phi_o}{h} = \dfrac{(3.04 \text{ eV})(1.60 \times 10^{-19} \text{ J/eV})}{6.63 \times 10^{-34} \text{ J·s}} = 7.34 \times 10^{14}$ Hz.

(c) The maximum kinetic energy is equal to the stopping potential, eV_o, expressed in eV.

$K_{max} = eV_o = 0.500$ eV.

---

3. Quantum "Particles": The Compton Effect

The **Compton effect** is the increase in wavelength of light scattered by electrons or other charged particles. When a photon collides with an electron a new photon of shorter wavelength (lower energy because the electron carries off some energy) emerges. Using the conservation of relativistic momentum and total energy, the shift is given by $\Delta\lambda = \lambda_1 - \lambda_2 = \lambda_C (1 - \cos\theta)$, where $\lambda_C = \frac{h}{m_e} = 2.43 \times 10^{-12}$ m $= 2.43 \times 10^{-3}$ nm is called the *Compton wavelength* of the electron. Since the Compton shift is very small, it is only significant for X-ray and gamma-ray scattering where the wavelengths are smaller.

Note: The Compton wavelength is *not* the wavelength of the electron. It is the characteristic wavelength when X-rays or gamma rays are scattered by an electron. If light is scattered by a proton, the Compton wavelength of the proton would be different.

Einstein's and Compton's successes in explaining electromagnetic phenomena in terms of quanta left scientists with two apparently competing theories of electromagnetic radiation, the wave theory and the photon theory. The two theories gave rise to a description that is called the **dual nature of light**. That is, light apparently behaves sometimes as a wave and at other times as photons or "particles".

Example 27.5 X-rays of wavelength of 2.00×10^{-10} m are scattered by a metal. The wavelength shift is observed to be 2.00×10^{-12} m at a certain scattering angle measured relative to the incoming X-ray.

(a) What is the scattering angle?

(b) What is the maximum shift possible for the Compton efffect?

Solution: Given: $\lambda = 2.00 \times 10^{-10}$ m, $\Delta\lambda = 2.00 \times 10^{-12}$ m.

Find: (a)θ (b) $\Delta\lambda_{max}$.

(a) From $\Delta\lambda = \lambda_C (1 - \cos\theta) = (2.43 \times 10^{-12}\text{ m})(1 - \cos\theta)$,

$$\cos\theta = 1 - \frac{\Delta\lambda}{\lambda_C} = 1 - \frac{2.00 \times 10^{-12}\text{ m}}{2.43 \times 10^{-12}\text{ m}} = 0.177, \quad \text{so} \quad \theta = 79.8°.$$

(b) Maximum shift occurs when $\theta = 180°$ (or $\cos\theta = -1$).

$\Delta\lambda_{max} = \lambda_C (1 - \cos 180°) = 2\lambda_C = 4.86 \times 10^{-12}$ m.

---

4. The Bohr Theory of the Hydrogen Atom

It is experimentally observed that a hydrogen atom can only emit and absorb light at certain wavelengths, not at all wavelengths (continuous spectrum). The small set of visible wavelengths the hydrogen atom can emit and absorb (the **Balmer series**) are given by an empirical equation $\frac{1}{\lambda} = R\left(\frac{1}{2^2} - \frac{1}{n^2}\right)$ for $n = 3, 4, 5, 6$, where $R = 1.097\times10^{-2}\ \text{nm}^{-1}$ is called the *Rydberg constant.* The **Bohr theory of the hydrogen atom** successfully explained the **emission spectrum** (a series bright lines) and **absorption spectrum** (a series dark lines superimposed on a continuous spectrum) of the hydrogen atom. In his theory, he assumed that

- the hydrogen electron orbits the nuclear proton in a circular orbit (analogous to planets orbiting the Sun)
- the angular momentum of the electron is quantized in integral multiples of Planck's constant, h
- the electron does not radiate energy when it is in certain discrete circular orbits
- the electron radiates or absorbs energy only when it makes a transition to another orbit.

From these assumptions, Bohr showed that the electron can have only certain sized orbits with certain energies. The energies and the radii of the orbits are given by $E_n = -\frac{13.6\ \text{eV}}{n^2}$ and $r_n = 0.0529n^2$ nm for $n = 1, 2, 3, 4, \ldots,$ where n is called the *quantum number*, specifically the **principal quantum number**. The $n = 1$ orbit is known as the **ground state** and orbits with $n > 1$ are called the **excited states**. The energy of the electron in any state is E_n, and the energy needed to completely free the electron from the atom is $-E_n$, which is called the **binding energy**.

An electron generally does not remain in an excited state for long. It decays, or makes a transition to a lower energy level, in a short time. The time an electron spends in an excited state is called the **lifetime** of the excited state. If an electron makes a transition from n_i state to n_f state, a photon is released and the energy of the photon is equal to the energy difference between the final and initial states, $\Delta E = E_f - E_i = 13.6\left(\frac{1}{n_f^2} - \frac{1}{n_i^2}\right)$ eV.

The wavelength of the photon is then $\lambda = \frac{hc}{\Delta E} = \frac{1.24\times10^3\ \text{eV}\cdot\text{nm}}{\Delta E}$. For transitions with $n_f = 1$, the spectrum series is called the *Lyman* series (ultraviolet), for $n_f = 2$, the *Balmer series* (visible if $n_i = 3, 4, 5,$ and 6), for $n_f = 3$, the *Paschen series* (infrared), etc.

In **fluorescence**, an electron that has been excited by absorbing a photon returns to the ground state in two or more steps. At each step a photon is emitted.

Example 27.6 What are the orbital radius and total energy of an electron for a hydrogen atom in (a) the ground state and (b) the second excited state?

Solution: Ground state corresponds to $n = 1$, and $n = 2$ and 3 are for the first and second excited states, respectively.

Given: (a) $n = 1$ (b) $n = 3$. Find: r and E.

(a) From $r_n = 0.0529n$ nm, $r_1 = 0.0529(1)$ nm $= 0.0529$ nm.

From $E_1 = -\dfrac{13.6 \text{ eV}}{n^2}$, $E_1 = -\dfrac{13.6 \text{ eV}}{1^2} = -13.6$ eV.

(b) $r_3 = 0.0529(3)$ nm $= 0.159$ nm, and $E_3 = -\dfrac{13.6 \text{ eV}}{3^2} = -1.51$ eV.

Example 27.7 The electron of a hydrogen atom makes a transition from the fourth excited state to the first excited state. What are the energy and wavelength of the emitted photon?

Solution: Since ground state corresponds to $n = 1$, $n = 2$ and 5 are for the first and fourth excited states.

Given: $n_i = 5$, $n_f = 2$. Find: (a) ΔE (b) λ.

(a) $\Delta E = 13.6\left(\dfrac{1}{n_f^2} - \dfrac{1}{n_i^2}\right)$ eV $= 13.6\left(\dfrac{1}{2^2} - \dfrac{1}{5^2}\right)$ eV $= 2.86$ eV.

(b) From $\Delta E = \dfrac{1.24 \times 10^3 \text{ eV}\cdot\text{nm}}{\lambda}$, we have $\lambda = \dfrac{1.24 \times 10^3 \text{ eV}\cdot\text{nm}}{\Delta E} = \dfrac{1.24 \times 10^3 \text{ eV}\cdot\text{nm}}{2.86 \text{ eV}} = 434$ nm.

The emitted photon is in the blue/violet region of the visible spectrum.

---

5. A Quantum Success: The Laser

The quantum theory of atomic structure gave rise to the development of the **laser**, an acronym which stands for *l*ight *a*mplification by *s*timulated *e*mission of *r*adiation. **Stimulated emission** is the process in which a photon with an energy equal to an allowed transition strikes an atom in an excited state, stimulating the atom to make a transition and emit a photon of the same energy, frequency, traveling direction, and phase as the incoming photon. In this process, one photon goes in and two come out (the incoming one and the one emitted by the atom), resulting in the amplification of that light. Another basic requirement for laser action is **population inversion**, that is, more electrons must occupy a **metastable state**, a state in which an excited electron remains for a relatively longer time, than an excited state of lower energy. (Normally, there are more electrons at lower energy levels.) This can be achieved by various "pumping" or energy input processes. Laser beams are very intense, highly directional, coherent (same phase), and monochromatic (same frequency). Depending on the lasing medium, lasers can be classified as gas lasers, solid lasers, semiconductor lasers, dye lasers, X-ray lasers, free-electron lasers, etc. Lasers have a wide variety of applications, ranging from compact disk players to welding and eye surgery.

An interesting application of laser light is the production of three-dimensional images in a process called **holography**. The key to holography is the coherent property of laser light, which gives the light waves a definite spatial relationship to each other. The process does not use lenses as ordinary image-forming processes do, yet it recreates the original scene in three dimensions. Holography has many potential applications in medicine, sciences, and even in our everyday life (holographic three-dimensional television).

IV. Mathematical Summary

Wien's Displacement Law	$\lambda_{max} T = 2.90 \times 10^{-3}$ m·K (27.1)	Relates the absolute temperature and the wavelength of maximum radiation
Photon Energy	$E = hf$ (27.4) $h = 6.63 \times 10^{-34}$ J·s	Defines the energy of a quantum in terms of its frequency
K_{max} and Stopping Potential in the Photoelectric Effect	$K_{max} = eV_o$ (27.5)	Relates the maximum electron kinetic energy and the stopping potential in the photoelectric effect
Energy Conservation in the Photoelectric Effect	$hf = K + \phi$ (27.6)	Applies energy conservation to the photoelectric effect
Work-Function and Threshold Frequency	$f_o = \frac{\phi_o}{f}$ (27.8)	Gives the minimum frequency (threshold frequency) to eject photo-electron
Compton Equation (scattering from free electrons)	$\Delta\lambda = \lambda_1 - \lambda_2 = \lambda_C (1 - \cos\theta)$ (27.9) $\lambda_C = h/(m_e c) = 2.43 \times 10^{-12}$ m $= 0.00243$ nm	Computes the wavelength shift (increase) of light scattered by a free electron or other charged particles
Bohr Theory Orbit Radius	$r_n = 0.0529 n^2$ nm $n = 1,2,3,\ldots$ (27.16)	Calculates the orbit radius of an electron in a hydrogen atom
Bohr Theory Electron Energy	$E_n = \frac{-13.6}{n^2}$ eV $n = 1, 2, 3, \ldots$ (27.17)	Calculates the total energy of an electron in a hydrogen atom
Bohr Theory Transition Energy (in eV)	$\Delta E = 13.6\left(\frac{1}{n_f^2} - \frac{1}{n_i^2}\right)$ (27.18)	Computes the energy difference (transition energy) between two orbits of an electron in a hydrogen atom
Bohr Theory Photon Wavelength (energies in eV)	$\lambda = \frac{1.24 \times 10^3}{\Delta E}$ nm (27.19)	Calculates the wavelength of a photon emitted when an electron makes a transition

V. Solutions of Selected Exercises and Paired Exercises

5. (a) From $\lambda_{max} T = 2.90\times10^{-3}$ m·K, we have $T = \dfrac{2.90\times10^{-3}\text{ m·K}}{\lambda_{max}} = \dfrac{2.90\times10^{-3}\text{ m·K}}{c} f \propto f.$

(b) $T = \dfrac{2.90\times10^{-3}\text{ m·K}}{\lambda_{max}} \propto \dfrac{1}{\lambda_{max}}.$

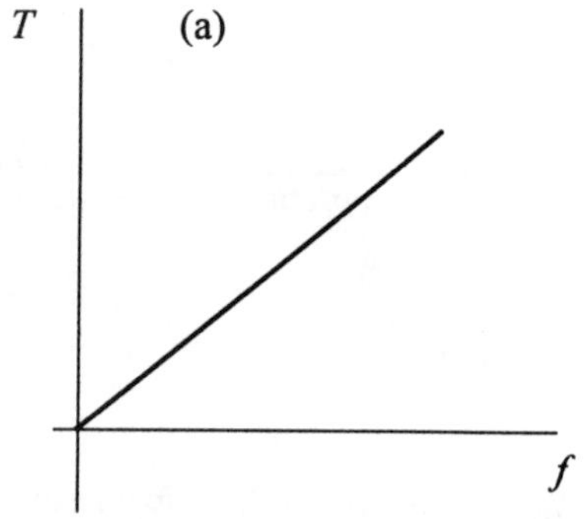

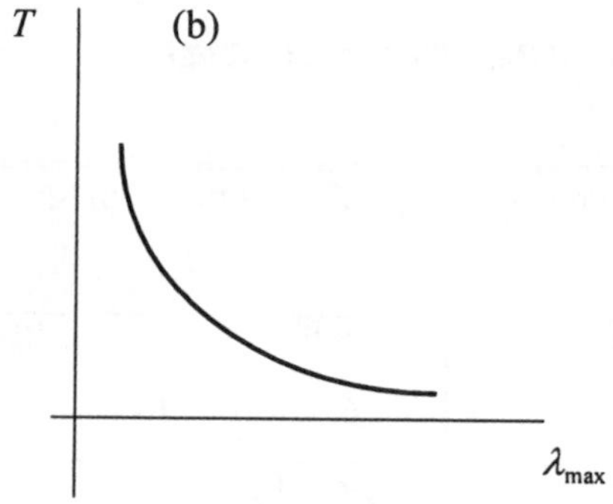

6. From $\lambda_{max} T = 2.90\times10^{-3}$ m·K, we have $T = \dfrac{2.90\times10^{-3}\text{ m·K}}{\lambda_{max}} = \dfrac{2.90\times10^{-3}\text{ m·K}}{700\times10^{-9}\text{ m}} = 4.1\times10^{3}$ K.

12. From $\lambda_{max} T = 2.90\times10^{-3}$ m·K, we have $\lambda_{max} = \dfrac{2.90\times10^{-3}\text{ m·K}}{T} = \dfrac{2.90\times10^{-3}\text{ m·K}}{1000\text{ K}} = 2.9\times10^{-6}$ m.

So $f = \dfrac{c}{\lambda_{max}} = \dfrac{3.00\times10^{8}\text{ m/s}}{2.9\times10^{-6}\text{ m}} = 1.03\times10^{14}$ Hz.

Therefore $E = hf = (6.63\times10^{-34}\text{ J·s})(1.03\times10^{14}\text{ Hz}) = 6.52\times10^{-20}$ J/quantum.

The number of quanta per square meter per second is

$n = \dfrac{I}{E} = \dfrac{3.0\text{ W/m}^2}{6.52\times10^{-20}\text{ J/quantum}} = 4.6\times10^{19}$ quanta/s·m².

18. From $E = hf$, we have $f = \dfrac{E}{h} = \dfrac{3.3\times10^{-15}\text{ J}}{6.63\times10^{-34}\text{ J·s}} = 4.98\times10^{18}$ Hz.

So $\lambda = \dfrac{c}{f} = \dfrac{3.00\times10^{8}\text{ m/s}}{4.98\times10^{19}\text{ Hz}} = 6.0\times10^{-11}$ m; X-ray.

24. $E = Pt = IAt = 0.20[(2.5\times10^{12}\text{ J/km}^2\text{·h})(1\text{ km}/1000\text{ m})^2\ (1\text{ h}/3600\text{ s})](1.0\times10^{-4}\text{ m}^2)(1.0\text{ s}) = 0.0139$ J.

$E_o = hf = \dfrac{hc}{\lambda} = \dfrac{(6.63\times10^{-34}\text{ J·s})(3.00\times10^{8}\text{ m/s})}{550\times10^{-9}\text{ m}} = 3.62\times10^{-19}$ J/quantum.

So $n = \dfrac{E}{E_o} = \dfrac{0.0139\text{ J}}{3.62\times10^{-19}\text{ J/quantum}} = 3.8\times10^{16}$ photons.

25. (a) Since $eV_o = hf + \phi_o$, the slope of the graph is equal to Planck's constant.

$$h = \frac{[(3-1)\text{ eV}](1.6\times10^{-19}\text{ J/eV})}{(115-67)\times10^{13}\text{ Hz}} = 6.7\times10^{-34}\text{ J.}$$

(b) $\phi_o = hf_o = (6.63\times10^{-34}\text{ J}\cdot\text{s})(43.9\times10^{13}\text{ Hz}) = 2.9\times10^{-19}\text{ J}$.

30. (a) $eV_o = K_{max} = hf - \phi_o = \dfrac{hc}{\lambda} - \phi_o = \dfrac{1.24\times10^3\text{ eV}\cdot\text{nm}}{300\text{ nm}} - 3.5\text{ eV} = 0.63\text{ eV}$.

So $V = 0.63$ V.

(b) $f_o = \dfrac{\phi_o}{h} = \dfrac{(3.5\text{ eV})(1.6\times10^{-19}\text{ J/eV})}{6.63\times10^{-34}\text{ J}\cdot\text{s}} = 8.4\times10^{14}\text{ Hz}$.

34. $K_{max} = hf - \phi_o = \dfrac{hc}{\lambda} - \phi_o$.

For red, $K_{red} = \dfrac{hc}{680\text{ nm}} - \phi_o$; for blue, $K_{blue} = \dfrac{hc}{420\text{ nm}} - \phi_o$.

Dividing the two equations yields $1.8 = \dfrac{K_{blue}}{K_{red}} = \dfrac{\dfrac{hc}{420\text{ nm}} - \phi_o}{\dfrac{hc}{680\text{ nm}} - \phi_o}$.

So $\phi_o = \dfrac{hc}{0.8}\left(\dfrac{1.8}{680\text{ nm}} - \dfrac{1}{420\text{ nm}}\right) = \dfrac{1.24\times10^3\text{ eV}\cdot\text{nm}}{0.8}\left(\dfrac{1.8}{680\text{ nm}} - \dfrac{1}{420\text{ nm}}\right) = 0.41\text{ eV}$.

39. For each scattering the wavelength shift is approximately equal to $\Delta\lambda = \lambda_c = 0.00243$ nm.

The wavelength change from X-ray to visible light is in the order of $550\text{ nm} - 0.01\text{ nm} = 550\text{ nm}$.

So it requires approximately $\dfrac{550\text{ nm}}{0.00243\text{ nm}} = 200{,}000$ scatterings.

46. $\Delta\lambda = \lambda_c(1-\cos\theta) = (2.43\times10^{-12}\text{ m})(1-\cos 45°) = 7.12\times10^{-13}\text{ m}$.

So $\lambda = \lambda_o + \Delta\lambda = 2.80\times10^{-10}\text{ m} + 7.12\times10^{-13}\text{ m} = 2.81\times10^{-10}\text{ m} = 0.281\text{ nm}$.

56. (a) $\Delta E = (13.6\text{ eV})\left(\dfrac{1}{n_f^2} - \dfrac{1}{n_i^2}\right) = (13.6\text{ eV})\left(\dfrac{1}{5^2} - \dfrac{1}{2^2}\right) = 2.856\text{ eV}$.

$f = \dfrac{\Delta E}{h} = \dfrac{(2.856\text{ eV})(1.60\times10^{-19}\text{ J/eV})}{6.63\times10^{-34}\text{ J}\cdot\text{s}} = 6.89\times10^{14}\text{ Hz}$.

(b) $\Delta E = (13.6\text{ eV})\left(\dfrac{1}{\infty} - \dfrac{1}{2^2}\right) = 3.40\text{ eV}$.

$$f = \frac{(3.40\text{ eV})(1.60\times10^{-19}\text{ J/eV})}{6.63\times10^{-34}\text{ J·s}} = 8.21\times10^{14}\text{ Hz}.$$

58. (a) $r_n = 0.0529\,n^2$ nm. $r_3 = (0.0529)(3)^2$ nm $= 0.476$ nm.

(b) $r_6 = 0.0529(6)^2 = 1.90$ nm.

(c) $r_{10} = 0.0529(10)^2$ nm $= 5.29$ nm.

64. (a) $\Delta E = (13.6\text{ eV})\left(\frac{1}{n_f^2} - \frac{1}{n_i^2}\right)$.

$$\Delta E_{52} = (13.6\text{ eV})\left(\frac{1}{2^2} - \frac{1}{5^2}\right) = 2.856\text{ eV}. \quad \lambda_{52} = \frac{1.24\times10^3}{\Delta E\text{ (in eV)}}\text{ nm} = \frac{1.24\times10^3}{2.856}\text{ nm} = 434\text{ nm}.$$

$$\Delta E_{21} = (13.6\text{ eV})\left(\frac{1}{1^2} - \frac{1}{2^2}\right) = 10.2\text{ eV}. \quad \lambda_{21} = \frac{1.24\times10^3}{10.2}\text{ nm} = 122\text{ nm}.$$

(b) Yes, 5 to 2 is in the visible region (violet).

68. $$E = -\frac{ke^2}{2r} = -\frac{(9.0\times10^9\text{ N·m}^2/\text{C}^2)(1.6\times10^{-19}\text{ C})^2}{2(0.0529\times10^{-9}\text{ m})} = -2.18\times10^{-18}\text{ J} = -13.6\text{ eV}.$$

71. (a) Since $E_{en} = -\frac{2\pi^2 k^2 e^4 m}{h^2}\frac{1}{n^2} = -\frac{13.6\text{ eV}}{n^2}$ for electron, $E_{mn} = (207)E_{en}$.

$$E_{m1} = (207)\frac{-13.6\text{ eV}}{1^2} = -2.82\times10^3\text{ eV}, \quad E_{m2} = (207)\frac{-13.6\text{ eV}}{2^2} = -7.04\times10^2\text{ eV}.$$

Since $r_{en} = \frac{h^2}{4\pi^2 ke^2 m}n^2 = (0.0529\text{ nm})n^2$ for electron, $r_{mn} = \frac{r_{en}}{207}$.

$$r_{m1} = \frac{0.0529\text{ nm}}{207}(1)^2 = 2.56\times10^{-4}\text{ nm}, \quad r_{m2} = \frac{0.0529\text{ nm}}{207}2^2 = 1.02\times10^{-3}\text{ nm}.$$

(b) $$\lambda = \frac{1.24\times10^3}{\Delta E\text{ (in eV)}}\text{ nm} = \frac{1.24\times10^3}{-7.04\times10^2\text{ eV} - (-2.82\times10^3\text{ eV})} = 0.586\text{ nm}.$$

78. The energy of a typical photon in the visible region is about

$$E_{min} = hf = \frac{hc}{\lambda} = \frac{1.24\times10^3}{700}\text{ eV} = 1.77\text{ eV} \quad \text{and} \quad E_{max} = \frac{1.24\times10^3}{400}\text{ eV} = 3.1\text{ eV}.$$

Since $\Delta E = (13.6\text{ eV})\left(\frac{1}{n_f^2} - \frac{1}{n_i^2}\right)$, the electron has to be in the $n = 2$ state, because $|\Delta E_{23}| = 1.89$ eV,

$|\Delta E_{24}| = 2.55$ eV, $|\Delta E_{25}| = 2.86$ eV, $|\Delta E_{26}| = 3.02$ eV.

However, most of the electrons are in the ground state at room temperature and so the visible light will not be absorbed. To excite the electrons from $n = 1$ to $n = 2$, an amount of energy $|\Delta E_{12}| = 10.2$ eV is needed. So a specific thermal energy (temperature) is required to excite the electrons to the $n = 2$ state. If the temperature is too high, the electrons will be excited to $n > 2$ state and the visible light will not be absorbed again.

85. The energy of a typical photon in the visible region is about

$$E_{\text{min}} = hf = \frac{hc}{\lambda} = \frac{1.24\times10^3}{700}\ \text{eV} = 1.77\ \text{eV} \quad \text{and} \quad E_{\text{max}} = \frac{1.24\times10^3}{400}\ \text{eV} = 3.1\ \text{eV}.$$

Since $\Delta E = (13.6\ \text{eV})\left(\frac{1}{n_\text{f}^2} - \frac{1}{n_\text{i}^2}\right)$, the electron has to be in the $n = 2$ state, because $|\Delta E_{23}| = 1.89$ eV, $|\Delta E_{24}| = 2.55$ eV, $|\Delta E_{25}| = 2.86$ eV, $|\Delta E_{26}| = 3.02$ eV. Therefore, there are four transitions.

VI. Practice Quiz

1. Which of the following colors is associated with the highest temperature?

(a) red (b) yellow (c) orange (d) green (e) blue

2. If the wavelength of a photon is halved, by what factor does the energy change?

(a) 4 (b) 2 (c) 1 (d) 1/2 (e) 1/4

3. A hypothetical atom has three excited states in addition to its ground state. How many different spectral lines are possible?

(a) 3 (b) 4 (c) 5 (d) 6 (e) 7

4. The kinetic energy of a photoelectron depends on which one of the following?

(a) intensity of light (b) duration of illumination (c) stopping potential
(d) wavelength of light (e) angle of illumination

5. If the scattering angle in Compton's scattering is 90°, what is the wavelength shift?

(a) 1.22×10^{-12} nm (b) 2.43×10^{-12} nm (c) 3.65×10^{-12} nm (d) 4.86×10^{-12} nm (e) 6.08×10^{-12} nm

6. What is the frequency of the most intense radiation from an object with temperature 100°C?

(a) 7.8×10^{-6} Hz (b) 2.9×10^{-5} Hz (c) 3.9×10^{13} Hz (d) 1.0×10^{13} Hz (e) 1.0×10^{11} Hz

7. A monochromatic light beam is incident on the surface of a metal with work function 2.50 eV. If a 1.0-V stopping potential is required to make the photocurrent zero, what is the wavelength of light?

(a) 354 nm (b) 423 nm (c) 497 nm (d) 744 nm (e) 1.42×10^3 nm

8. The binding energy of the hydrogen atom in its ground state is 13.6 eV. What is the energy of the atom when it is in the $n = 5$ state?

   (a) 2.72 eV (b) −2.72 eV (c) 0.544 eV (d) −0.544 eV (e) −13.6 eV

9. A hydrogen atom in ground state absorbs a photon of energy 12.09 eV. To which state will the electron make a transition?

   (a) $n = 1$ (b) $n = 2$ (c) $n = 3$ (d) $n = 4$ (e) $n = 5$

10. The wavelength of a ruby laser is 694.3 nm. What is the energy difference between the two energy states involved in laser action?

    (a) 1.537 eV (b) 1.646 eV (c) 1.786 eV (d) 1.812 eV (e) 1.936 eV

Answers to Practice Quiz:

1.e 2.b 3.d 4.d 5.b 6.c 7.a 8.d 9.c 10.c

CHAPTER 28

Quantum Mechanics

I. Chapter Objectives

Upon completion of this chapter, you should be able to:

1. explain the de Broglie hypothesis, calculate the "wavelength" of a matter wave, and specify under what circumstances the wave nature of matter will be observable.
2. qualitatively understand the reasoning that underlies the Schrödinger wave equation and its use in finding particle wave functions.
3. understand the structure of the periodic table in terms of quantum mechanical electron orbits and the Pauli exclusion principle.
4. understand the inherent quantum mechanical limits on the accuracy of physical observations -- in particular, those of location and speed, and energy and time.
5. understand the relationship between particles and antiparticles and the energy requirements for pair production.

II. Key Terms

Upon completion of this chapter, you should be able to define and/or explain the following key terms:

quantum mechanics
de Broglie hypothesis
de Broglie (matter) wave
wave function
Schrödinger's wave equation
probability density (ψ^2)
orbital quantum number (ℓ)
magnetic quantum number (m_ℓ)
spin quantum number (m_s)
shell
subshell
Pauli exclusion principle
electron configuration
electron period
periodic table of elements
period
group
Heisenberg uncertainty principle
positron
antiparticle
pair production
pair annihilation
antimatter

The definitions and/or explanations of the most important key terms can be found in the following section: **III. Chapter Summary and Discussion.**

III. Chapter Summary and Discussion

1. Matter Waves: The de Broglie Hypothesis

The **de Broglie hypothesis** assigns a wavelength to moving material particles, by reverse analogy with the assignment of particle nature of light. From relativity, the momentum of a massless (no rest mass) photon is $p = \frac{E}{c} = \frac{hf}{c} = \frac{h}{\lambda}$. The de Broglie hypothesis states that whenever a particle has momentum of magnitude, p, there is a wave is associated with it. That wave has a wavelength of $\lambda = \frac{h}{p} = \frac{h}{mv}$. These waves associated with moving particles are called **matter waves** or, more commonly, **de Broglie waves**. Matter waves of particles have been experimentally verified by the *Davisson-Germer* experiment (electron diffraction) and the *Thomson* experiment (also electron diffraction).

Note: For convenience, the wavelength of the matter wave of an electron accelerated from rest through a potential V (in volts) can be calculated with $\lambda = \sqrt{\frac{1.50}{V}} \times 10^{-9}\text{ m} = \sqrt{\frac{1.50}{V}}\text{ nm}$.

Example 28.1 What is the wavelength of the matter wave associated with
(a) a ball of mass 0.50 kg moving with a speed of 25 m/s?
(b) an electron moving with a speed of 2.5×10^7 m/s?

Solution: Given: (a) $m = 0.50$ kg, $v = 25$ m/s. (b) $m = 9.11 \times 10^{-31}$ kg, $v = 2.5 \times 10^7$ m/s.

Find: λ (a) and (b).

(a) $\lambda = \frac{h}{p} = \frac{h}{mv} = \frac{6.63 \times 10^{-34}\text{ J·s}}{(0.50\text{ kg})(25\text{ m/s})} = 5.3 \times 10^{-35}$ m.

(b) $\lambda = \frac{6.63 \times 10^{-34}\text{ J·s}}{(9.11 \times 10^{-31}\text{ kg})(2.5 \times 10^7\text{ m/s})} = 2.9 \times 10^{-11}\text{ m} = 290$ nm.

The wavelength of the electron is much smaller than that of everyday objects, however, on the order of atomic distances. That is why matter waves are important for particles like electrons since their wavelengths are comparable to the sizes of the objects they interacts with. It is easier to observe interference and diffraction of electrons than the ball and all everyday objects.

Example 28.2 An electron is accelerated by a potential difference of 100 V.

(a) What is the wavelength of the matter wave associated with the electron?

(b) What is the momentum of the electron?

(c) What is the kinetic energy of the electron?

Solution: Given: $m = 9.11\times10^{-31}$ kg, $V = 100$ V. Find: (a) λ (b) p (c) K.

(a) $\lambda = \sqrt{\dfrac{1.50}{V}}$ nm $= \sqrt{\dfrac{1.50}{100}}$ nm $= 0.122$ nm.

(b) From $\lambda = \dfrac{h}{p}$, we have $p = \dfrac{h}{\lambda} = \dfrac{6.63\times10^{-34}\ \text{J}\cdot\text{s}}{0.122\times10^{-9}\ \text{m}} = 5.43\times10^{-24}\ \text{kg}\cdot\text{m/s}$.

(c) $K = \frac{1}{2} m v^2 = \dfrac{p^2}{2m} = \dfrac{(5.43\times10^{-24}\ \text{kg}\cdot\text{m/s})^2}{2(9.11\times10^{-31}\ \text{kg})} = 1.62\times10^{-17}$ J.

2. The Schrödinger Wave Equation

The **Schrödinger wave equation** is an equation that enables us to calculate matter waves for particles in various structures. The function satisfying the wave equation is called the **wave function**, ψ, which describes the wave as a function of time and space. The general form of the wave equation can be written as $(K + U)\ \psi = E\ \psi$, where K, U, and E are the kinetic, potential, and total energy of the particle, respectively. The square of the wave function (ψ^2) is called the **probability density**, which represents the relative probability of finding a particle in space and time. The interpretation of ψ^2 as the probability of finding a particle at a particular place altered the idea that particles are found in certain definitive locations. Now we can only say a particle has a certain probability of being at some location and can not predict exactly where it is. In some instances, a particle has a finite probability of being even in a classically forbidden region. The classical locations of particles such as the electron orbits in a hydrogen atom are only the *most probable* locations the particles can be.

The solutions to the Schrödinger wave equation lead to many experimentally proven quantum effects (classically forbidden) such as *tunneling* though a *potential energy barrier*. A classical analogy to quantum tunneling would be that a person could run though a concrete wall if he/she tries enough jumps. It takes a lot of energy, say E, to go through the concrete wall and the kinetic energy of a human being is certainly smaller than E. Therefore, classically the person cannot run through the concrete wall because he/she does not have enough energy to overcome the energy barrier of the wall. However, there is a small probability for the person to run though the wall. It is possible for the impossible (in classical sense) to happen if it is tried enough times. The electron tunneling microscope is an application of this quantum effect.

3. Atomic Quantum Numbers and the Periodic Table

When the Schrödinger wave equation is solved for the hydrogen atom, the solution gives three quantum numbers: the principal quantum number n, the orbital quantum number ℓ, the magnetic quantum number m_ℓ. Later, the spin quantum number m_s was added to explain experimental observations. The **principal quantum number** solely determines the energy levels of the electron of a hydrogen atom as predicted by the Bohr theory, in the absence of external magnetic field. The **orbital quantum number** determines the allowed value of angular momentum for an electron orbit. The **magnetic quantum number** determines the orientation of the electron orbit with respect to a given axis. The **spin quantum number** is a purely quantum-mechanical concept, that, in a classical analogy, indicates whether the spin angular momentum of an electron is up or down.

Numerically, the principal quantum number can be any positive integer, $n = 1, 2, 3, \ldots, \infty$. For a given n, the orbital quantum number can be $\ell = 0, 1, 2, 3, \ldots, (n-1)$, or n different values. For example if $n = 2$, $\ell = 0, 1$, or two different values; if $n = 5$, $\ell = 0, 1, 2, 3$, or four different values. For a given ℓ, the magnetic quantum number can be $m_\ell = 0, \pm1, \pm2, \pm3, \ldots, \pm\ell$, or $2\ell + 1$ different values. The spin quantum number can be either $+\frac{1}{2}$ (spin up) or $-\frac{1}{2}$ (spin down). Each set of unique quantum numbers, (n, ℓ, m_ℓ, m_s), defines a quantum state.

Example 28.3 Write down the possible sets of quantum numbers for each individual quantum state in the $n = 4$ shell.

Solution: For each n, $\ell = 0, 1, 2, 3, \ldots, n-1$. For each ℓ, $m_\ell = 0, \pm1, \pm2, \pm3, \ldots, \pm\ell$. m_s can be $\pm\frac{1}{2}$.

For $n = 4$, $\ell = 0, 1, 2$, and 3.

For $\ell = 0$, $m_\ell = 0$, and $m_s = +\frac{1}{2}$ or $-\frac{1}{2}$.

So there are two states, $(4, 0, 0, +\frac{1}{2})$ and $(4, 0, 0, -\frac{1}{2})$, for $\ell = 0$.

For $\ell = 1$, $m_\ell = 0$, and ±1.

For $m_\ell = 0$, $m_s = +\frac{1}{2}$ or $-\frac{1}{2}$, so there are two states, $(4, 1, 0, +\frac{1}{2})$ and $(4, 1, 0, -\frac{1}{2})$.

For $m_\ell = +1$, $m_s = +\frac{1}{2}$ or $-\frac{1}{2}$, so there are two states, $(4, 1, 1, +\frac{1}{2})$ and $(4, 1, 1, -\frac{1}{2})$.

For $m_\ell = -1$, $m_s = +\frac{1}{2}$ or $-\frac{1}{2}$, so there are two states, $(4, 1, -1, +\frac{1}{2})$ and $(4, 1, -1, -\frac{1}{2})$.

Therefore, there are six states for $\ell = 1$.

For $\ell = 2$, $m_\ell = 0, \pm 1, \pm 2$.

For $m_\ell = 0$, $m_s = +\frac{1}{2}$ or $-\frac{1}{2}$, so there are two states, $(4, 2, 0, +\frac{1}{2})$ and $(4, 2, 0, -\frac{1}{2})$.

For $m_\ell = +1$, $m_s = +\frac{1}{2}$ or $-\frac{1}{2}$, so there are two states, $(4, 2, 1, +\frac{1}{2})$ and $(4, 2, 1, -\frac{1}{2})$.

For $m_\ell = -1$, $m_s = +\frac{1}{2}$ or $-\frac{1}{2}$, so there are two states, $(4, 2, -1, +\frac{1}{2})$ and $(4, 2, -1, -\frac{1}{2})$.

For $m_\ell = +2$, $m_s = +\frac{1}{2}$ or $-\frac{1}{2}$, so there are two states, $(4, 2, 2, +\frac{1}{2})$ and $(4, 2, 2, -\frac{1}{2})$.

For $m_\ell = -2$, $m_s = +\frac{1}{2}$ or $-\frac{1}{2}$, so there are two states, $(4, -2, -1, +\frac{1}{2})$ and $(4, -2, -1, -\frac{1}{2})$.

Therefore, there are ten states for $\ell = 2$.

For $\ell = 3$, $m_\ell = 0, \pm 1, \pm 2, \pm 3$.

For $m_\ell = 0$, $m_s = +\frac{1}{2}$ or $-\frac{1}{2}$, so there are two states, $(4, 3, 0, +\frac{1}{2})$ and $(4, 3, 0, -\frac{1}{2})$.

For $m_\ell = +1$, $m_s = +\frac{1}{2}$ or $-\frac{1}{2}$, so there are two states, $(4, 3, 1, +\frac{1}{2})$ and $(4, 3, 1, -\frac{1}{2})$.

For $m_\ell = -1$, $m_s = +\frac{1}{2}$ or $-\frac{1}{2}$, so there are two states, $(4, 3, -1, +\frac{1}{2})$ and $(4, 2, -1, -\frac{1}{2})$.

For $m_\ell = +2$, $m_s = +\frac{1}{2}$ or $-\frac{1}{2}$, so there are two states, $(4, 3, 2, +\frac{1}{2})$ and $(4, 3, 2, -\frac{1}{2})$.

For $m_\ell = -2$, $m_s = +\frac{1}{2}$ or $-\frac{1}{2}$, so there are two states, $(4, 3, -2, +\frac{1}{2})$ and $(4, 3, -2, -\frac{1}{2})$.

For $m_\ell = +3$, $m_s = +\frac{1}{2}$ or $-\frac{1}{2}$, so there are two states, $(4, 3, 3, +\frac{1}{2})$ and $(4, 3, 3, -\frac{1}{2})$.

For $m_\ell = -3$, $m_s = +\frac{1}{2}$ or $-\frac{1}{2}$, so there are two states, $(4, 3, -3, +\frac{1}{2})$ and $(4, 3, -3, -\frac{1}{2})$.

Therefore, there are fourteen states for $\ell = 3$.

Hence, there are a total of $2 + 6 + 10 + 14 = 32$ states.

Generally, there are $2n^2$ states for a given value of n. Here $2n^2 = 2(4)^2 = 32$.

It is common to refer to the total of all the states in a given n as forming a **shell**, and to ℓ levels of that shell as **subshells**. That is, atomic electrons with the same n value are said to be in the same shell. Electrons with the same ℓ value are said to be in the same subshell. The subshells are designated by the letters *s*, *p*, *d*, *f*, *g*, . . . for ℓ = 0, 1, 2, 3, 4, . . ., respectively.

The **Pauli exclusion principle** states that no two electrons in a multielectron atom can have the same set of quantum numbers (n, ℓ, m_ℓ, m_s). That is, no two electrons can be in the same quantum state.

Example 28.4 How many possible sets of quantum numbers or electron states are there in the 5*f* subshell? What is the maximum number of electrons that can occupy this subshell?

Solution: For the $5f$ subshell, $n = 5$ and $\ell = 3$.

For $\ell = 3$, m_ℓ can be 0, ±1, ±2, ±3, or seven different values, $(2\ell + 1) = (2 \times 3 + 1) = 7$ values.

For each m_ℓ, m_s can be either $+\frac{1}{2}$ or $-\frac{1}{2}$, 2 values.

So there are $2 \times 7 = 14$ sets of quantum numbers or electron states.

According to the Pauli exclusion principle, only one electron can occupy an electron state, so the maximum number of electrons that can occupy the $5f$ subshell is 14.

In a shorthand notation called **electron configuration**, we write the quantum states in increasing order of energy, and designate the numbers of electrons in each level with a subscript. For example, $4p^5$ means that the $4p$ subshell contains five electrons. The electron configuration for the ground state of the element neon (Ne), for example, is $1s^2 2s^2 2p^6$ (both shells $n = 1$, $n = 2$ completely full).

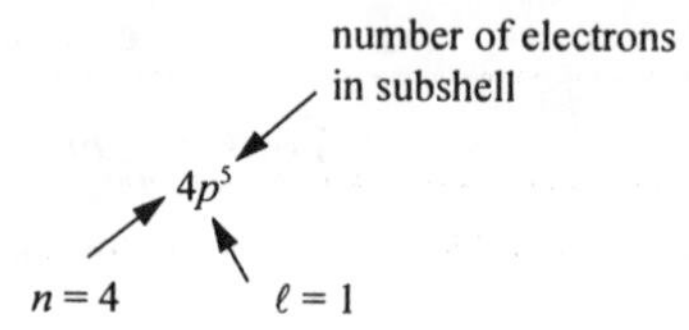

Some energy levels, such as $2s$ and $2p$, have similar energies. We refer to such sets of energy levels that have about the same energy as **electron period**. These electron period are the basis of the **periodic table of elements**. In a periodic table, elements are arranged in horizontal rows, which are called **periods**, in order of increasing atomic mass. Elements of similar chemical properties are arranged in vertical columns called **groups**, or families of elements with similar properties.

4. The Heisenberg Uncertainty Principle

The **Heisenberg uncertainty principle** as applied to position (x) and momentum (magnitude p) may be stated as follows: it is impossible to know simultaneously an object's exact position and momentum. This principle overthrows the *deterministic* view of nature in classical physics. The minimum uncertainty of the product of position and momentum is $(\Delta p)(\Delta x) \geq \frac{h}{2\pi}$, expressed alternatively in energy and time is $(\Delta E)(\Delta t) \geq \frac{h}{2\pi}$.

Note: The value $h/2\pi$ is the minimum uncertainty of the product of position and momentum, that is, *at the very best*, the uncertainty of the product is $h/2\pi$, that is, never can it ever be zero.

Example 28.5 A measurement of an electron's speed is 2.0×10^6 m/s and has an uncertainty of ±10%. What is the minimum uncertainty in its position?

Solution: Given: $m = 9.11 \times 10^{-31}$ kg, $v = 2.0 \times 10^6$ m/s, % uncertainty in $v = \pm 10\%$.

Find: Δx.

$v_{max} = v + 0.10v = 2.0 \times 10^6 \text{ m/s} + (0.10)(2.0 \times 10^6 \text{ m/s}) = 2.2 \times 10^6$ m/s. $v_{min} = 1.8 \times 10^6$ m/s.

So the uncertainty in momentum is

$\Delta p = p_{max} - p_{min} = m\,v_{max} - m\,v_{min} = (9.11 \times 10^{-31} \text{ kg})(2.2 \times 10^6 \text{ m/s} - 1.8 \times 10^6 \text{ m/s}) = 3.6 \times 10^{-25}$ kg·m/s.

From the uncertainty principle, $(\Delta p)(\Delta x) \geq \frac{h}{2\pi}$,

we have $\Delta x \geq \frac{h}{2\pi(\Delta p)} = \frac{6.63 \times 10^{-34} \text{ J·s}}{2\pi(3.6 \times 10^{-25} \text{ kg·m/s})} = 2.9 \times 10^{-10} \text{ m} = 0.29 \text{ nm} \approx$ atomic dimensions.

Example 28.6 The energy of an electron state has an uncertainty of about ±0.0500 eV. What is the uncertainty in the life-time (the time the electron remains in that level before making a transition to another level) of that level?

Solution: Given: $\Delta E = 2 \times 0.0500 \text{ eV} = 0.100 \text{ eV} = 1.60 \times 10^{-20}$ J.

Find: Δt.

From $(\Delta E)(\Delta t) \geq \frac{h}{2\pi}$, we have $\Delta t \geq \frac{h}{2\pi(\Delta E)} = \frac{6.63 \times 10^{-34} \text{ J·s}}{2\pi(1.60 \times 10^{-20} \text{ J})} = 4.14 \times 10^{-14}$ s.

---

5. Particles and Antiparticles

A **positron** is a particle that has the same mass as an electron, but carries a *positive* electronic charge. The oppositely-charged positron is said to be the **antiparticle** of the electron. All subatomic particles have been found to have antiparticles. A positron can only be created with the simultaneous creation of an electron in a process called **pair production**. In this process, the energy of a high-energy photon is converted to a positron and an electron and usually they have kinetic energies. By the mass-energy equivalence of relativity, the *threshold energy for pair production* (the minimum energy of the proton) required is $E_{min} = hf = 2\,m_e c^2 = 1.022$ MeV, that is, $E \geq 1.022$ MeV. A positron and an electron can also be "destroyed" or annihilated in a process called **pair annihilation**. It is the direct conversion of rest mass into electromagnetic energy (a photon), the inverse of pair production so to speak. It is conceivable that antiparticles predominate in some parts of the universe. If so, the atoms of the **antimatter** in this region would consist of negatively charged nuclei (composed of antiprotons and antineutrons), surrounded by positively charged positrons (antielectrons). It would be impossible to distinguish between antimatter and ordinary matter, except, however, if antimatter and ordinary matter should come into contact, that they would annihilate each other with an explosive release of energy.

IV. Mathematical Summary

Momentum of a Photon	$p = \frac{E}{c} = \frac{hf}{c} = \frac{h}{\lambda}$ (28.1)	Defines the momentum of a photon
de Broglie wavelength of a Moving Particle:	$\lambda = \frac{h}{p} = \frac{h}{mv}$ (28.2)	Defines the momentum of a moving particle (matter wave)
Electron Wavelength When Accelerated through Potential V (nonrelativistic)	$\lambda = \sqrt{\frac{1.50}{V}} \times 10^{-9}$ m $= \sqrt{\frac{1.50}{V}}$ nm (28.3)	Calculates the wavelength of an electron when it is accelerated through a potential V (nonrelativistic)
Heisenberg Uncertainty Principle	$(\Delta p)(\Delta x) \geq \frac{h}{2\pi}$ (28.5) $(\Delta E)(\Delta t) \geq \frac{h}{2\pi}$ (28.6)	Expresses the Heisenberg uncertainty principle for momentum-position and energy-time
Condition for Electron Pair Production	$hf = 2m_ec^2 = 1.022$ MeV (28.7)	Gives the minimum energy (frequency) required to produce an electron-positron pair

V. Solutions of Selected Exercises and Paired Exercises

8. From Eq. 28.3: $\lambda = \sqrt{\frac{1.50}{V}}$ nm, so $\lambda \propto \frac{1}{\sqrt{V}}$.

Therefore $\frac{\lambda_2}{\lambda_1} = \sqrt{\frac{V_1}{V_2}} = \sqrt{\frac{250 \text{ kV}}{600 \text{ kV}}} = 0.645.$

14. (a) $\lambda = \frac{h}{m\,v} = \frac{6.63 \times 10^{-34} \text{ J}\cdot\text{s}}{(9.11 \times 10^{-31} \text{ kg})(2.19 \times 10^{6} \text{ m/s})} = 3.32 \times 10^{-10}$ m.

(b) $2\pi r_1 = 2\pi(0.529 \times 10^{-10} \text{ m}) = \lambda$. So they are the same,

18. (a) The following standing waves can be set up in the well: $L = \frac{n\lambda}{2}$, $n = 1, 2, 3, \ldots$

or $\lambda = \frac{2L}{n}$. Also the wave function must be zero at $x = 0$ and $x = L$.

So $\psi_n = A \sin \frac{2\pi}{\lambda} x = A \sin \frac{n\pi x}{L}$ for $n = 1, 2, 3, \ldots$

(b) $K_n = \frac{p^2}{2m} = \frac{(h/\lambda)^2}{2m} = \frac{h^2}{2m\lambda^2} = \frac{h^2}{2m\,(2L/n)^2} = n^2 \frac{h^2}{8mL^2}$.

25. The periodical table groups elements according to the values of quantum numbers n and ℓ. Within a group, the elements have the same or very similar electronic configurations for the outmost electrons.

26. (a) $2n^2 = 2(2)^2 = 8$ and $2(3)^2 = 18$.

(b) For $n = 2$: (2, 1, 1, +1/2); (2, 1, 1, –1/2); (2, 1, 0, +1/2); (2, 1, 0, –1/2); (2, 1, –1, +1/2); (2, 1, –1, –1/2); (2, 0, 0, +1); (2, 0, 0, –1/2).

For $n = 3$: (3, 2, 2, +1/2); (3, 2, 2, –1/2); (3, 2, 1, +1/2); (3, 2, 1, –1/2); (3, 2, 0, +1/2); (3, 2, 0, –1/2); (3, 2, –1, +1/2); (3, 2, –1, –1/2); (3, 2, –2, +1/2); (3, 2, –2, –1/2); (3, 1, 1, +1); (3, 1, 1, –1/2); (3, 1, –1, +1/2); (3, 1, –1, –1/2); (3, 1, 0, +1/2); (3, 1, 0, –1/2); (3, 0, 0, +1/2); (3, 0, 0, –1/2).

30.

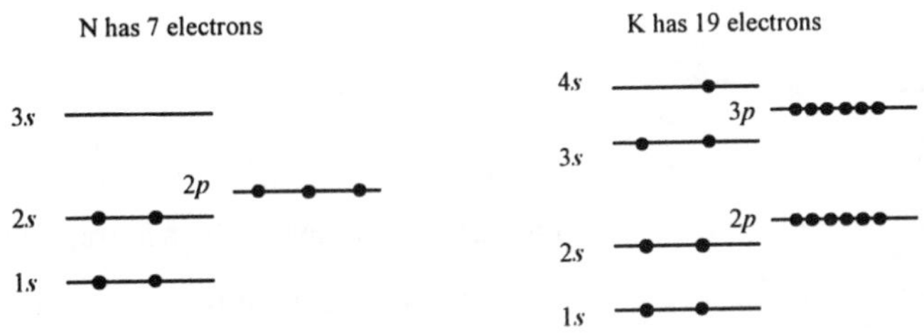

38. From $\Delta p\,\Delta x \geq \dfrac{h}{2\pi}$,

$$\Delta v \geq \frac{\Delta p}{m} \geq \frac{h}{2\pi\,\Delta x\,m} = \frac{6.63\times10^{-34}\ \text{J·s}}{2\pi[(0.10-0.050)\times10^{-9}\ \text{m}]\,(9.11\times10^{-31}\ \text{kg})} = 2.3\times10^{6}\ \text{m/s}.$$

44. From $\Delta E\,\Delta t \geq \dfrac{h}{2\pi}$, $\Delta t \geq \dfrac{h}{2\pi\,\Delta E} = \dfrac{6.63\times10^{-34}\ \text{J·s}}{2\pi(2\times0.0003\ \text{eV})(1.6\times10^{-9}\ \text{J/eV})} = 1.1\times10^{-12}\ \text{s}.$

53. Since $d\sin\theta = m\lambda$, $\lambda = \dfrac{d\sin\theta}{m} = \dfrac{(0.190\times10^{-9}\ \text{m})\sin 50^\circ}{1} = 1.46\times10^{-10}\ \text{m} = 0.146\ \text{nm}.$

From Eq. 28.3, $\lambda = \sqrt{\dfrac{1.50}{V}}$ nm, we have $V = \dfrac{1.50}{\lambda^2} = \dfrac{1.50}{(0.146)^2} = 70\ \text{V}.$

54. From $\Delta p\,\Delta x \geq \dfrac{h}{2\pi}$, $\Delta x \geq \dfrac{h}{2\pi\,\Delta p} = \dfrac{h}{2\pi\,m\,\Delta v} = \dfrac{6.63\times10^{-34}\ \text{J·s}}{2\pi(9.11\times10^{-31}\ \text{kg})(10^{-5}\ \text{m/s})} = 11.6\ \text{m}.$

55. (a) $p = \dfrac{h}{\lambda} = \dfrac{hf}{c} = \dfrac{E}{c} = \dfrac{(7.5\times10^{6}\ \text{eV})(1.6\times10^{-19}\ \text{J/eV})}{3.00\times10^{8}\ \text{m/s}} = 4.0\times10^{-21}\ \text{kg·m/s}.$

(b) $\lambda = \dfrac{h}{p} = \dfrac{6.63\times10^{-34}\ \text{J·s}}{4.0\times10^{-21}\ \text{kg·m/s}} = 1.7\times10^{-13}\ \text{m}.$

62. (a) $2E = 2hf = 2m_pc^2 = 2(1.67\times10^{-27}\text{ kg})(3.00\times10^8\text{ m/s})^2 = 3.01\times10^{-10}\text{ J} = 1.88\text{ GeV}$.

So $E = \dfrac{1.88\text{ GeV}}{2} = 0.94\text{ GeV}$.

(b) $f = \dfrac{3.01\times10^{-10}\text{ J}}{2(6.63\times10^{-34}\text{ J·s})} = 2.3\times10^{23}\text{ Hz}$.

64. They do not have to go off back to back according to momentum conservation.

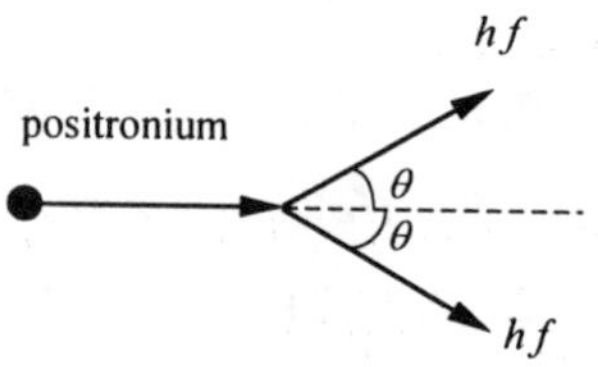

VI. Practice Quiz

1. If the accuracy in measuring the position of a particle increases, the accuracy in measuring its velocity will:
(a) increase (b) decrease (c) remain the same (d) be uncertain (e) be exact

2. What is the electron configuration for Li which has three electrons?
(a) $1s^3$ (b) $1s^12s^2$ (c) $1s^22s^1$ (d) $1s^21p^1$ (e) $1s^11p^2$

3. An electron is accelerated through a potential difference of 150 V. What is the de Broglie wavelength associated with the matter wave of the electron?
(a) 0.1 nm (b) 1 nm (c) 1 mm (d) 1 cm (e) 1 m

4. What is the maximum number of electrons that can occupy the d subshell?
(a) 2 (b) 6 (c) 10 (d) 14 (e) 18

5. The probability of finding an electron in a hydrogen atom is directly proportional to its
(a) energy (b) momentum (c) angular momentum
(d) the wave function (e) the square of the wave function

6. The element krypton has 36 electrons. How many electrons are in the $n = 5$ shell?
(a) 2 (b) 4 (c) 8 (d) 10 (e) 18

7. If the principal quantum number is 5, which of the following is not an allowed magnetic quantum number?

(a) 0 (b) −2 (c) 3 (d) −4 (e) 5

8. An antiproton is a particle which has

(a) the same mass as an electron (b) the same mass as a positron (c) negative electronic charge (d) positive electronic charge (e) no electronic charge

9. A proton has a kinetic energy of 1.0 MeV. If its momentum is measured with an uncertainty of ±1.0%, what is the uncertainty in its position?

(a) 2.8×10^{-14} m (b) 5.6×10^{-14} m (c) 9.1×10^{-14} m (d) 2.3×10^{-13} m (e) 9.1×10^{-13} m

10. Which one of the following particles would tend to have the smallest de Broglie wavelength (assuming they travel at the same speed)?

(a) electron (b) proton (c) neutron (d) baseball (e) bowling ball

Answers to Practice Quiz:

1. b 2. c 3. a 4. c 5. e 6. b 7. e 8. c 9. d 10. e

CHAPTER 29

The Nucleus

I. Chapter Objectives

Upon completion of this chapter, you should be able to:

1. distinguish between the Thompson and Rutherford-Bohr models of the atom, specify some of the basic properties of the strong nuclear force, and understand nuclear notation.
2. define the term "radioactivity," distinguish among alpha, beta, and gamma decay, and write nuclear decay equations.
3. explain the concepts of activity, decay constant, and half life of a radioactive sample, and use radioactive decay to find the age of objects.
4. state which proton and neutron number combinations result in stable nuclei, explain the pairing effect and magic numbers in relation to nuclear stability, and calculate nuclear binding energy.
5. gain insight into the operating principle of various nuclear radiation detectors, investigate the medical and biological effects of radiation exposure, and study some of the practical uses and applications of radiation.

II. Key Terms

Upon completion of this chapter, you should be able to define and/or explain the following key terms:

Rutherford-Bohr model
nucleon
strong nuclear force
atomic number (*Z*)
proton (atomic) number
mass number
neutron number
isotope
nuclide
radioactivity
alpha particle
beta particle
gamma ray
alpha decay
gamma decay
activity
decay constant
half-life
curie (Ci)
becquerel (Bq)
carbon-14 dating
pairing effect
atomic mass unit (u)
mass defect
total binding energy
magic number
radiation detector
roentgen (R)

conservation of nucleons

conservation of charge

tunneling

barrier penetration

beta (β^-) decay

beta (β^+) decay

electron capture

rad

gray (Gy)

rem

relative biological effectiveness (RBE)

sievert (Sv)

neutron activation analysis

The definitions and/or explanations of the most important key terms can be found in the following section: **III. Chapter Summary and Discussion**.

III. Chapter Summary and Discussion

1. Nuclear Structure and the Nuclear Force

The **Rutherford-Bohr model** of the atom is essentially the planetary model, with negative electrons orbiting the positively charged nucleus in which almost all the mass of the atom is located. The atomic nucleus is composed of two types of particles -- protons and neutrons -- which are collectively referred to as **nucleons**. Using scattering experiments, the *upper limit* for the nuclear radius is found to be on the order of 10^{-12} m.

The **strong nuclear force**, or simply the nuclear force, is the short range attractive force between nucleons which is responsible for holding the nucleus together against the repulsive electric force of its protons. This force possesses the following properties:

- It is strongly attractive and much larger in relative magnitude than the electrostatic and magnetic forces.
- It is very short-ranged; that is, a nucleon interacts only with its nearest neighbors, over a distance on the order of 10^{-15} m.
- It acts between any two nucleons within a short range, that is, between two protons, a proton and a neutron, or two neutrons.

In a nuclear notation ${}^{A}_{Z}X_{N}$, Z is the **atomic number** or **proton number** (the number of protons), N is the **neutron number** (the number of neutrons), A is the **mass number** (the number of nucleons), and X is the chemical symbol of the element of the nucleus. The neutron number is often omitted because it is determined from $N = A - Z$. For example, the ${}^{13}_{6}C$ is for carbon-13, which has 6 protons, $13 - 6 = 7$ neutrons, and 13 nucleons.

Isotopes are nuclei with the same number of protons (atomic number) but different number of neutrons (neutron numbers or mass numbers), such as ${}^{12}_{6}C$, ${}^{13}_{6}C$, and ${}^{14}_{6}C$. A particular nuclear species or isotope of any element is called a **nuclide**, for example ${}^{11}_{6}C$, ${}^{12}_{6}C$, ${}^{13}_{6}C$, ${}^{14}_{6}C$, ${}^{15}_{6}C$, and ${}^{16}_{6}C$, are all nuclides of carbon.

Example 29.1 The nuclear notation for calcium-39 is ${}^{39}_{20}Ca$. If the atom is neutral,

(a) how many protons are present?

(b) how many neutrons are present?

(c) how many nucleons are present?

(d) how many electrons are present?

Solution: $A = 39$, $Z = 20$.

(a) The number of protons is equal to the atomic number, $Z = 20$ protons.

(b) The number of neutrons is equal to the neutron number, $N = A - Z = 39 - 20 = 19$ neutrons.

(c) The number of nucleons is equal to the mass number, $A = 39$ nucleons.

(d) In a neutral atom, the number of electrons is equal to the number of protons, so there are 20 electrons.

Example 29.2 Which one of the following is an isotope of ${}^{39}_{20}X_{19}$?

(a) ${}^{40}_{20}X_{20}$ (b) ${}^{39}_{19}Y_{20}$ (c) ${}^{39}_{21}Z_{18}$.

Solution: Isotopes are nuclei with the same number of protons and a different number of neutrons or nucleons, so the answer is (a).

---

2. Radioactivity

The nuclei of some isotopes are not stable and they decay spontaneously with the emission of particles. Such isotopes are said to be *radioactive* or to exhibit **radioactivity**. Five decay modes have been found experimentally and they are:

- **Alpha decay,** in which the nucleus emits an **alpha particle**, a doubly charged ($+2e$) particle containing two protons and two neutrons (helium nucleus, ${}^{4}_{2}He$).
- **Beta decay** (β^-) in which the nucleus emits a **beta particle** or an electron (${}^{0}_{-1}e$).
- **Gamma decay,** in which an excited nucleus emits a **gamma ray**, or a "particle" or quantum, or a photon, of electromagnetic energy.

- β^+ **decay**, in which the nucleus emits a positron ($^{0}_{+1}e$).
- **Electron capture**, in which the nucleus *absorbs* one of the orbital electrons.

Alpha decay involves the quantum mechanical process of **tunneling** or **barrier penetration**, that is, there is a finite probability of finding the alpha particle outside the nucleus. In basic beta β^- decay, a neutron decays to a proton and an electron, $^{1}_{0}n \rightarrow {}^{1}_{1}p + {}^{0}_{-1}e$; and in basic beta β^+ decay, a proton decays to a neutron and a positron, $^{1}_{1}p \rightarrow {}^{1}_{0}p + {}^{0}_{+1}e$. (Neutrinos are also involved in beta decays. They will be discussed later.)

Two laws apply to all nuclear reactions: the **conservation of nucleons** (the total number of nucleons remains constant) and the **conservation of charge** (the total charge remains constant).

Example 29.3 Write the nuclear equation for the following nuclear reactions:

(a) Alpha decay of $^{214}_{84}Po$.

(b) Beta (β^-) decay of $^{14}_{6}C$.

(c) Gamma decay of $^{61}_{28}Ni^*$ (the * represents an excited nucleus).

Solution: For every nuclear reaction, the total number of nucleons and the total charge remain constant.

(a) If $^{214}_{84}Po$ emits an alpha particle ($^{4}_{2}He$), the daughter nucleus must have $214 - 4 = 210$ nucleons (conservation of nucleons) and $84 - 2 = 82$ protons (conservation of charge). From the periodic table, the nucleus that has 82 protons is lead (Pb). Therefore, the nuclear equation is $^{214}_{84}Po \rightarrow {}^{210}_{82}Pb + {}^{4}_{2}He$.

(b) If $^{14}_{6}C$ emits a beta particle or electron ($^{0}_{-1}e$), the daughter nucleus must have the same number of nucleons (14) and $6 + 1 = 7$ protons to conserve nucleons and charge. The element with 7 protons is nitrogen (N). Therefore, the nuclear equation is $^{14}_{6}C \rightarrow {}^{14}_{7}N + {}^{0}_{-1}e$.

(c) Since a gamma ray (γ) is a photon, the daughter nucleus is still Ni, but at a lower energy level. The nuclear equation is $^{61}_{28}Ni^* \rightarrow {}^{61}_{28}Ni + \gamma$.

The absorption or degree of penetration of nuclear radiation is an important consideration in applications such as radioisotope treatment of cancer and nuclear shielding around a nuclear reactor. Alpha particles have large mass and are doubly charged, and generally move slowly. A few centimeters of air or a sheet of paper will usually completely stop or absorb them. Beta particles can travel a few meters in air or a few centimeters in aluminum before being stopped. Gamma rays are more penetrating. They can penetrate a centimeter or more of a dense material such as lead.

3. Decay Rate and Half-Life

The **activity** of a radioactive isotope is defined as the number of disintegrations or decays (ΔN) per unit time (Δt), that is, $\frac{\Delta N}{\Delta t}$. This activity is proportional to the number of undecayed nuclei present (N), that is, $\frac{\Delta N}{\Delta t} = -\lambda N$, where λ is the **decay constant**, which is different for different isotopes. The negative sign indicates the decrease in the number of undecayed nuclei present in a decay process. Solving the activity equation yields the expression for the undecayed nuclei as a function of time, $N = N_o\, e^{-\lambda t}$, where N_o is the initial number of nuclei at $t = 0$. The **half-life** ($t_{1/2}$) of an isotope is the time required for the number of undecayed nuclei in a sample (hence the radioactive decay rate or activity of the sample) to fall to one-half of its original value. It is related to the decay constant through $t_{1/2} = \frac{0.693}{\lambda}$. The SI unit of activity is the **becqerel** (Bq), and 1 Bq = 1 decay/s. Another common unit is the **curie** (Ci), and 1 Ci = 3.70×10^{10} decays/s = 3.70×10^{10} Bq.

Note: When using $N = N_o\, e^{-\lambda t}$, it is not necessary to change time to seconds. You can use any unit (min, h, or y) as long as λ is in the same units (inverse), for example, ($d^{-1} \times d$) = unitless.

Radioactive isotopes can be used as nuclear clocks. A common radioactive dating method used on materials that were once part of living things is **carbon-14 dating**. This method uses the fact that the carbon-14 content decreases with time once a living organism dies, so measurements of the concentration of carbon-14 in dead matter relative to that in living things can then be used to establish when the organism died.

Example 29.4 A radioactive isotope sample has an half-life of 10 min. What percentage of the sample is left after 55 min?

Solution: Given: $t_{1/2} = 10$ min, $t = 55$ min. Find: $\frac{N}{N_o}$.

$$\lambda = \frac{0.693}{t_{1/2}} = \frac{0.693}{10\text{ min}} = 0.0693\text{ min}^{-1}, \quad \text{so} \quad \lambda t = (0.0693\text{ min}^{-1})(55\text{ min}) = 3.81.$$

From $N = N_o\, e^{-\lambda t}$, we have $\frac{N}{N_o} = e^{-\lambda t} = e^{-3.81} = 2.2\times10^{-2} = 2.2\%$.

The problem can also be solved with half-life directly. After one half-life, the fraction of nuclei remaining is $\frac{N}{N_o} = \frac{1}{2} = \frac{1}{2^1}$; after two half-lives, the fraction of nuclei remaining is $\frac{N}{N_o} = \frac{1}{2} \times \frac{1}{2} = \frac{1}{2^2}$; so after *n* half-lives, the fraction of nuclei remaining is $\frac{N}{N_o} = \frac{1}{2^n}$. $n = \frac{55\text{ min}}{10\text{ min}} = 5.5$ half-lives, so $\frac{N}{N_o} = \frac{1}{2^{5.5}} = 2.2\%$.

Example 29.5 A radioactive sample with a half-life of 5.00 min has 1.50×10^{10} nuclei.

(a) What is the decay constant?

(b) What is the initial activity in Ci?

(c) What is the activity in Ci after 2.00 min.

Solution:

Given: $N_o = 1.50\times10^{10}$, $t_{1/2} = 5.00$ min.

Find: (a) λ (b) $\left|\dfrac{\Delta N}{\Delta t}\right|_o$ (c) $\left|\dfrac{\Delta N}{\Delta t}\right|$.

(a) $\lambda = \dfrac{0.693}{t_{1/2}} = \dfrac{0.693}{5.00\text{ min}} = 0.128\text{ min}^{-1} = 2.13\times10^{-3}\text{ s}^{-1}$.

(b) $1\text{ Ci} = 3.70\times10^{10}$ decays/s.

$$\left|\frac{\Delta N}{\Delta t}\right|_o = \lambda N_o = (2.13\times10^{-3}\text{ s}^{-1})(1.50\times10^{10}) = 3.195\times10^7\text{ decays/s} = 8.64\times10^{-4}\text{ Ci}.$$

(c) $\lambda t = (0.128\text{ min}^{-1})(2.00\text{ min}) = 0.256$, so $N = N_o\, e^{-\lambda t} = (1.50\times10^{10})\, e^{-0.256} = 1.16\times10^{10}$ nuclei.

Therefore, $\left|\dfrac{\Delta N}{\Delta t}\right| = \lambda N = (2.13\times10^{-3}\text{ s}^{-1})(1.16\times10^{10}) = 2.471\times10^7\text{ decays/s} = 6.68\times10^{-4}\text{ Ci}.$

4. Nuclear Stability and Binding Energy

Experiments show that to lower their total energy, two protons will pair up and as will two neutrons. This so-called **pairing effect** gives the following criteria for nuclear stability:

(1) All isotopes with more than 83 protons ($Z > 83$) are unstable.

(2) Most even-even (even number of protons and even number of neutrons) are stable. Many odd-even or even-odd nuclei are stable. Only four odd-odd nuclei are stable, ^{2}H, ^{6}Li, ^{10}Be, and ^{14}N.

(3) Stable nuclei with less than 40 nucleons ($A < 40$) have approximately the same number of protons and neutrons. Stable nuclei with $A > 40$ have more neutrons than protons.

Example 29.6 Is the sodium isotope $^{22}_{11}$Na likely to be stable?

Solution:

(1) *Satisfied.* With $Z = 8$, the criterion is satisfied.

(2) *Not satisfied.* The isotope $^{22}_{11}$Na has a odd-odd nucleus and so it is unstable.

Since the masses of nuclei are so small, another unit, the **atomic mass unit (u)** is used to measure nuclear masses. $1\ \text{u} = 1.6606\times10^{-27}$ kg, and it is referenced from a neutral atom of ^{12}C, which is taken to have an exact mass of 12.000000 u. According to mass-energy equivalence, 1 u = 931.5 MeV. It is found that the masses of the nucleons in a nucleus (bound particles) is less than that when the nucleons are separated as free particles. This mass difference, or **mass defect** (Δm), has an energy equivalence called the total **binding energy** (E_b), which is the minimum amount of energy needed to separate a given nucleus apart into its constituent nucleons, $E_b = (\Delta m)c^2$. The *average binding energy per nucleon, E_b/A*, gives an indication of nuclear stability. The greater the E_b/A, the more tightly bound are the nucleons in a nucleus, hence the more stable it is.

Example 29.7 What is the total binding energy and average binding energy per nucleon for sodium $^{22}_{11}$Na. The mass of $^{22}_{11}$Na is 21.994435 u.

Solution: Sodium $^{22}_{11}$Na has 11 protons and 11 neutrons.

Given: $^{22}_{11}$Na mass = 21.994435 u, $^{1}_{1}$p mass = 1.007825 u, $^{1}_{0}$n mass = 1.008665 u.

Find: E_b and E_b/A.

The mass of the separated nucleons is $m = 11(1.007825\ \text{u}) + 11(1.008665\ \text{u}) = 22.18139\ \text{u}$.

So the mass defect is $\Delta m = 22.18139\ \text{u} - 21.994435\ \text{u} = 0.18696\ \text{u}$.

Therefore the total binding energy is $E_b = (0.18696\ \text{u})(931.5\ \text{MeV/u}) = 174.15\ \text{MeV}$.

The average binding energy per nucleon is $\dfrac{E_b}{A} = \dfrac{174.15\ \text{MeV}}{22\ \text{nucleons}} = 7.916\ \text{MeV/nucleon}$.

---

5. Radiation Detection and Applications

A *Geiger counter*, a *scintillation counter*, a *solid state (semiconductor) detector*, a *cloud chamber*, a *bubble chamber*, and a *spark chamber* are all examples of **radiation detectors**.

An important consideration in radiation therapy and radiation safety is the amount, or *dose*, of radiation. Several quantities are used to describe this in terms of *exposure*, *absorbed dose*, and *equivalent dose*. The earliest unit of dosage, the **roentgen** (R) was based on exposure. The roentgen has been largely replaced by the rad (*r*adiation *a*bsorbed *d*ose), which is an absorbed dose unit. The SI unit for absorbed dose is the **gray** (Gy): 1 Gy = 1 J/kg = 100 rad. The effective dose is measured in terms of the **rem** unit (*r*oentgen or *r*ad *e*quivalent *man*). The different degrees of effectiveness of different particles are characterized by the **relative biological effectiveness (RBE)**. Then, effective dose (in rem) = dose (in rad) × RBE. Another unit of effective dose is the **sievert** (Sv), and effective dose (in Sv) = dose (in Gy) × RBE. Since 1 Gy = 100 rad, it follows that 1 Sv = 100 rem.

Some other applications of radiation in domestic and industrial use are the smoke detector, radioactive tracer (to detect leaks), and in **neutron activation analysis** (to activate compounds).

IV. Mathematical Summary

Decay Rate of a Radioisotope	$\frac{\Delta N}{\Delta t} = -\lambda N$ (29.2) $1\ \text{Ci} = 3.70\times 10^{10}$ decays/s $= 3.70\times 10^{10}$ Bq	Defines the decay rate of an radioisotope in terms of its decay constant and number of nuclei
Activity of a Radioisotope	$\left\|\frac{\Delta N}{\Delta t}\right\| = \lambda N$	Defines the activity of a radioisotope
Number of Undecayed Nuclei	$N = N_o\, e^{-\lambda t}$ (29.3)	Calculates the number of undecayed nuclei as a function of time
Half-life and Decay Constant	$t_{1/2} = \frac{0.693}{\lambda}$ (29.4)	Relates the half-life and the decay constant of a radioisotope
Total Binding Energy	$E_b = (\Delta m)c^2$ (29.5)	Computes the binding energy of a nucleus from the mass defect between the nucleus and the constituent particles
Effective Dose	Dose (rem) = does (rad) × RBE (29.6) Dose (Sv) = dose (Gy) × RBE (29.7)	Calculates the effective dose by including the relative biological effectiveness (RBE)

V. Solutions of Selected Exercises and Paired Exercises

8. The mass number is 8 + 8 = 16, 8 + 9 = 17, and 8 + 10 = 18, respectively. So they are $^{16}_{8}O$, $^{17}_{8}O$, $^{18}_{8}O$.

14. (a) $^{60}_{27}Co \rightarrow {}^{60}_{28}Ni + {}^{0}_{-1}e$.

(b) $^{226}_{88}Ra \rightarrow {}^{222}_{86}Rn + {}^{4}_{2}He$.

18. (a) $^{222}_{86}Rn \rightarrow {}^{218}_{84}Po + {}^{4}_{2}He$.

(b) α–decay: $^{218}_{84}Po \rightarrow {}^{214}_{82}Pb + {}^{4}_{2}He$; β–decay: $^{218}_{84}Po \rightarrow {}^{218}_{85}At + {}^{0}_{-1}e$.

20. (a) ${}^{4}_{2}\text{He}$. (b) $4{}^{1}_{0}\text{n}$. (c) γ. (d) ${}^{29}_{12}\text{Mg}$.

30. (a) $40 \text{ mCi} = (40\times10^{-3} \text{ Ci})\dfrac{3.70\times10^{10} \text{ decays/s}}{1 \text{ Ci}} = 1.5\times10^{9} \text{ decays/s}.$

(b) $1.5\times10^{9} \text{ decays/s} = (1.5\times10^{9} \text{ betas/s})\dfrac{60 \text{ s}}{1 \text{ min}} = 8.9\times10^{10} \text{ betas/min}.$

32. $\lambda = \dfrac{0.693}{t_{1/2}} = \dfrac{0.693}{18\times60 \text{ s}} = 6.42\times10^{-4} \text{ s}^{-1}; \quad \lambda t = (5.92\times10^{-4} \text{ s}^{-1})(3600 \text{ s}) = 2.31.$

So $\dfrac{\Delta N}{\Delta t} = \lambda N = \lambda N_o\, e^{-\lambda t} = \dfrac{\Delta N_o}{\Delta t}\, e^{-\lambda t} = (10 \text{ mCi})\, e^{-2.31} = 1 \text{ mCi}.$

40. From $\dfrac{\Delta N}{\Delta t} = \lambda N = \lambda N_o\, e^{-\lambda t} = \dfrac{\Delta N_o}{\Delta t} e^{-\lambda t}$, we have $\dfrac{\Delta N/\Delta t}{\Delta N_o/\Delta t} = 0.10 = e^{-\lambda t}$.

So $t = -\dfrac{\ln 0.10}{\lambda} = -\dfrac{t_{1/2} \ln 0.10}{0.693} = -\dfrac{(5.3 \text{ y}) \ln 0.10}{0.693} = 18 \text{ y}.$

46. The mass of U_3O_8 is $3(238 \text{ u}) + 8(16 \text{ u}) = 942 \text{ u}$.

So the fraction of U by mass is $\dfrac{3\times238}{942} = 0.758.$

In 500,000 tons of U_3O_8, there are $(0.758)(500{,}000 \text{ tons})(10 \text{ kg/ton}) = 3.8\times10^{6} \text{ kg}$ of U today.

4.6 billion years is about 1 half-life.

Therefore, there were $2(3.8\times10^{6} \text{ kg}) = 7.6\times10^{6} \text{ kg}$ of U.

53. (a) ${}^{17}_{8}\text{O}$ because of an unpaired neutron beyond a magic number.

(b) ${}^{42}_{20}\text{Ca}$ because of a magic number difference making ${}^{40}_{20}\text{Ca}$ more stable (both are paired).

(c) ${}^{10}_{5}\text{B}$ because of lack of paring.

(d) Approximately the same as both have 126 neutrons (paired and magic) and the different number of protons does not affect neutrons.

60. $E_b = \Delta m\, c^2 = (4m_H + 5m_n - m_{Be})\, c^2 = [4(1.007825 \text{ u}) + 5(1.008665 \text{ u}) - 9.012183 \text{ u}](931.5 \text{ MeV/u})$

$= 58.2 \text{ MeV}. \quad \text{So} \quad \dfrac{E_b}{A} = \dfrac{58.2 \text{ MeV}}{9 \text{ nucleon}} = 6.46 \text{ MeV/nucleon}.$

66. For Al: $E_b = (13m_H + 14m_n - m_{Al})\, c^2 = [13(1.007825 \text{ u}) + 14(1.008665 \text{ u}) - 26.981541 \text{ u}](931.5 \text{ MeV/u})$

$= 225.0 \text{ MeV}. \quad \text{So} \quad \dfrac{E_b}{A} = \dfrac{225.0 \text{ MeV}}{27 \text{ nucleon}} = 8.33 \text{ MeV/nucleon}.$

For Na: $E_b = (11m_H + 12m_n - m_{Na})\, c^2 = [11(1.007825\text{ u}) + 12(1.008665\text{ u}) - 22.989770\text{ u}](931.5\text{ MeV/u})$

$= 186.6\text{ MeV}$. So $\dfrac{E_b}{A} = \dfrac{186.6\text{ MeV}}{23\text{ nucleon}} = 8.11\text{ MeV/nucleon}$.

So nucleons are more tightly bound in Al.

74. (a) $E = \text{m} \times \text{Dose} = (0.20\text{ kg})(1.25\text{ rad}) = 0.25\text{ J}$.

(b) Dose (in rem) = Dose (in rad) × RBE = (1.25 rad)(4) = 5.0 rem, which exceeds the maximum permissible radiation dosage with background radiation included. So the answer is yes.

80. The first α decay to ${}^{230}_{90}\text{Th}$; the second α decay to ${}^{226}_{88}\text{Ra}$; the third α decay to ${}^{222}_{86}\text{Rn}$; the fourth α decay to ${}^{218}_{84}\text{Po}$; the first β decay to ${}^{222}_{85}\text{At}$; the second β decay to ${}^{218}_{86}\text{Rn}$.

88. (a) ${}^{229}_{90}\text{Th} \to {}^{225}_{88}\text{Ra} + {}^{4}_{2}\text{He}$. (b) ${}^{225}_{88}\text{Ra} \to {}^{225}_{89}\text{Ac} + {}^{0}_{-1}\text{e}$.

94. (a) Since ${}^{7}\text{Be}$ suffers from a severe shortage of neutrons compared to its proton number, we expect either positron decay, or electron capture, or both.

(b) It is electron capture. So ${}^{7}_{4}\text{Be} + {}^{0}_{-1}\text{e} \to {}^{7}_{3}\text{Li}$. Yes ${}^{7}\text{Li}$ is stable (probably after some γ decay).

VI. Practice Quiz

1. How many neutrons are there in ${}^{15}_{8}\text{O}$?

(a) 7 (b) 8 (c) 15 (d) 22 (e) 23

2. Which particle has the most mass?

(a) alpha (b) beta (β^-) (c) beta (β^+) (d) electron (e) gamma

3. A radioactive sample has a half-life of 5.0 min. What fraction of the sample is left after 20 min?

(a) 0 (b) 1/2 (c) 1/4 (d) 1/8 (e) 1/16

4. An atom has 98 protons and 249 nucleons. If it undergoes alpha decay, what are the numbers of protons and nucleons, respectively, in the daughter nucleus?

(a) 100, 245 (b) 94, 247 (c) 96, 245 (d) 100, 249 (e) 98, 249

5. The half-life of radioactive iodine-137 is 8.0 days. How many iodine nuclei are necessary to produce an activity of 1.0 μCi?

(a) 2.9×10^{9} (b) 4.6×10^{9} (c) 3.7×10^{10} (d) 7.6×10^{12} (e) 8.1×10^{13}

6. What is the average binding energy per nucleon for ${}^{197}_{79}\text{Au}$? The mass of ${}^{197}_{79}\text{Au}$ is 196.96656 u.

(a) 6.8 MeV (b) 7.3 MeV (c) 7.7 MeV (d) 7.9 MeV (e) 8.3 MeV

7. The nuclear equation, ${}^{14}_{6}\text{C} \rightarrow {}^{14}_{7}\text{N} + {}^{0}_{-1}\text{e}$, is for

(a) alpha decay. (b) beta (β^-) decay. (c) gamma decay (d) electron capture. (e) beta (β^+) decay.

8. An X-ray technician takes an average of ten X-rays per day and receives 2.5×10^{-3} rad per X-ray. What is the total dose in rem the technician receives in 250 working days?

(a) 2.50 rem (b) 5.00 rem (c) 6.25 rem (d) 7.75 rem (e) 9.00 rem

9. Which one of the following nuclei is likely to be unstable?

(a) ${}^{4}_{2}\text{He}$ (b) ${}^{27}_{13}\text{Al}$ (c) ${}^{13}_{6}\text{C}$ (d) ${}^{24}_{11}\text{Na}$ (e) ${}^{2}_{1}\text{H}$

10. The binding energy of a nucleus is contributed to

(a) radioactivity. (b) alpha decay. (c) too many neutrons.

(d) conservation of nucleons. (e) mass defect.

Answers to Practice Quiz:

1. a 2. a 3. e 4. c 5. c 6. d 7. b 8. c 9. d 10. e

CHAPTER 30

Nuclear Reactions and Elementary Particles

I. Chapter Objectives

Upon completion of this chapter, you should be able to:

1. write nuclear reaction equations using charge and nucleon conservation, and understand the concepts of *Q*-value and threshold energy and utilize them in the analysis of nuclear reactions.
2. understand what determines which nuclei are fissionable, the nature and cause of a nuclear chain reaction, and the basic principles involved in the operation of nuclear reactors.
3. explain the fundamental difference between fusion and fission, calculate energy releases in fusion reactions, and understand how fusion might eventually produce useful electric energy for the world.
4. explain why the neutrino is necessary to account for observed beta decay data, specify some of the physical properties of neutrinos, and write complete beta decay equations.
5. understand the quantum mechanical picture of forces, and classify the various forces according to their strengths, properties, ranges, and virtual particles.
6. classify the various elementary particles into families and understand the different properties of the various families of elementary particles.

II. Key Terms

Upon completion of this chapter, you should be able to define and/or explain the following key terms:

nuclear reaction
particle accelerators
***Q*-value**
endoergic
exoergic
threshold energy
cross section
fission reaction
chain reaction
critical mass
nuclear reactor
fuel rods
inertia confinement
neutrino
fundamental forces
virtual particle
exchange particle
photon
meson
muon
pion
weak nuclear force
W particle
graviton

coolant

control rods

moderator

breeder reactor

LOCA

meltdown

fusion reaction

plasma

magnetic confinement

elementary particles

hadrons

leptons

tauon

quarks

baryons

gluons

electroweak force

The definitions and/or explanations of the most important key terms can be found in the following section: **III. Chapter Summary and Discussion**.

III. Chapter Summary and Discussion

1. Nuclear Reactions

In **nuclear reactions**, nuclei are converted into the nuclei of other isotopes, which in general are nuclides of completely different elements. Some reactions are produced by energetic particles from radioactive sources or from **particle accelerators**, in which particles are accelerated to high speeds. The general form of a nuclear reaction equation is $\text{A} + \text{a} \rightarrow \text{B} + \text{b}$, or A(a, b)B, where upper case letters represent the nuclei and the lowercase letters represent the particles, A and a are the reactants and B and b are the products.

The ***Q*-value** of a reaction or decay represents the energy released ($Q > 0$) or absorbed ($Q < 0$) in the process; it appears as a change in the total rest mass of the system, $Q = (m_\text{A} + m_\text{a} - m_\text{B} - m_\text{b})c^2 = (\Delta m)c^2$. When the Q-value of a reaction is positive ($Q > 0$), the mass of the reactants is greater than the mass of the products, energy is released, and the reaction is said to be **exoergic**. When the Q-value of a reaction is negative ($Q < 0$), the mass of the products is greater than the mass of the reactants, energy is absorbed, and the reaction is said to be **endoergic**. All naturally occurring radioactive decay processes are exoergic.

In an endoergic reaction, the minimum energy required for the reaction to go forward is called the **threshold energy** and is given by $K_\text{min} = \left(1 + \frac{m_a}{m_A}\right)|Q|$. A measure of the probability that a particular reaction will occur is called the **cross section** of the reaction.

Note: The mass-energy equivalence of 1 u = 931.5 MeV is used in calculating Q-value and threshold energy.

Example 30.1 Is the following reaction exoergic or endoergic?

$$^{7}\text{Li} \quad + \quad ^{1}\text{p} \quad \rightarrow \quad ^{4}\text{He} \quad + \quad ^{4}\text{He}$$

(7.016005 u) (1.007825 u) (4.002603 u) (4.002603 u)

Solution: Given: $m_A = 7.016005$ u, $m_a = 1.007825$ u, $m_B = 4.002603$ u, $m_b = 4.002603$ u.

Find: If exoergic or endoergic.

$Q = (\Delta m)c^2 = (m_A + m_a - m_B - m_b)c^2$

$= (7.016005 \text{ u} + 1.007825 \text{ u} - 4.002603 \text{ u} - 4.002603 \text{ u})(931.5 \text{ MeV/u}) = +17.35 \text{ MeV}.$

Since the Q-value is positive, the reaction is exoergic.

Example 30.2 Find the threshold energy for the following reaction.

$$^{13}\text{C} \quad + \quad ^{1}\text{p} \quad \rightarrow \quad ^{13}\text{N} \quad + \quad ^{1}\text{n}$$

(13.003355 u) (1.007825 u) (13.005739 u) (1.008665 u)

Solution: Given: $m_A = 13.003355$ u, $m_a = 1.007825$ u, $m_B = 13.005739$ u, $m_b = 1.008665$ u.

Find: K_{min}.

$Q = (\Delta m)c^2 = (m_A + m_a - m_B - m_b)c^2$

$= (13.003355 \text{ u} + 1.007825 \text{ u} - 13.005739 \text{ u} - 1.008665 \text{ u})(931.5 \text{ MeV/u}) = -3.003 \text{ MeV}.$

So the threshold energy is $K_{\text{min}} = \left(1 + \frac{m_a}{m_A}\right)|Q| = \left(1 + \frac{1.007825 \text{ u}}{13.003355 \text{ u}}\right)(3.003 \text{ MeV}) = 3.236 \text{ MeV}.$

The threshold energy is more than 3.003 MeV because products have kinetic energies.

2. Nuclear Fission

In a nuclear **fission reaction**, a heavy nucleus divides into two lighter nuclei whose total mass is less than that of the heavy nuclei, with the emission of two or more neutrons. The energy liberated in a fission reaction is about 1 MeV per nucleon in fission products. Since there are billions and billions of nuclei in even a tiny sample of fissionable material, the amount of energy released in nuclear fission is enormous. A sustained release of nuclear energy can be accomplished by a **chain reaction**, in which neutrons from one fission reaction initiate more fission reactions, and the process multiplies, the number of neutrons doubling or tripling with each generation. When this occurs uniformly with time, we have exponential growth (or a bomb). To have a sustained chain reaction, there must be a minimum, or **critical mass** of the fissionable material. Basically, enough mass is needed so the neutrons from the previous reaction do not escape without causing more fission reactions.

Example 30.3 Estimate the energy released in the reaction, ${}^{235}_{92}\text{U} + {}^{1}_{0}\text{n} \rightarrow {}^{140}_{54}\text{Xe} + {}^{94}_{38}\text{Sr} + 2({}^{1}_{0}\text{n})$.

Solution: There are 140 + 94 = 234 nucleons in the fission products, so the accompanying energy release is approximately (1 MeV/nucleon)(234 nucleons) ≈ 234 MeV. This is the energy released per initial ${}^{235}\text{U}$ nucleus!

Example 30.4 Find the energy released in the following fission reaction.

${}^{1}\text{n}$	+	${}^{235}\text{U}$	$\rightarrow$	${}^{141}\text{Ba}$	+	${}^{92}\text{Kr}$	+	$3({}^{1}\text{n})$
(1.008665 u)		(235.043925 u)		(140.91420 u)		(91.9252 u)		(1.008665 u)

Solution: The mass difference between the reactants and the products is

$$\Delta m = 1.008665\text{ u} + 235.043925\text{ u} - 140.91420\text{ u} - 91.9252\text{ u} \; -3(1.008665\text{ u}) = 0.1872\text{ u}.$$

So the energy released is $(0.1872\text{ u})(931.5\text{ MeV/u}) = 174.4\text{ MeV}$.

Currently, the only type of practical **nuclear reactor** is based on the fission chain reaction. There are four key elements to a reactor: fuel rods, coolant, control rods, and a moderator. **Fuel rods** are tubes packed with pellets of enriched uranium oxide (or other fissionable material) located in the reactor core. A **coolant** is usually ordinary water (H_2O) or heavy water (D_2O) flowing around the fuel rods to remove the energy released from the fission chain reaction. The chain reaction and energy output of a reactor are controlled by means of boron or cadmium **control rods**, which can be inserted into or withdrawn from the reactor core. (Boon and cadmium are very effective in absorbing neutrons.) The water flowing around the fuel rods acts not only as a coolant, but also as a **moderator**, which slows down the speed of the neutrons from fission before they initiate more fission reactions. Other materials, such as graphite (carbon), may be used as moderators.

A **breeder reactor** produces more fissionable materials than it consumes. U-235 is consumed and fissionable Pu-239 is produced from a neutron reaction with U-238.

Should the coolant be lost or stop flowing through the core of a reactor, the reactor will overheat. This could result in a **LOCA** (loss-of-coolant accident) and the **meltdown** of the core. The huge amount of energy released can cause explosions and the release of radioactive materials to the environment. Therefore, a safety shutdown must be started when a loss of coolant is detected and emergency cooling water must be provided to keep the radioactive fragments from overheating.

3. Nuclear Fusion

In a nuclear **fusion reaction**, light nuclei fuse together to form a heavier nucleus with less total mass than the original nuclei, with the release of energy. At the centers of some stars, four hydrogen nuclei (^1H) fuse to form helium (^4He). The energy released per fusion is about 24.7 MeV. Due to the light mass of H and He, fusion releases more energy than fission on a per unit mass basis.

Fusion requires high temperatures that ionize the fusion material into a gas of positive ions and negative electrons called **plasma**. Due to the high temperature of plasma, plasma confinement is a major problem. The problem of plasma confinement is being approached in several different ways. Among these are magnetic confinement and inertial confinement. In **magnetic confinement**, magnetic fields are used to hold the plasma in a confined space, a so-called magnetic bottle. In **inertial confinement**, pulsed laser, electron, or ion beams would be used to implode hydrogen fuel pellets, producing compression and high temperatures.

Example 30.5 Find the energy released in the following fusion reaction.

$$^2\text{H} + {}^2\text{H} \rightarrow {}^3\text{He} + {}^1\text{n}$$

(2.014102 u) (2.014102 u) (3.016029 u) (1.008665 u)

Solution: The mass difference between the reactants and the products is

$$\Delta m = 2.014102\text{ u} + 2.014102\text{ u} - 3.016029\text{ u} - 1.008665\text{ u} = 3.510\times10^{-3}\text{ u}.$$

So the energy released is $(3.510\times10^{-3}\text{ u})(931.5\text{ MeV/u}) = 3.270\text{ MeV}$.

---

4. Beta Decay and the Neutrino

In a beta decay, there is an apparent violation of conservation of energy, linear momentum, and angular momentum (spin). To account for this apparent problem, an additional particle called the **neutrino** (ν_e) was proposed. Neutrons have zero rest mass and interact with matter by the weak nuclear force (to be discussed later). Neutrinos (ν_e) are produced in β^+ decays and antineutrinos ($\bar{\nu}_e$) are produced in β^- decays. The complete basic reactions are

β^- decay: $\text{n} \rightarrow \text{p} + \text{e}^- + \bar{\nu}_e$ $\qquad$ β^+ decay: $\text{p} \rightarrow \text{n} + \text{e}^+ + \nu_e$.

5. Fundamental Forces and Exchange Particles

There are only four known **fundamental forces**: the *gravitational force*, the *electromagnetic force*, the *strong nuclear force*, and the *weak nuclear force*.

In quantum mechanics, forces are transmitted through the exchange of particles. These particles are created within the uncertainty principle and are called **virtual particles**. The fundamental forces are considered to be carried by virtual **exchange particles.** The greater the mass of an exchange particle, the more energy required to create it, and the shorter the range of the particle. Therefore, the range of the particle is inversely proportional to its mass.

The exchange particle for the electromagnetic force is a (virtual) **photon**. Since a photon has zero mass, its range is infinite. The strong nuclear force is associated with the **meson**. The relationship between the mass of the meson and the range of the strong nuclear force is given by $R = c\Delta t = \dfrac{h}{2\pi m_\text{m} c}$, where m_m is the mass of the exchange particle. The mass of π mesons **(pions)** is found to be from 247 to 264 times that of an electron, so its range is about 10^{-15} m, which indicates a very short-ranged force. The **W particle** carries the **weak nuclear force.** The weak force has a mass of about 100 times that of a proton, which explains the extremely short range ($\approx \times 10^{-17}$ m) and the weakness of the force. The weak force is the only force that acts on neutrinos, which explains why they are so difficult to detect. **Graviton** is the massless exchange particle for the gravitational force and interacts very weakly with matter, making it very elusive to detect. There is still no firm evidence of the existence of this massless particle. A summary of the relative strength, range, and exchange particles of the four fundamental forces are given in the following table:

Force	Relative Strength	Action Distance	Exchange Particle
Strong Nuclear	1	Short range ($\approx 10^{-15}$ m)	Pion (π meson)
Electromagnetic	10^{-3}	Infinite	photon
Weak Nuclear	10^{-8}	Extremely short range ($\approx 10^{-17}$ m)	W particle
Gravitational	10^{-45}	infinite	Graviton

Example 30.6 The π^o meson has a mass of 264 times that of an electron. What is the range of the force mediated by this particle?

Solution: Given: m_m (meson) $= 264 m_\text{e} = 264(9.11\times10^{-31}\text{ kg}) = 2.41\times10^{-28}$ kg.

Find: R.

$$R = \frac{h}{2\pi m_\text{m} c} = \frac{6.63\times10^{-34}\text{ J·s}}{2\pi(2.41\times10^{-28}\text{ kg})(3.00\times10^{8}\text{ m/s})} = 1.46\times10^{-15}\text{ m}.$$

6. Elementary Particles

Elementary particles are fundamental particles, or the building blocks, of atoms. **Hadrons** are particles that interact by the strong nuclear force, for example, protons, neutrons, and pions. **Leptons** are particles that interact by the weak nuclear force, for example, electrons, muons, and neutrinos. **Quarks** are elementary particles that make up hadrons. Quarks combine only in two ways, either in three or in quark-antiquark pairs. Three quark combinations are called **baryons**, and quark-antiquark combinations are called mesons.

There are six *flavors* or types of quarks: up (u), down (d), strange (s), charm (c), top (t), and bottom (b). Quarks have fractional electronic charges of either $-e/3$ or $+2e/3$. Quarks interact by the strong nuclear force, but they are also subject to the weak force. A weak force acting on a quark changes its flavor, and gives rise to the decay of hadrons.

The exchange particle for quarks is the **gluon**. To give the strong force a field representation, each quark is said to possesses an analog of electric charge which is the source of the "gluon field". Instead of charge, their property is called *color* (no relationship with ordinary color). Each quark can come in one of the three possible colors: red, blue, and green. There are corresponding anticolors and antiquarks. When a quark emits or absorbs a gluon, it changes its color.

The electromagnetic force and the weak force are two parts of a single **electroweak force**. A theory that would merge the strong nuclear force and the electroweak force is called the *grand unified theory* (GUT). Perhaps, all forces are part of a single *superforce*, which is a real theoretical challenge.

IV. Mathematical Summary

Q Value	$Q = (m_A + m_a - m_B - m_b)c^2$ $= (\Delta m)c^2$ (30.4)	Defines the Q value of a nuclear reaction in terms of the mass defect
Threshold Energy	$K_{min} = \left(1 + \frac{m_a}{m_A}\right)\lvert Q\rvert$ (30.5)	Calculates the minimum energy required (threshold energy) for an endoergic reaction
Range of Exchange Particle	$R = c\Delta t = \frac{h}{2\pi m_m c}$ (30.6)	Defines the range of an exchange particle

V. Solutions of Selected Exercises and Paired Exercises

6. (a) ${}^{14}_{7}\text{N}$. (b) ${}^{2}_{1}\text{H}$. (c) ${}^{30}_{15}\text{P}$. (d) ${}^{17}_{8}\text{O}$. (e) ${}^{10}_{5}\text{B}$.

8. ${}^{211}\text{Pb} \rightarrow {}^{211}\text{Bi} + {}^{0}_{-1}\text{e}$. So the energy released is

$$Q = (m_{\text{Pb}} - m_{\text{Bi}} - m_{\text{e}})\, c^2 = (210.98874\ \text{u} - 210.98726\ \text{u})(931.5\ \text{MeV/u}) - 0.511\ \text{MeV} = 0.868\ \text{MeV}.$$

12. $Q = (15.994915\ \text{u} + 1.008665\ \text{u} - 13.003355\ \text{u} - 4.002603\ \text{u})(931.5\ \text{MeV/u}) = -2.215\ \text{MeV}.$

$$K_{\text{min}} = \left(1 + \frac{m_{\text{a}}}{M_{\text{A}}}\right)|Q| = \left(1 + \frac{1.008665}{15.994915}\right)(2.215\ \text{MeV}) = 2.35\ \text{MeV}.$$

14. $Q = (199.968321\ \text{u} + 1.007825\ \text{u} - 196.96656\ \text{u} - 4.002603\ \text{u})(931.5\ \text{MeV}) = +6.50\ \text{MeV}.$

So it is exoergic.

20. ${}^{226}\text{Ra} \rightarrow {}^{222}\text{Rn} + {}^{4}\text{He}$. $Q = (226.025406\ \text{u} - 222.017574\ \text{u} - 4.002603\ \text{u})(931.5\ \text{MeV/u}) = 4.871\ \text{MeV}.$

So the kinetic energy of the recoiling daughter nucleus is $K = 4.871\ \text{MeV} - 4.706\ \text{MeV} = 0.165\ \text{MeV}$, according to conservation of energy.

23. (a) $\pi R^2 = \pi R_{\text{o}}^2 (A)^{2/3} = \pi(1.2\times10^{15}\ \text{m})^2 (12)^{2/3} = 2.37\times10^{29}\ \text{m}^2 = (2.37\times10^{29}\ \text{m}^2)\dfrac{1\ \text{b}}{10^{28}\ \text{m}^2} = 0.24\ \text{b}.$

(b) $\pi R^2 = \pi R_{\text{o}}^2 (A)^{2/3} = \pi(1.2\times10^{15}\ \text{m})^2 (56)^{2/3} = 6.62\times10^{29}\ \text{m}^2 = 0.66\ \text{b}.$

(c) $\pi R^2 = \pi R_{\text{o}}^2 (A)^{2/3} = \pi(1.2\times10^{15}\ \text{m})^2 (208)^{2/3} = 1.59\times10^{28}\ \text{m}^2 = 1.6\ \text{b}.$

(d) $\pi R^2 = \pi R_{\text{o}}^2 (A)^{2/3} = \pi(1.2\times10^{15}\ \text{m})^2 (238)^{2/3} = 1.74\times10^{28}\ \text{m}^2 = 1.7\ \text{b}.$

30. (a) 231 nucleons are involved in the fission process.

So the energy released is (1 MeV/nucleon)(231 nucleon) = 231 MeV.

(b) 233 nucleons are involved in the fission process.

So the energy released is (1 MeV/nucleon)(233 nucleon) = 233 MeV.

38. According to momentum conservation, the momentum of the beta particle plus the daughter nucleus is equal in magnitude and opposite in direction to the momentum of the neutrino.

$$p = \frac{E}{c} = \frac{(2.65\times10^{6}\ \text{eV})(1.6\times10^{19}\ \text{J/eV})}{3.00\times10^{8}\ \text{m/s}} = 1.41\times10^{-21}\ \text{kg·m/s opposite to neutrino's.}$$

40. In a β^- decay: ${}^{\text{A}}_{\text{Z}}\text{p} \rightarrow {}^{\text{A}}_{\text{Z+1}}\text{d} + {}^{0}_{-1}\text{e}$.

So $Q = (m_{\text{P}} - m_{\text{D}} - m_{\text{e}})\, c^2 = \{(m_{\text{P}} + Zm_{\text{e}}) - [m_{\text{D}} + (Z+1)m_{\text{e}}]\}\, c^2 = (M_{\text{P}} - M_{\text{D}})\, c^2$.

43. In a β^{+} decay: ${}^{A}_{Z}\text{p} \rightarrow {}_{Z-1}^{\;\;A}\text{d} + {}_{+1}^{\;\;0}\text{e}$.

So $Q = (m_{\text{P}} - m_{\text{D}} - m_{\text{e}})\,c^2 = \{(m_{\text{P}} + Zm_{\text{e}}) - [m_{\text{D}} + (Z-1)m_{\text{e}}] - 2m_{\text{e}}\}\,c^2 = (M_{\text{P}} - M_{\text{D}} - 2m_{\text{e}})\,c^2$.

50.

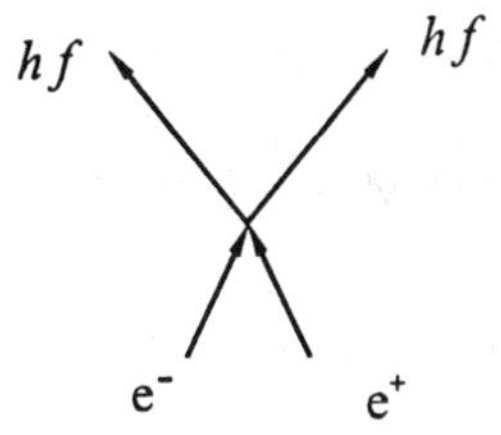

55. $$R = \frac{h}{2\pi\, m\, c} = \frac{hc}{2\pi\, m\, c^2} = \frac{hc}{2\pi\, E_o} = \frac{(6.63\times 10^{34}\ \text{J·s})(3.00\times 10^{8}\ \text{m/s})}{2\pi(100\times 10^{6}\ \text{eV})(1.6\times 10^{19}\ \text{J/eV})} = 1.98\times 10^{-15}\ \text{m}..$$

64. ${}^{210}_{84}\text{Po} \rightarrow {}^{206}_{82}\text{Pb} + {}^{4}_{2}\text{He}$.

So $Q = (209.98286\ \text{u} - 205.97446\ \text{u} - 4.002603\ \text{u})(931.5\ \text{MeV/u}) = 5.40\ \text{MeV}$.

VI. Practice Quiz

1. What is the Q-value of the following reaction?

^{14}N + ^{4}He → ^{1}p + ^{17}O

(14.003074 u) (4.002603 u) (1.007825 u) (16.999131 u)

(a) 0 (b) -1.279×10^{-3} MeV (c) 1.279×10^{3} MeV (d) -1.191 MeV (e) 1.191 MeV

2. The meson particle is associated with which of the following forces?

(a) strong nuclear (b) weak nuclear (c) electromagnetic (d) gravitational (e) weak

3. The fuel for nuclear fusion is

(a) hydrogen. (b) helium. (c) uranium (d) plutonium (e) any radioactive material.

4. What is the energy released (positive) or absorbed (negative) in the following reaction.

^{3}H + ^{3}H → ^{4}He + $2(^{1}\text{n})$

(3.016049 u) (3.016049 u) (4.002603 u) (1.008665 u)

(a) 0 (b) -0.0122 MeV (c) 0.0122 MeV (d) -11.3 MeV (e) 11.3 MeV

5. Which one of the following is an example of a hadron?

(a) electron (b) proton (c) neutrino (d) W particle (e) graviton

6. Calculate the range of a hypothetical force with the proton as the virtual exchange particle.

(a) 6.7×10^{-25} m (b) 2.1×10^{-16} m (c) 1.5×10^{-15} m (d) 6.0×10^{-8} m (e) 1.5×10^{-5} m

7. The charge of some quarks and antiquarks is

(a) 0. (b) $e/4$. (c) $e/3$. (d) $e/2$. (e) e.

8. In a nuclear fission, the mass of the products compared with the mass of the reactants is

(a) zero. (b) greater. (c) smaller. (d) the same. (c) infinite.

9. A chain reaction can occur

(a) in any uranium core. (b) when critical mass is reached (c) in the center of the Sun

(d) when the control rods are fully inserted (e) when the coolant is too hot.

10. What is the threshold energy for the following reaction?

$$^{14}\text{N} \quad + \quad ^{4}\text{He} \quad \rightarrow \quad ^{1}\text{p} \quad + \quad ^{17}\text{O}$$

(14.003074 u) (4.002603 u) (1.007825 u) (16.999133 u)

(a) 0 (b) −1.193 MeV (c) 1.193 MeV (d) −1.534 MeV (e) 1.534 MeV

Answers to Practice Quiz:

1. d 2. a 3. a 4. e 5. b 6. b 7. c 8. c 9. b 10. e